Logical Forms

Logical Forms

An Introduction to Philosophical Logic

Mark Sainsbury

Basil Blackwell

© Mark Sainsbury 1991

First published 1991

Basil Blackwell Ltd
108 Cowley Road, Oxford OX4 1JF, UK

Basil Blackwell, Inc.
3 Cambridge Center
Cambridge, Massachusetts 02142, USA

British Library Cataloguing in Publication Data
A CIP catalogue record for this book is available from the British Library.

Library of Congress Cataloging in Publication Data
Sainsbury, R. M. (Richard Mark)
   Logical forms: an introduction to philosophical logic/p. cm.
   Includes bibliographical references and index.
   ISBN 0-631-17777-9 – ISBN 0-631-17778-7 (pbk.)
   1. Logic, Symbolic and mathematical.   I. Title.
BC135.S14   1991
160 – dc20
90-40876
CIP

Typeset by the Alden Press, Osney Mead, Oxford
Printed in Great Britain by T.J. Press Ltd, Padstow, Cornwall

Contents

Acknowledgements

An early draft of the first three chapters of this book circulated in manuscript in 1980: I am very grateful to the many people – colleagues, friends and students – who commented upon that draft.

Most of the final version was written in Belize, Central America, in the summer months of 1987. I am grateful to King's College London for making this possible by a grant of sabbatical leave; and to the Government of Belize for kindly granting me exemption from import duty on generating equipment required to run my word processor.

I would like to thank Danial Bonevac, who read the entire manuscript in a near-final draft, and whose comments led to many improvements; Marianne Talbot, who provided invaluable help with references; and Stephen Read, whose very penetrating and knowledgeable criticisms of what had been intended as the final draft led to some considerable rewriting, and saved me from many mistakes.

Introduction

Some kind of knowledge of logical forms, though with most people it is not explicit, is involved in all understanding of discourse. It is the business of philosophical logic to extract this knowledge from its concrete integuments, and to render it explicit and pure.

Bertrand Russell, *Our Knowledge of the External World*

This book is an introduction to philosophical logic. It is primarily intended for people who have some acquaintance with deductive methods in elementary formal logic, but who have yet to study associated philosophical problems. However, I do not presuppose knowledge of deductive methods, so the book could be used as a way of embarking on philosophical logic from scratch.

Russell coined the phrase "philosophical logic" to describe a programme in philosophy: that of tackling philosophical problems by formalizing problematic sentences in what appeared to Russell to be *the* language of logic: the formal language of *Principia Mathematica*. My use of the term "philosophical logic" is close to Russell's. Most of this book is devoted to discussions of problems of formalizing English in formal logical languages.

I take validity to be the central concept in logic. In the first chapter I raise the question of why logicians study this property in connection with artificial languages, which no one speaks, rather than in connection with some natural language like English. In chapters 2–5 I indicate some of the possibilities and problems for formalizing English in three artificial logical languages: that of propositional logic (chapter 2), of first order quantificational logic (chapter 4) and of modal logic (chapter 5). The final chapter takes up the purely philosophical discussion, and, using what has been learned on the way, addresses such questions as whether there was any point in those efforts at formalizing, what can be meant by *the* logical form of an English sentence, what is the domain of logic, and what is a logical constant.

In this approach, one inevitably encounters not only questions in the philosophy of logic, but also questions in the philosophy of language, as when one considers how best to formalize English sentences containing empty names, or definite descriptions, or adverbs, or verbs of propositional attitude.

My own preference in teaching logic is to begin with the elementary formal part, keeping the students' eyes blinkered to philosophical questions, which are dauntingly hard. In introducing the philosophical issues later on, I am conscious of the width of the gap I expect students to leap, from the drill of truth tables and proofs, to discussions of the semantics of names or conditionals. This is the gap this book is designed to fill.

Logic raises a host of problems that call for philosophical discussion, like the nature of truth, the relation between logical rules and psychological processes, the nature of logical knowledge, the question of what exists. I have turned aside from as many of these as I could, limiting myself to the single theme of the nature of logical form, together with whatever tributaries I seemed absolutely compelled to navigate.

There are exercises throughout, and these are required for completeness. Various issues that could well be regarded as essential to the main business are relegated to exercises, so that the reader can, in effect, contribute to the development of the argument. For example, expert readers of the section on Russell's theory of descriptions might be aghast to find that the body of the text makes no mention of Russell's claim that "The present King of France is not bald" is ambiguous, and in a way which resolves a certain puzzle. The reader is invited to discover the ambiguity, and its relation to the puzzle, in an exercise. Exercises start on page 337. There are bibliographical Notes at the end of each chapter which suggest further reading and provide some information about the sources of some of the points in the text.

The reader will find it useful to look at the section which immediately follows ("Organization," p. 2), which explains the system of numbering displayed material, the arrangement of glossary entries, and related matters.

The book is progressive. The first chapter is written with the complete novice in mind; the last chapter addresses a considerably more knowledgeable audience. My hope is that the intervening material will help beginners progress from the one state to the other.

Organization

Optional material is indicated by an asterisk in the title of the section or chapter. It is optional in that subsequent discussions do not presuppose it. Chapter **3**, on conditionals, is the only optional chapter. It could be viewed as an appendix to chapter **2**.

Most displayed material is numbered by a bold numeral followed by a right parenthesis, thus:

1)   This is how displayed material appears.

Subsequent references to displayed material use the numeral enclosed in parentheses, so the above is referred to as (1). Numbering begins afresh with each section, and a reference like "(1)" refers to the displayed sentence labelled "1)" within the current section.

Sections and chapters are also numbered. A reference like "(2.3)" refers to the displayed sentence labelled "3)" in the second section within the current chapter. A reference like "(4.2.3)" refers to the displayed sentence labelled "3)" in the second section of chapter **4**.

Reference to a section within the current chapter takes the form: §3. Reference to a section outside the current chapter takes the form: chapter **3.2** (that is, chapter **3**, section 2).

Footnotes refer the reader to exercises at the back of the book.

Left-hand page headings give the current chapter number and title. Right-hand page headings give the current section number and title.

A full understanding of the matters discussed requires further reading. Suggestions will be found in the bibliographical notes at the end of each of chapters **2–6**. Works are referred to by the author's name followed by a date in square brackets. The bibliography (p. 383) must be used to obtain full publication details.

There is a glossary (p. 369) which lists definitions of technical terms in the order in which they appear in the text, and in most cases the page number of their introduction into the text. Glossary entries are numbered consecutively by chapter. This is a different numbering system from that of the displayed material (since not all terms are introduced by displayed definitions). Each term occurring in the glossary also occurs in the Index, where its glossary entry number is given, preceded by "G", for example "G4.12". This gives alphabetical access to glossary entries.

Glossary entries include one or two set-theoretical terms (e.g. "sequence") that are used though not explained in the text.

There is a list of symbols (p. 380), with a brief definition of each. They are arranged in order of occurrence, with a page indicating where they were first introduced.

I have not been very particular about my use of inverted commas. In many cases, strict accuracy requires corner quotes,[1] but I decided to dispense with this additional complexity, in the belief that no confusion is likely to result.

Notes

1 See Quine [1940], §6, "quasi quotation".

1

Validity

1 What is logic about?

The philosophy of logic gives an account of what logic is, of the concepts that it uses, and of how it relates to other disciplines and to our ordinary thought and talk.

Logic is about reasons and reasoning. There are reasons for *acting*: wanting to keep thin is a reason for avoiding fatty foods. There are reasons for *believing*: that the potatoes have been boiling for twenty minutes is a reason for believing that they are ready to eat. Historically, logic has primarily concerned itself with reasons for believing.

If we ask what a reason for believing is, we might be tempted to say: we give a reason for believing when we answer the question "Why does so-and-so believe such-and-such?" But we must be careful, for such a question can be answered in two different kinds of way.

Suppose we ask of an orthodox Hindu: why does he suppose that one should not eat meat? One kind of answer is: this belief was instilled in him by his family at an early age, and has been sustained by a variety of social and personal pressures. This kind of answer may *explain* the origin of the belief. But it does not give a *reason* for the belief, in the sense of "reason" in which logic is concerned with reasons. Explanations of this kind belong to psychology or sociology. They are quite foreign to logic.

Suppose we answer the question in a different way, saying: the Hindu believes that killing, and everything which requires killing, is wrong; and that eating meat requires killing. This answer *may* explain the origin of the belief. But it also does, or purports to do, something else: it *justifies* the belief. Understood in this way, as attempting to provide a justification, the answer shows a concern with reasons in the sense in which logic can be said to be the study of *reasons*.

Logic is a *normative* discipline. It aims to say what reasons are *good* reasons. It does not merely describe the reasons that in fact move people. It lays down standards. It says what reasons *ought* to move one. Even so, the starting-point has to be what we generally think of as good reasons. Logic starts with an intuitive commonsensical and pretheoretical distinction between good and bad reasons, a distinction made by people pursuing their ordinary daily concerns. From this the logician hopes to fashion an articulate and defensible distinction between good reasons and bad. One would expect a large measure of agreement between the logician's technical distinction and the commonsensical one. But one should not turn one's back on the possibility of a divergence: common sense may need to be corrected in the light of reflection.

Here are some examples, of an everyday kind, of the commonsensical distinction at work. Most people would agree that:

> James is a banker and all bankers are rich

is a good reason for

> James is rich.

By contrast, most people would agree that:

> Henry is a playwright and some playwrights are poor

is not a good reason for

> Henry is poor.

There is (or until recently was) no general agreement about whether

> James and Henry lead pretty similar lives except that James is a non-smoker and Henry smokes twenty cigarettes a day

is or is not a good reason for

> Henry is more likely to die of heart disease than James.

We can regard a reason as a good reason without having to believe it ourselves. We do not have to believe that all bankers are rich to see that "James is a banker and all bankers are rich" constitutes a good reason for

"James is rich". This is extremely important to logic, as I shall illustrate towards the end of this section.

When we talk about reasons, we do not have to talk about particular people and what they believe. Even if no one had ever had the beliefs we attributed to the Hindu, we could still say that

>Killing, and everything which requires killing, is wrong;

>Eating meat requires killing

together form good reasons for

>One should not eat meat.

What one means can be partly understood in this way: if anyone were to believe that killing, and everything which requires killing, is wrong, and were also to believe that eating meat requires killing, he would thereby be right (rational, reasonable, logical, justified) also to believe that one should not eat meat.

When we want to consider something like

>Killing is wrong

in abstraction from whether anyone believes it or not, we shall call it a *proposition*. A proposition is the sort of thing that *can* be believed, or asserted, or denied, but it does not have to be: it can be disbelieved, or merely entertained, or not even thought of at all. Perhaps no one believes, or has even until just this moment supposed or entertained, the proposition that Julius Caesar built New York single-handed in a day. We shall, nevertheless, say that there is such a proposition.

The most general question which confronts the logician can now be expressed as follows: what makes one proposition (or collection of propositions) a good reason for a proposition?

We shall call the propositions offered as reasons *premises*, and the proposition which the premises are supposed to support the *conclusion*. When some premises and a conclusion are assembled together, we shall call the result an *argument*. The technical use of these terms, as just introduced, differs in some ways from the ordinary use. In particular, as used in logic, the term "argument" does not imply any kind of disagreement or dispute.

We have already considered various arguments:

1) *Premises*:
Killing, and anything which requires killing, is wrong;
Eating meat requires killing.
Conclusion:
One should not eat meat.

2) *Premise*:
James is a banker and all bankers are rich.
Conclusion:
James is rich.

3) *Premise*:
Henry is a playwright and some playwrights are poor.
Conclusion:
Henry is poor.

4) *Premise*:
James and Henry lead pretty similar lives except that James is a non-smoker and Henry smokes twenty cigarettes a day.
Conclusion:
Henry is more likely than James to die of heart disease.

Common sense pronounces that in (**1**) and (**2**) the premises constitute good reasons for the conclusion, that this is definitely not so for (**3**), and that (**4**) is a debatable case.

In (**1**) and (**2**) the conclusion *follows from* the premise(s). The branch of logic with which we shall be mainly concerned – *deductive logic* – investigates the contrast between arguments in which the conclusion follows from the premises, and those in which it does not. One way in which premises can give good reason for a conclusion is for the conclusion to follow from the premises.

We ask whether an argument's premises constitute good reason for the conclusion, or whether the conclusion follows from the premises. In either case we are asking about a *relation* between two lots of propositions: on the one hand the premises, on the other hand the conclusion. In the case of arguments like (**1**) and (**2**), in which the conclusion follows from the premises, the point is not that the premises make the conclusion true, or even likely to be true. Perhaps the premises themselves are false, or likely to be false. Rather, *if* the premises were true, so would be the conclusion. That is why it would be rational to believe the conclusion if one believed the premises. Whether or not the relation obtains is something that can often, though not always, be detected *a priori*, that is, without any appeal to experience or experiment.[1] Our knowledge that the premises in (**1**) and (**2**) constitute good reasons for their respective conclusions is quite independent of knowing whether bankers are rich, or whether James is a banker (who is James, anyway?), or whether killing is wrong, or whether eating meat requires killing. Traditionally, logic has been held to be wholly *a priori*, and this has been used as a mark to distinguish it from other disciplines. This is a controversial view. But it would not be tenable even for a moment if the logician was called upon to pronounce on whether or not the premises of arguments are true, or probable. The view owes its attraction to the thought that the *relational* fact – whether the premises are so related to the conclusion that they constitute good reason for it – can be known *a priori*, even if the propositions involved in the relation cannot be known *a priori*.[†]

2   Inductive versus deductive logic

An old tradition has it that there are two branches of logic: deductive logic and inductive logic. More recently, the differences between these disciplines have become so marked that most people nowadays use "logic" to mean deductive logic, reserving terms like "confirmation theory" for at least some of what used to be called inductive logic. I shall follow the more recent practice, and shall construe "philosophy of logic" as "philosophy of deductive logic". In this section, I try to set out the differences between the two disciplines, and to indicate very briefly why some people think that inductive logic is not logic at all.

In §**1**, we saw that one way in which an argument's premises can constitute good reasons for its conclusion is for the conclusion to follow from the premises. Let us say that any argument whose conclusion follows from its premises is *valid*. An initial test for validity is this. We ask: is it possible for the premises to be true, yet the conclusion false? In the case of (**1.3**), about poverty and playwrights, the answer is "Yes". Even if *some* playwrights are poor, it is possible that others, perhaps even the vast majority, are rich, and that Henry is among the rich ones. In general, an argument is valid just on condition that it is impossible for the premises all to be true yet the conclusion false. Could one hope to distinguish deductive from inductive logic by saying that the former, but not the latter, is concerned with validity?

Consider two arguments which occur in hundreds of text-books:

1)   All men are mortal, Socrates is a man; so Socrates is mortal.

† See Exercise 1, page 337.

    2)   The sun has risen every morning so far; so (probably) it will rise tomorrow.

The first is a standard example of an argument classified as valid by deductive logic. The second is an argument which is not classified as valid by deductive logic. However, the inductive logician is supposed to accord it some reasonably favourable status. Certainly, the reasons which the premises of (**2**) give for its conclusion are better by far than those given by the same premises for the opposite conclusion:

    3)   The sun has risen every morning so far; so (probably) it will *not* rise tomorrow.

This may seem a silly argument, but apparently something quite like it moves some gamblers. The "Monte Carlo fallacy" consists in the belief that if there has been a long run of reds on the roulette wheel, it is more likely to come up black next time.[†]

The deductive logician contrasts (**1**) and (**2**) by saying that the first but not the second is valid. The inductive logician will make a contrast between (**2**) and (**3**) – probably not by using the word "valid", but perhaps by saying that (**2**), unlike (**3**), is "inductively strong". The premises of (**2**), but not those of (**3**), provide strong reasons for the conclusion.

The premises of (**1**) also provide strong reasons for its conclusion. How are we to distinguish strong deductive reasons from strong inductive ones? We have a suggestion before us: the truth of the premises of a valid deductive argument makes the falsity of its conclusion impossible, but this is not so in the case of inductively strong arguments. Another way of putting this is: the reasons given by a deductively valid argument are *conclusive*: the truth of the premises guarantees the truth of the conclusion. This way of making the contrast fits (**1**) and (**2**). The truth of the premises of (**2**) may make the conclusion *probable*, but it does not guarantee it: it does not make it *certain*.

Inductive logic, as the terminology of inductive strength suggests, must be concerned with a relation which holds to a greater or lesser *degree*. Some non-conclusive reasons are stronger than others. So unlike deductive logic, which will make a sharp dichotomy between valid and invalid arguments, inductive logic will discern a continuum of cases, along which (**2**), perhaps, registers fairly high, whereas (**3**) registers very low indeed.

† See Exercise 2, page 337.

Deductive validity is, as logicians say, *monotonic*. That is, if you start with a deductively valid argument, then, no matter what you *add* to the premises, you will end up with a deductively valid argument. Inductive strength is not monotonic: adding premises to an inductively strong argument can turn it into an inductively weak one. Consider (2), which is supposed to be a paradigm of inductive strength. Suppose we add the premises: there is a very large meteor travelling towards us; by tonight it will have entered the solar system and will be in stable orbit around the sun; it will lie between the sun and the earth, so that the earth will be in permanent shadow. When these premises are added, the resulting argument is far from strong. (I have assumed one particular interpretation of what it is for the sun to "rise". However one interprets this phrase, it is easy enough to find premises adding which would weaken the argument.)

Much everyday reasoning is non-monotonic, and there are endless much simpler and more realistic cases than the one just given. At the start of the investigation, Robinson's confessing to the crime gives you a powerful reason for believing him guilty. But you may rightly change your mind about his guilt, without changing your mind about whether he confessed, when a dozen independent witnesses testify to his being a hundred miles away at the time of the crime. This is a typical case in which adding information can weaken reasons which, on their own, were strong.

Table 1.1 summarizes the differences between deductive and inductive logic that we have so far mentioned.

Table 1.1

	Valid deductive reasoning	Strong inductive reasoning
Truth of premises gives good reason for truth of conclusion	√	√
Truth of premises makes falsity of conclusion impossible	√	×
Premises are conclusive reasons	√	×
Monotonic	√	×
Degrees of goodness of reasons	×	√

I said that not everyone would agree that there is any such thing as inductive logic. A famous proponent of an extreme version of this view is Karl Popper ([1959], chapter 1, §1). He has argued that the only sort of good reason is a deductively valid one. A consequence of his view is that there is nothing to choose between (2) and (3), considered simply as arguments: both are equally bad, being alike deductively invalid. He would therefore reject the ticks in the first and last rows of the right hand column of Table 1.1. For Popper, there is no such subject matter as the one I have tried to demarcate by the phrase "inductive logic"; no inductive argument gives a good reason; and there is no difference of degree among the goodness of "inductive reasons", all being equally bad.

A less radical sceptic about inductive logic may allow that there are good reasons which are not deductively valid, but deny that there is any systematic discipline worthy of the name "inductive logic". Reflection on the role of background knowledge in what are called inductively strong arguments, like (2), may ground such a scepticism. Inductive strength, as we have seen, is non-monotonic. Hence an argument cannot be assessed as inductively strong *absolutely*: for some possible background information will contain information which would greatly weaken the conclusion. This means that every assessment of inductive strength must be relativized to a body of background knowledge. However, it is far from obvious how the project of inductive logic should attempt to accommodate this point, for it is quite unclear how the background knowledge could be specified in a way which is neither question begging (for example, saying that such-and-such an argument is inductively strong relative to any bodies of background knowledge not containing any information which would weaken the conclusion), nor quite unsystematic (for example, listing various bodies of background knowledge). There is thus a genuine (I do not say decisive) ground for doubting whether inductive logic could aspire to the kind of system and generality attained by deductive logic.

A still less radical scepticism about the possibility of inductive logic takes the form: there is such a subject matter, but it does not deserve to be called *logic*. Here is one reason why a person might hold this view. It may be said that anything worthy of the name of logic must be *formal*: the property of arguments with which it is concerned must arise wholly from the form or pattern or structure of the propositions involved. Whatever exactly "formal" means (see below, §10), it certainly seems to be the case that no formal question is at issue between those who do, and those who do not, think that the evidence shows that smoking increases the risk of heart disease.[2]

Table 1.2

	Deductive logic	*Inductive "logic"*
Truth of premises gives good reason for truth of conclusion	√	?
Systematic	√	?
Formal	√	?
A priori	√	?

Another form of this kind of scepticism is as follows. Logic is *a priori*, but inductive "logic" is not, so it is not really logic. Consider the assessment of (**1.4**), the argument about smoking and heart disease.[†] No doubt the interpretation of statistical evidence would be important, and *perhaps* there is an *a priori* discipline of statistics. But even conceding this, it seems at least arguable that some non-*a priori* material is involved. If so, if, that is, it is not a purely *a priori* matter whether or not some argument is inductively strong, then inductive "logic" would not be an *a priori* discipline, and this would make it very unlike deductive logic.

Table 1.2 summarizes the various kinds of scepticism about the possibility of inductive logic.

I offer no assessment of the sceptical claims. However, from now on I shall discuss only deductive logic – logic, for short – and deductive validity – validity, for short.

3 Possibility: logical and physical

Consider the following two arguments:

1) This young tomato plant has all the moisture, nutrients, warmth and light that it needs; so it will grow good tomatoes.

2) This person is an adult male and has never married; so he is a bachelor.

Tradition has it that the first is invalid (i.e. not valid) and the second is valid. We suggested in §2 that a valid argument is one whose premises

† See Exercise 3, page 337.

cannot be true without the conclusion being true also. But is there not a sense of "impossible" in which it is impossible for the premises of (1) to be true without the conclusion also being true?

Perhaps. But there is also a sense of "possible" in which this does not hold. The plant might be attacked by wireworms or destroyed by a meteorite before the tomatoes grow, even though it has all the moisture etc., that it needs.

The following two claims will help us distinguish two kinds of possibility: *physical possibility* and *logical possibility*.[3]

> 3)   It is impossible for an internal combustion engine, used on level roads on the earth's surface, to return 5000 mpg.

> 4)   It is impossible for there to be a car which, at a given time, both has exactly three and exactly four wheels.

(3) is probably true, if the kind of impossibility involved is physical: the laws of nature being what they are, no ICE could be as efficient as that. But it is not true if the kind of impossibility involved is logical. What is logically impossible involves some kind of contradiction or inconsistency, as illustrated in (4). Logical impossibility typically issues from the very nature of the concepts involved, and is not beholden to the laws of nature. It is logically possible for the laws of nature to be very different from what they actually are.[†]

A definition of validity needs to draw upon the notion of logical rather than physical possibility, if it is to give a correct account of logician's usage. Consider the following example:

> 5)   This creature has the form of a finch. So it will not discourse intelligently about Virginia Woolf.

As the word "valid" is standardly used in logic, this is not a valid argument. But consider the following objection: intelligent discourse requires suitable musculature and thorax, and suitable complexity of brain; but it is *impossible* that a creature having the form of a finch should have such a thorax etc., and a sufficiently large brain. So it is impossible for the premise of the argument to be true without the conclusion also being true. So the argument is valid.

This objection uses the notion of physical rather than logical possibi-

† See Exercise 4, page 337.

lity. The laws of nature that we actually have rule out there being a brain of sufficient complexity for discourse in a finch-sized skull. So it is physically impossible for the premise of (5) to be true yet its conclusion false. But it is not logically impossible. We *might* have had different laws of nature. There is no *logical* guarantee that discourse requires a larger than finch-sized brain. So it is logically possible for the premise of (5) to be true yet the conclusion false. So the argument is not valid.[†]

We can now set out our preliminary definition of validity:

> An argument is VALID if and only if it is logically impossible for all the premises to be true yet the conclusion false.[‡]

This definition has some merits. For one thing, it suggests an answer to why we should use valid arguments: valid arguments are necessarily truth-preserving. So long as you start out with truth, you will never depart from the truth if you keep to valid arguments. Moreover, it is rather surprising how much can be extracted about the nature of validity from even this preliminary definition (see §6).

However, the definition has many defects. We characterized logic as the study of validity. But now, in defining validity, we have used the notion of *logical* impossibility. If we fully understood what *logical* impossibility is, presumably we would already know what *logic* itself is. So our characterizations have run in a circle.

We mentioned a connection between logical impossibility and inconsistency and contradictoriness. But these terms themselves were left unexplained.

This unsatisfactory state of affairs will have to persist for some time. One feature of definitions of validity for the formal languages to be considered in later chapters is that they can entirely avoid such notions as logical possibility and inconsistency (in the ordinary sense). For the moment, we shall see how far the ordinary notions can take us.

4   Validity, inconsistency and negation

A collection of propositions is *inconsistent* if and only if it is logically impossible for all of them to be true. Here logical impossibility is used to explain inconsistency, whereas in §3 inconsistency was used to explain

---

† See Exercise 5, page 338.   ‡ See Exercise 6, page 338.

logical impossibility. This shows that the two notions are closely related. It also shows that we could reasonably hope for a further elucidation of both notions, one which takes us out of the circle. For the moment, we shall simply take for granted the notion of logical impossibility.

Consider the propositions:

1) The earth is spheroid.

2) The earth is not spheroid.

It is logically impossible for both these propositions to be true and logically impossible for both of them to be false. In short, (1) and (2) are *contradictories*: each is a *contradictory* of the other.[†] Moreover, (2) is the *negation* of (1). You get the negation of a proposition if you insert "not" (or some equivalent expression) into it in such a way as to form a contradictory of it.

Being the negation of a proposition is one way, but not the only way, of being a contradictory of it. Being contradictories is one way, but not the only way, for two propositions to be inconsistent. I shall amplify these points, and then connect the notions of *contradictory*, *inconsistency*, and *negation* with that of validity.

If one proposition is the negation of another, it follows trivially from the definition that the two propositions are contradictories. The converse does not hold. Two propositions can be contradictories without either being the negation of the other. For example:

3) John is more than six feet tall

and

4) John is either exactly six feet tall or else less than six feet tall

are contradictories, but neither is the negation of the other. Negation is one way, but not the only way, of forming a contradictory.

Inserting "not" into a proposition does not always yield the negation of it, for inserting "not" does not always yield a contradictory. Consider:

5) Some men are happy.

6) Some men are not happy.

† See Exercise 7, page 339.

The second results from the first by inserting a "not", but the two propositions are not contradictories, since both could be – and presumably actually are – true. So (6) is not the negation of (5).

Similarly,

7)   Reagan believes that Shakespeare was a genius

8)   Reagan believes that Shakespeare was not a genius

are not contradictories, since both could be false. They would be false if Reagan had no view one way or the other about Shakespeare's qualities. Hence (8) is not the negation of (7).[†]

Any collection of propositions containing a contradictory pair is inconsistent. It is impossible for both of two contradictory propositions to be true, so it is impossible for all the propositions in a collection containing a contradictory pair to be true. The converse does not hold: there are inconsistent collections containing no contradictory pair. For example:

9)   John is over six feet tall. John is under six feet tall

is an inconsistent collection, for it cannot be that both propositions are true. Since both could be false (and would be, if John were exactly six feet tall), they are not contradictories.[‡]

Figure 1.1 summarizes the relationships mentioned. All pairs of propositions of which one is a negation of the other are contradictories, and all contradictories are inconsistent. However, there are inconsistencies which are not contradictories, and contradictories of which neither is a negation of the other.

Part of the link between validity and inconsistency, mediated by the notion of contradictoriness, consists in the following fact:

10)   If an argument is valid, a collection of propositions consisting of its premises together with a contradictory of its conclusion is inconsistent.

To illustrate, consider the following argument:

† See Exercise 8, page 340.   ‡ See Exercise 9, page 340.

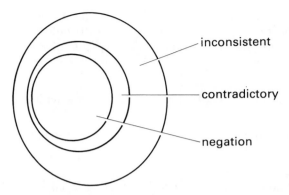

Figure 1.1

11)   Anyone who drives a car risks death. Anyone who risks death is brave. So all car drivers are brave.†

The following collection contains the premises of (**11**) as (i) and (ii) and the negation of its conclusion as (iii):

12)   (i) Anyone who drives a car risks death.
          (ii) Anyone who risks death is brave.
          (iii) Not all car drivers are brave.

Take any two of these propositions. We can see that, if these two are true, then the third cannot be. So the collection is inconsistent.

We can argue quite generally for (**10**), using the definition of validity given in §**3**. Take any valid argument. By definition, it is logically impossible for its premises to be true yet its conclusion false. In other words, it is logically necessary that if all the premises are true, so is the conclusion. But if the conclusion is true, then, necessarily, any contradictory of it is false. So, necessarily, if the premises are true, any contradictory of the conclusion is false. So it is logically impossible for the premises and a contradictory of the conclusion all to be true. So this collection is inconsistent.

The link between validity and inconsistency also runs in the other direction:

13)   If a collection of propositions is inconsistent, any argument whose premises consist of all but one of the collection, and

† See Exercise 10, page 340.

whose conclusion is a contradictory of the remaining proposition, is valid.

The argument for this is rather like the one just given for **(10)**.[†] Taking **(10)** and **(13)** together shows that we could have defined validity in terms of inconsistency, rather than in terms of logical impossibility.[‡]

5   Arguments and argument-claims

We use "argument" to refer to any collection of propositions, one of which is singled out as the conclusion. It is useful to have a standard pattern for writing out arguments. We adopt the convention that where there is no contrary indication, the conclusion of an argument is the last proposition in a list, and is marked off from its predecessors by being preceded by a semicolon. Thus if an argument has two premises, A and B, and a conclusion, C, we shall write it:

$A, B; C.$

More generally, where the argument has n premises, and its conclusion is C, we write:

$A_1, \ldots A_n; C$

(n may be equal to or greater than zero. For the case in which $n = 0$ (there are no premises) see §**6**.)
    A useful abbreviation is "⊨", short for "is valid". It works like this:

1) $A_1, \ldots A_n \vDash C$ abbreviates "$A_1, \ldots A_n; C$" is valid.
    $A_1, \ldots A_n \nvDash C$ abbreviates "$A_1, \ldots A_n; C$" is not valid.

The symbol "⊨" is pronounced "(double) turnstile".
    It is important to distinguish between an argument ($A_1, \ldots A_n; C$) and what I shall call an *argument-claim*: $A_1, \ldots A_n \vDash C$, or $A_1, \ldots A_n \nvDash C$. The component propositions in an argument are true or false, but the argument itself cannot significantly be said to be true or false. One correct dimension of assessment for an argument is whether it is valid or not;

† See Exercise 11, page 340.   ‡ See Exercise 12, page 340.

another is whether it is persuasive or not; but truth and falsehood do not provide a proper dimension of assessment. By contrast, an argument-claim *is* true or false: true if it is a positive argument-claim (⊨) and the argument in question is valid, or if it is a negative argument-claim (⊭) and the argument in question is not valid; and otherwise false.

In an argument-claim, "⊨" appears in the very place in which, in an expression of the argument in ordinary English, one would find a word like "so", "therefore" or "hence": a word used to show that one has come to the conclusion which is being drawn from the previous propositions. This gives rise to a tendency to confuse the role of "⊨" with that of conjunctions like "therefore". But the roles are really very different.

First, "⊨" and "therefore" belong to different grammatical categories. "⊨" is a predicate, the sort of expression which can be used to attribute a property to something. "Therefore" is not a predicate, but rather a word used to join sentences together. To see the force of this point, consider the fact that we can meaningfully say "Some arguments are valid but others are not" (bearing in mind that "⊨" abbreviates "is valid"), though we cannot meaningfully say "Some arguments are therefore, but some are not".

Secondly, something of the form "$A_1, \ldots A_n$, therefore C" is an argument, about which the question arises whether or not it is valid. By contrast, something like "$A_1, \ldots A_n \vDash C$" is not itself an argument, but rather a claim *about* an argument, the claim that it is valid.

Thirdly, in ordinary circumstances, one who propounds an argument, $A_1 \ldots A_n$, therefore C, is thereby committing himself to the truth of all of $A_1, \ldots A_n$. But one who makes the claim that $A_1, \ldots A_n \vDash C$ makes no such commitment, since there are valid arguments whose premises are not true.[4][†]

6 Some important properties of validity

Although our definition of validity in §3 is not as illuminating as one might wish, it none the less enables us to discover some important general features of validity.

The key property of validity is that it logically guarantees the preservation of truth.[‡] If you start with truth and argue validly then you are bound to end up with truth. That is why it is a good thing to argue

† See Exercise 13, page 340.   ‡ See Exercise 14, page 341.

validly. But validity does not always generate truth (see (**1**)), nor does truth always generate validity (see (**3**)).

1) There are valid arguments with false conclusions.

Example:

2) All heavenly bodies revolve around the earth. The sun is a heavenly body. Therefore the sun revolves around the earth.

Moreover:

3) There are invalid arguments with true premises and true conclusions.

Example:

4) Petroleum can be used as a fuel. More people live in Paris than in Boston. Therefore, the first man on the moon was an American.

We have already seen that deductive validity is monotonic. Using the notation of §5, this can be expressed:

5) If $[A_1, \ldots A_n \vDash C]$ then

$[A_1, \ldots A_n, B \vDash C$, whatever B may be].[5]

In other words, you cannot turn a valid argument into an invalid one by adding to the premises. This elaborates what is meant by saying that deductive logic aims to pick out arguments in which the premises give *conclusive* reasons for the conclusion.

Another important property of validity, as classically conceived, and as defined in §3, is a kind of *transitivity*. Chaining arguments together will preserve validity:

6) If $[A_1, \ldots A_n \vDash C]$ and $[B_1, \ldots B_k, C \vDash D]$, then

$[A_1, \ldots A_n, B_1, \ldots B_k \vDash D]$.

The intermediate conclusion, C, can be cut out, since the premises which establish it can establish anything it can establish.†

Validity has a property akin to *reflexivity*:

7)   If C is among the $A_1, \ldots A_n$, then

$$[A_1, \ldots A_n \vDash C].‡$$

This shows that circular arguments are valid. (Of course, they are not normally *useful*: see (**7.4**) below.)

A new piece of terminology: we shall express the claim that a collection of propositions $(A_1, \ldots A_n)$ is inconsistent by writing:

$$(A_1, \ldots A_n) \vDash.$$

The terminology is justified by the fact that if an argument's premises are inconsistent, it is valid; and this is suggested by the blank after the turnstile, accepting any completion. More formally:

8)   If $[(A_1, \ldots A_n) \vDash]$, then

$$[A_1, \ldots A_n \vDash B, \text{ whatever } B \text{ may be}].¶$$

Like all the other properties of validity described in this section, this one follows from the definition given in §3. If premises are inconsistent, they cannot all be true. If premises cannot all be true, then, in particular, the following case cannot arise: that all the premises are true, and also some arbitrary proposition, B, is false. So it is impossible for all the premises to be true yet B false. So an argument with inconsistent premises is valid, whatever its conclusion.

(**8**) should not be read as saying that you can infer anything from an inconsistency. This makes it sound as if one might be in a position to make such an inference. But one can rationally make an inference only from what one takes to be true, and one cannot rationally take an inconsistency to be true.

A further piece of terminology. Let us write

$$\vDash A$$

† See Exercise 15, page 341.   ‡ See Exercise 16, page 341.
¶ See Exercise 17, page 341.

to abbreviate: "it is logically impossible for *A* to be false". The terminology is justified by the fact that if an argument's conclusion cannot be false, then it is valid; and this is suggested by the blank before the turnstile, accepting any completion. This claim can be expressed as follows:

9)   If [⊨ *A*], then [*B*₁, ... *B*ₙ ⊨ *A*], whatever *B*₁, ... *B*ₙ may be.

This shows how we can extend the notion of an argument to include the case in which there are zero premises. This does not reflect anything in ordinary usage, but it is convenient for logic.†

The properties of validity mentioned in this section are properties of the traditional notion. In various ways, some of which we shall discuss in §7, the traditional notion may seem to fall short of what we want. This has prompted the development of various "non-classical" conceptions of validity. Our concern is confined to the classical notion.‡

7   Validity and usefulness: "sound", "relevant", "persuasive"

It is important to realize that even if an argument is valid, it may not be *useful*: it may not be a good one to use, either to discover what is true or to persuade an audience of something. For example, consider:

1)   Some circles are square. Therefore there will be no third world war.

Since it is logically impossible for any circles to be square, (1) is valid (its validity follows from (**6.8**)). But the argument would not be a good one to use for any purpose, and certainly not to convince someone that there will be no third world war. Normally, a good argument is not merely valid. In addition, it has true premises. An argument which has true premises and is valid is called *sound*.

The last remark is qualified by "normally" since there is at least one circumstance in which it is useful to propound a valid argument with a false premise. This is when one hopes that one's hearer will recognize that the conclusion is false and that the argument is valid, and so will be

† See Exercise 18, page 342.   ‡ See Exercise 19, page 342.

persuaded that at least one premise is false. This mode of argument is called *reductio ad absurdum*.

Suppose your hearer believes that Harry is a merchant seaman, but you disagree. Suppose also that you both know that Harry's arms are not tattooed. Then you might say:

> 2) Suppose Harry is a merchant seaman. All merchant seamen have tattoos on their arms. So Harry must have tattoos on his arms.

One intends one's hearer to recognize the validity of the argument, and, persisting in his belief that the conclusion is false, to come to infer that at least one premise is false. One has to hope that he will be more firmly persuaded of the truth of "All merchant seamen have tattoos on their arms" than of "Harry is a merchant seaman", so that he will retain the former belief and abandon the latter.

A sound argument may fail to persuade an audience if the audience does not realize that the premises are true, or does not realize that the argument is valid. Here the fault lies with the audience, not with the argument. But a sound argument can still be defective, in that it may not be useful. Consider:

> 3) Washington is the capital of the USA. Therefore all dogs are dogs.

Since it is logically impossible for "All dogs are dogs" to be false, (3) is valid (its validity follows from (**6.9**)). Since the premise is true, it is also sound. But the argument is not useful. Part of the reason is that no argument is needed in order to persuade someone of something so trivial as the conclusion of (3). Another part of the reason is that the premise has no proper relevance to the conclusion. For an argument to be useful, it must, normally, be sound, and must, always, be relevant. Logicians have tried to devise special logics to reflect the concept of relevance. But this is one more topic we shall not pursue.[6]

Consider:

> 4) The whale will become extinct unless active measures are taken to protect it. Therefore the whale will become extinct unless active measures are taken to preserve it.

This is valid (its validity follows from (**6.7**)). It is sound, since its premise is true (if you disagree, select your own example). It is intuitively relevant, for whatever precise account we give of this notion it appears that nothing could be more relevant to whether a proposition is true than whether *that very proposition* is true. But the argument is plainly useless. It could not persuade anyone of anything, and it could not help in the discovery of truth.

For an argument to be *persuasive* for a person he must be willing to accept each of the premises but, before the argument is propounded to him, be unwilling to accept the conclusion.† When the premise *is* the conclusion, he cannot be in this state. This is the general reason for the uselessness of circular arguments.

How could a valid argument ever be persuasive? It is possible because we do not always acknowledge or take explicit note of all the logical consequences of our beliefs. If we did explicitly hold before our minds all the logical consequences of our beliefs, seeing them *as* consequences, we would already have accepted the conclusion of any valid argument whose premises we have accepted. Hence no valid argument could be persuasive. This is how things would be with a perfectly rational being. The utility of valid arguments is a monument to our frailty: to the fact that we are not completely rational beings.

To sum up this section: validity is not the only desirable property in an argument. But it is the only one which normally concerns logicians.

8   Sentences and propositions

An argument consists of propositions. A proposition is what is believed, asserted, denied, and so on. This section elaborates this idea.

We can start with the relatively straightforward idea of a meaningful sentence. A sentence is a series of words, arranged in accordance with the grammatical rules of the language in question. So "The cat sat on the mat" is a sentence, but "cat sat mat on the the", though composed of just the same words, is not a sentence. Properly speaking, we should say: not an *English* sentence, for a sentence is defined relative to a language. The same series of words could be a sentence in one language but not in another. Thus:

1)   Plus Robert court, plus Juliette change

† See Exercise 20, page 342.

is not an English sentence, despite being composed only of English words. But it is (or so I am told) a French sentence.

It is a disputed question whether every sentence – that is, every *grammatical* sentence – is meaningful. A standard example is:

2)   Green ideas sleep furiously together.

The suggestion is that this is a genuine sentence – it is grammatically correct – but has no meaning. We will not enter this dispute. But we do need to help ourselves to the notion of a meaningful sentence, whether or not this coincides with the notion of a grammatically correct one.

A preliminary definition of a *proposition* might run as follows:

3)   A proposition is what is expressed, in a given context, by a meaningful, declarative, indicative sentence.

Various aspects of this definition require comment. A *declarative* sentence is one that could be used to make an assertion, to affirm that something is or is not the case. Thus:

4)   The King is in his counting house

is a declarative sentence. By contrast

5)   Is the King in his counting house?

is not a declarative sentence, but rather an interrogative one. It cannot be used to *affirm* how things are, but only to *ask* how things are.

6)   Put the King in his counting house

is not a declarative sentence, but rather an imperative one. It cannot be used to affirm that things are so-and-so, but only, rather, to order that they be so-and-so.

An *indicative* sentence, one in what grammarians call the indicative *mood*, contrasts with a *subjunctive* sentence. Corresponding to the indicative (4) is the subjunctive:

7)   Were the King in his counting house.

Subjunctive sentences are not used by themselves to affirm anything, but they may occur in sentences usable to affirm things. One common use is in subjunctive conditionals, for example:

   8)   Were the King in his counting house, the Queen would be content.

Two distinct sentences can express the same proposition. The English sentence

   9)   Snow is white

expresses the same proposition as the French sentence

   10)   La neige est blanche.

Even within a single language, distinct sentences can express the same proposition if they have the same meaning.

   11)   Blackie is a puppy

expresses the same proposition as

   12)   Blackie is a young dog.

Two sentences with different meanings can express the same proposition if they are used in different contexts, which is why the definition (3) mentions contexts. Suppose you are my only audience, and I address the following remark to you:

   13)   You are hungry.

Suppose that you then utter the sentence:

   14)   I am hungry.

We both express the same proposition. The first sentence, in the context of being *directed at* you, expresses the same proposition as the second, in the context of being *uttered by* you. Had you uttered the same sentence as me, (13), you would not have expressed the same proposition. This shows that the same sentence can, with respect to different contexts, express different propositions.†

† See Exercise 21, page 342.

There is another way in which this can occur. A sentence may be ambiguous. For example, "There's that crane again" may refer to a lifting-device or a bird. There is no such thing in general as *the* proposition such a sentence expresses, though relative to a context it may express just one rather than the other of the propositions.

I have simply stipulated certain features of propositions, and their relations to sentences. What makes this an appropriate definition to adopt in logic? The simplest way to answer the question is this: validity is defined in terms of truth conditions, and so one should identify a proposition by truth conditions. This answer relies on ideas that will be introduced in §9. However, we can see at once something of the motivation for the notion of a proposition, as used by logicians.

Consider the argument:

15)   I am hungry; therefore I am hungry.

Intuitively this should count as valid. But suppose we thought of the components of arguments as sentences, and suppose we imagine the context shifting between the utterance of the premise and the utterance of the conclusion. Suppose you are hungry and utter the premise, and I am not hungry and utter the conclusion. Then we would have a true premise and a false conclusion, so the argument could not be valid. Clearly we need to avoid such problems, and introducing the notion of a proposition, in the style of this section, is one way of doing so.

We still *could* have defined an argument as a collection of sentences, but we would have had to say something about the context being held constant over all the sentences of an argument. The upshot would have been the same. On some occasions, it is easiest to think of arguments as composed of propositions, on others it is easiest to think of them as composed of sentences, with a background assumption of constancy of context. We will help ourselves to both notions, as convenient.

9   Validity and truth conditions

A sentence like "Snow is white" is true under some but not all logically possible circumstances. There are logically possible circumstances, including those which actually obtain, under which the sentence is true, and logically possible circumstances (in which, say, snow is black) under which the sentence is false. A circumstance is one *under which* a sentence is true just on condition that if the circumstance actually obtained, then the sentence would be true.

Some sentences, for example, "Snow is snow", are true under all logically possible circumstances. Some sentences, for example "7 is less than 5", are true under no logically possible circumstances.

We shall say that a sentence's *truth conditions* are the circumstances under which it is true. We can think of these circumstances as bundled together in a collection or set – a set with no members, in the case of sentences like "7 is less than 5". Using this notion, we can give yet another definition of validity:

1) $[A_1, \ldots A_n \vDash C]$ if and only if the truth conditions of C include those of $(A_1, \ldots A_n)$.

We could put it another way: every circumstance under which all of $A_1, \ldots A_n$ are true is one under which C is true. Equivalently: the truth conditions of $A_1, \ldots A_n$ are included in those of C. It is not difficult to work out that (1) and these variants define the same notion as that defined in §**3**.†

The importance of this definition is that it shows that only the truth conditions of a sentence matter to the validity or otherwise of any argument in whose expression it occurs. A consequence is that if a sentence occurs in the expression of an argument and you replace it by one having the same truth conditions, the argument will remain valid, if it was valid before, or invalid, if it was invalid before. This will bear importantly on questions of formalization, to be considered later.

10   Formal validity

Logic, or at any rate formal logic, is not primarily concerned merely with the very general notion of validity which we have so far discussed. It is concerned with a particular species: *formal validity*. Formal validity, being a kind of validity, has all the properties of validity; but it has some additional distinctive features.

We could try to define formal validity by saying that an argument is formally valid if and only if it is *valid in virtue of its form or pattern*. This

† See Exercise 22, page 343.

captures part of what is intended, though it will need supplementation.
Examples will help. Consider:

> 1) Frank will marry Mary only if she loves him. But Mary does
> not love Frank. So he will not marry her.

> 2) The whale will be saved from extinction only if active measures
> are taken. But active measures will not be taken. So the whale
> will not be saved from extinction.

These have a common form or pattern, which we could distil out as
follows:

> 3) ... only if —— , it is not the case that ——, so it is not the case
> that ...

Here the dots are meant to be filled on both occurrences by the same
sentence, and likewise the dash. It is more convenient to use letters,
rather than dots and dashes, thus:

> 4) A only if B. It is not the case that B. So it is not the case that A.

This can be called an *argument-form*. It is an argument-form of each of (1)
and (2), since these both result from it by making suitable replacements
for A and B.[†]
   The logician wants to say that (1) and (2) are valid in virtue of their
pattern or form, the same in each case. This represents among other
things an attempt to attain generality. It would be hopeless to try going
through each argument in turn, picking out the valid ones. But if we are
granted the idea of an argument-form we can say: not only is this specific
argument valid; so are all of the same form.
   One way to elaborate this a little is to define a notion of validity for
argument-forms:

> 5) An argument-form is valid if and only if, necessarily,[‡] each of
> its instances is valid.

So (4) is an example of a valid argument-form.[7] Now we can say that an
argument is valid *in virtue of* its form just on condition that it is an
instance of a valid argument-form.

† See Exercise 23, page 343.   ‡ See Exercise 24, page 343.

This goes some way towards saying what formal validity is, and we can reinforce the idea with two more examples before showing that an important ingredient is missing.

Consider:

> **6)** All camels are herbivores. All herbivores are pacific. Therefore all camels are pacific.

This is an instance of the argument-form: "All ... are ——, all —— are ***, therefore all ... are ***." "Camel" and "herbivore" are traditionally classified as nouns, "pacific" as an adjective. We shall call both of these, as they occur in (**6**), *predicates*. We shall also include among predicates relational expressions like "loves" and "is bigger by ... than". To mark the sort of gap that a predicate can fill, we shall use capital letters starting with "*F*", so we shall write out the argument-form:

> **7)** All *F* are *G*. All *G* are *H*. Therefore, all *F* are *H*.

Assuming that this is a valid argument-form, it follows that (**6**) is not merely valid, but also formally valid.[†]

> **8)** Ian is Scottish. All Scots are prudent. So Ian is prudent.

This is an instance of the argument-form:

> **9)** α is *F*. All *F* are *G*. So α is *G*.

Here we use "α" (and if needed, "β", "γ", ...) to mark the position which a name can occupy. Assuming that (**9**) is valid, (**8**) is formally valid.[‡]

This discussion has already begun to include some controversial elements. Highlighting these must wait for later chapters. For the moment, I want to bring out what is missing from the account so far: a gap which makes it inadequate as a presentation of the traditional idea of formal validity.

The idea was that formal validity should be a special kind of validity. But as presented so far, nothing has been said to rule out formal validity coinciding with validity. For nothing has been said which prevents an argument itself counting as an argument-form. This is no accident of the particular way in which I have presented the idea. The general problem is

---

† See Exercise 25, page 343. ‡ See Exercise 26, page 343.

this: what is the difference between pattern or form, and what fills it: substance or content? In (**4**), for example, the remaining English words correspond to the pattern or form, the letters *A* and *B* to the places where one could insert content or substance to yield a genuine argument. But what is the basis for this distinction?

To bring out the problem, consider the sort of example that would standardly be given in a logic text of an argument which is valid but is not formally valid:

10)   Tom is a bachelor. Therefore, Tom is unmarried.

This is certainly valid (reading "bachelor" in a familiar way). A case for saying that it is not formally valid might start by pointing out that (**10**) is an instance of the invalid argument-form:

11) α is *F*. Therefore α is *G*.†

How could the case be pressed further? What is needed for formal validity is that there be *some* valid argument-form of which the argument is an instance. So to establish failure of formal validity, it is not enough to cite one invalid argument-form of which the argument is an instance.[8] You have to show that it is not an instance of *any* valid argument-form. But who is to say that (**10**) itself is not an argument-form? If it is, then, since (**10**) itself is its only instance, it is an instance of a valid argument-form, and so formally valid, contrary to the intention.

We might try to block this difficulty by stipulating that every argument-form must have some gaps (marked by dots and dashes, or letters). So every argument-form will have more than one instance. But (**10**) would still come out as formally valid, in virtue of being an instance of the valid

12) α is a bachelor. Therefore, α is unmarried.

If the concept of formal validity is to be narrower than that of validity, as the logician intends, the concept of form will have to be made more restrictive. The logician will stipulate that in an argument-form the only expressions we may use, other than the dots or dashes or letters which mark the gaps for the "content", are the *logical constants*. "Bachelor" and "unmarried", which occur in (**12**), are not logical constants, so (**12**) is not an argument-form, and so does not establish the formal validity of (**10**).

† See Exercise 27, page 344.

11   The logical constants

I shall begin by giving a list of expressions that are generally held to be logical constants:

1)   it is not the case that
      and
      or
      if ... then ...
      if and only if
      some
      a
      everything
      all
      is
      are
      is the same as
      [plus any expression definable just in terms of the above].

It is a matter for philosophical debate whether the list should be extended to include, for example, "necessarily" or "is a member of" (as used in set theory). The debates arise from the fact that there is no general agreement about what *makes* an expression a logical constant. A list like the above fails to speak to this issue.[9]

A widely held view, which certainly captures part of the truth, is that an essential feature of a logical constant is that it introduces no special subject matter. It should be "topic-neutral". This is because, in logic, we are concerned with reasoning in general, and not with this or that special area of knowledge. It is all very well for an anthropologist concerned with kinship to take a particular interest in what is signified by such words as "bachelor" and "married". The logician aims at greater generality. He will concern himself only with expressions which can occur in an argument on any subject whatsoever. The expressions in the list, but not expressions like "bachelor", satisfy this requirement.

Bearing in mind the suggestion that argument-forms should contain, apart from devices for marking gaps, only logical constants, we can verify that (**10.4**), (**10.7**), (**10.9**) are argument-forms, but (**10.12**) is not one. This is consistent with the formal validity of (**10.1**), (**10.2**), (**10.6**) and (**10.8**), but is hostile to the formal validity of (**10.10**). If our list of logical constants can be assumed to be complete, or at least to exclude expressions like "bachelor" and "unmarried", then it seems that the only argument-forms for (**10.10**) would be either the invalid (**10.11**) or the equally invalid:

  2) $A; B.$

If so, (**10.10**) is not formally valid.

In chapter **6.5** we will ask whether there is any illuminating and general account of *logical constant*. If there is, then we can use it to give an account of formal validity: it will amount to validity in virtue of the meaning of the logical constants, and in abstraction from other than structural features of premises and conclusion. By *structural features* I mean facts about the recurrence of certain non-logical elements, for example the fact that "camels" occurs in the two places it does in (**10.6**). For the moment, we will make do with a relativized notion: given some list of constants, we will say that an argument is formally valid if and only if it is valid in virtue of the meanings of the constants on the list, and in abstraction from other than structural features of premises and conclusion.

So much for preliminary notions. Much remains to be done: an illuminating account of what makes an expression deserve to be called a logical constant, an illuminating account of logical possibility, a deeper investigation of the notion of form. Some of these tasks will be taken up later. The next project arises from the following fact: the investigation of formal validity in practice proceeds by turning away from ordinary English, and studying instead some artificial, "logical", language, like the language of the *propositional calculus* or the language of the *predicate calculus*. Why? We can approach the question from two sides. Why turn away from English? And why consider precisely these artificial languages? The next section covers some aspects of the first approach.

12   The project of formalization

If logicians really aim to study validity, as it occurs in our everyday thought and talk, why do they study artificial languages, which no one speaks? Why not stick to English, or French, or some other natural language?

One way to begin an answer is this: an argument is a collection of propositions. But the *sentences* of natural languages like English do not adequately reflect the logical properties of the *propositions* they express.

Formal logic *is* concerned with the very arguments we use in daily life, but it has to express these arguments in a different way.

This introduces the crucial idea of the *logical form* of a sentence. A sentence's logical form lays bare the logical features of the proposition which it expresses.

A traditional hope is that logic should provide a *mechanical* means of testing for validity.[10] But how could you present a machine with arguments? If arguments are composed of propositions, then you cannot present a machine with arguments in any direct way, for propositions are too abstract. What you would have to feed into the machine are sentences. If the machine is to test the validity of the argument the sentences express, every logically relevant feature of the propositions must be correlated with some property of the physical make-up of the sentences.

It has been held that such a correlation does not obtain, or at least does not obtain in any readily statable fashion, between sentences of natural languages and the propositions they express. Hence the need for artificial languages. The idea is that these will supply the logical forms of sentences in natural languages. By translating a natural sentence into an artificial one, the hidden logical features of the proposition expressed will be brought to the surface.

Let us now consider some ways in which natural sentences may be inadequate for logical purposes: inadequate as vehicles for bringing out the logical features of arguments.

(1) Lexical ambiguity

As a special case of (**6.7**) it holds quite generally that:

$$C \vDash C.$$

That is, this holds whatever *proposition* C may be. But, as we saw in connection with (**8.15**), it does not hold for arbitrary *sentences* of natural languages. Here we note the following reason for its failure: many sentences are ambiguous, they have more than one meaning, express more than one proposition. When this is due to the sentence containing a word with more than one meaning, we shall call the ambiguity *lexical*.

    1)   John cut the painter. Therefore John cut the painter

is not valid. If we interpret "painter" in the first sentence to mean an artist, and interpret this word in the second sentence to mean a rope used to secure a boat, the first sentence may be true while the second is false. The argument that results from this interpretation is not valid.

An obvious way to deal with this problem is to distinguish two words, say "painter$_1$" and "painter$_2$", one for each of the meanings, and throw away the ambiguous word. Then the proposed interpretation of (1) would look like this:

2)   John cut the painter$_1$. Therefore John cut the painter$_2$

and no one would be particularly tempted to think that this expressed an argument of the form: *C*; *C*. (No doubt one would also have to distinguish "cut$_1$" and "cut$_2$".) This strategy, of eliminating ambiguous words, already involves departing from natural languages, in which ambiguous words are rife. But the proponent of artificial languages envisages altogether more radical departures.

(2) *Structural ambiguity*

Some sentences are ambiguous, yet the ambiguity cannot be attributed to the ambiguity of one or more words in the sentence: the ambiguity is not lexical but *structural*. Here are some examples of structurally ambiguous sentences, with alternative interpretations added in brackets.

3)   Harry is a dirty window cleaner. [(a) Harry is a dirty cleaner of windows; (b) Harry is a cleaner of dirty windows.]

4)   Tom and Mary are visiting friends. [(a) Tom and Mary are visiting some people, and they are friends with these people; (b) Tom and Mary are friends with one another, and they are visiting some people; (c) Tom and Mary are visiting some people, and these people are friends with one another.]

5)   Receipts from this source are not liable to income tax under section iv, paragraph 19. [(a) Section iv, paragraph 19, exempts receipts from this source from liability to income tax; (b) Section iv, paragraph 19, does not impose a liability to income tax on receipts from this source.]

6)   I thought you were someone else. [(a) I thought you were someone other than the person you in fact are; (b) I thought you were someone who is not identical to himself.]

7) *First speaker*: "I ought to send flowers."
   Second speaker: "No you ought not."
   [(a) You are not under an obligation to send flowers; (b) you are under an obligation not to send flowers.]

8) Nicholas has written a book about everything. [(a) Nicholas has written a book, and it treats every subject; (b) for every subject, Nicholas has devoted at least one whole book exclusively to it.]

In none of these cases can the ambiguity be attributed to any word. One way to test for this is to verify that each of the words can occur in a variety of sentences lacking the corresponding ambiguities, and this is not to be expected if the words are ambiguous.[†]

The existence of structural ambiguity shows that the elimination of lexical ambiguity is not enough. Some more radical approach is required.

(7) and (8) bring to light a particularly worrying problem. Structural ambiguity seems to affect even logical constants, here "not" – (7) – and "a" and "everything" – (8). On one reading of (7) we think of "not" as dominating the sentence, to form its negation, a reading which we might write as "Not: you ought to send flowers". On the other reading, we think of "not" as governing just the description of the action, a reading which we might write as "You ought to do this: not send flowers". On one reading of (8) we think of "a" as dominating the sentence, a reading which we might write as: "A book by Nicholas is like this: it is about everything". In the other reading, we think of "everything" as dominating the sentence, a reading which we might write as: "Everything has this property: Nicholas has written a book about it". The logical constants determine formal validity (see §10 and §11). If structural ambiguity can affect the logical constants, then the hope of giving a general characterization of formal validity for English as it stands is undermined. Consider:

9) Logic, epistemology and metaphysics are all the philosophical subjects there are. Nicholas has written a book about logic. Nicholas has written a book about epistemology. Nicholas has written a book about metaphysics. Therefore, Nicholas has written a book about every philosophical subject.

If (9) is valid, we are intuitively inclined to think that it is formally valid. But there is no straightforward answer to the question whether it is valid.

† See Exercise 28, page 344.

It all depends on how we understand the conclusion, which is structurally ambiguous in the same fashion as (**8**).

The problem for the account of formal validity is as follows. We said that a formally valid argument is valid in virtue of its form, and that this in turn is a matter of it being an instance of a form all of whose instances are valid. However, (**9**), read as invalid, is an instance of every argument-form of which (**9**), read as valid, is an instance. Hence (**9**) is not an instance of a valid argument-form. The problem of structural ambiguity threatens to deprive even the apparently formally valid reading of (**9**) of its formal validity.[†]

It is theoretically possible that structural validity could be filtered out of natural languages. After all, in (**3**)–(**8**) unambiguous paraphrases in English were given. But it is unclear whether precise rules can be given which would effect this filtering. One can therefore see why logicians should prefer artificial languages, so constructed that structural ambiguity is impossible.

(3) Syntactic irregularity

The syntax or grammar of a language is a set of rules which determine how sentences are constructed from the language's vocabulary. A syntactic distinction is one which we have to make in order to devise such rules.

As we have seen, there are two possible answers to the question: what ought to be picked out as sentences? One answer is: just the series of words which constitute *grammatical* sentences, where it is supposed that we have some antecedent grasp on what it is for a series of words to be a grammatical sentence. The other answer is: just the series of words which constitute *meaningful* sentences. It might be that the class of grammatical sentences is wider than the class of meaningful sentences.

Without prejudice to this debate, I shall in this section mean by syntactic rules ones which determine the class of meaningful sentences.

We have already been obliged to take note of various syntactic distinctions in English, for example, between sentences, predicates and names. We used $A_1, \ldots A_n, B, C$, as letters marking the sort of position that can be occupied by a sentence; F, G, H as letters marking the sort of position that can be occupied by a predicate; and α as a letter marking the sort of position that can be occupied by a name. We have attempted no definition of these categories. Rather, we have simply picked out examples, and gestured towards the category as a whole.

† See Exercise 29, page 344.

The gesture determines the category from the example in the following way. Anything belongs to the category if it can replace the example (at least in the context under consideration) without turning sense into nonsense. Thus, given that "Tom" belongs to the category of names, we can infer that "Harry" does too, since it can replace "Tom" without turning sense into nonsense. But neither "herbivores" nor "2 + 2 = 4" are names, by this test, since replacing "Tom" in "Tom is a bachelor" yields the nonsensical "Herbivores is a bachelor" and "2 + 2 = 4 is a bachelor". I call this way of determining syntactic categories the *naive syntactic test*. The test is woefully inadequate in the study of validity. It places expressions with similar logical powers in different categories; and it places expressions with dissimilar logical powers in the same categories.

The former point is the less important. The expressions "Mount Everest" and "Ronald Reagan" have logically similar powers. Each serves to pick out an object. Yet it is at least arguable that replacing the latter by the former in "Ronald Reagan is thinking of Vienna" turns sense into nonsense. If so, these names fall into different categories, according to the naive syntactic test. The uncertainty reveals the vagueness of the distinction between sense and nonsense.

I shall illustrate the second point by some examples. First, by the naive syntactic test, the category of names would contain not only expressions like "Tom" and "Harry" but also what logicians call *quantifier phrases*, like "everyone", "no one", "someone". For example, all of "Everyone is a bachelor", "No one is a bachelor" and "Someone is a bachelor" make perfectly good sense. But the logical powers of, say, "Harry" and "no one" are very different, as is brought out by the fact that (**10**) is valid but (**11**) is not:

10)   Harry is a bachelor. So someone is a bachelor.

11)   No one is a bachelor. So someone is a bachelor.

The contrast also emerges in the following passage from Lewis Carroll [1872]:

12)   "Who did you pass on the road?" the King went on, holding out his hand to the Messenger for some more hay.
"Nobody", said the Messenger.
"Quite right", said the King: "this young lady saw him too. So of course Nobody walks slower than you."

"I do my best", the Messenger said in a sullen tone. "I'm sure nobody walks much faster than I do!"

"He can't do that", said the King, "or else he'd have been here first." (pp. 143–4)

Despite its vagueness, the naive syntactic test at least doesn't definitely rule out counting both the expressions "is sensitive to pain" and "is evenly distributed over the earth's surface" as predicates. Since "Harry is sensitive to pain" is clearly sense, this means allowing that "Harry is evenly distributed over the earth's surface" is sense too. To convince ourselves that it is, we could imagine Harry being chopped into small pieces, which were then dropped at regular intervals from an aeroplane. If both expressions belong to the same category, however, we run into a problem.

13)   Human beings are sensitive to pain. Harry is a human being. So Harry is sensitive to pain

is, intuitively, formally valid. Does not its validity turn only on the logical constants it contains? It is an instance of the argument-form

14) F are G. α is an F.   So α is G.

It is tempting to believe that (14) is valid, and explains the formal validity of (13). But the temptation must be resisted, as the invalidity of (at least one reading of)[†] the following shows:

15)   Human beings are evenly distributed over the earth's surface. Harry is a human being. So Harry is evenly distributed over the earth's surface.

The invalidity of (15) establishes the invalidity of (14).[‡] We need a more refined notion of a predicate, if we are to attain interesting generalizations about valid forms of argument.

Intuitively, the following argument is formally valid:

16)   Every candidate is a clever or industrious person. Every clever or industrious person is worthy of praise. So every candidate is worthy of praise.

† See Exercise 30, page 344.   ‡ See Exercise 31, page 344.

But the argument-form we would reach for to sustain this judgement is invalid:

17)   Every F is a G. Every G is H. So every F is H.

For the following instance of (17) is invalid:

18)   Every number is a number or its successor. Every number or its successor is even. So every number is even.[11]

Expressions which look similar, at least to the naive eye, can contribute in very different ways to the meanings of sentences in which they occur. This is the phenomenon I refer to by the phrase *syntactic irregularity*. A closely allied phenomenon is that in natural languages it seems to be impossible, or at least difficult, to characterize properties which are of logical importance in the way which would make mechanical testing possible: that is, on the basis of the physical make-up of sentences.

This can be illustrated by the relation of *negation*, which is clearly important to logic, as its connection with inconsistency and validity has already made plain. The obvious initial idea is to say that if one sentence results from another by prefixing it with "It is not the case that", the former is the negation of the latter. This rule works well enough for some cases. For example, it says, correctly, that "It is not the case that the earth is flat" is the negation of "The earth is flat". But it does not work for all. For example, prefixing:

19)   I will marry you, if you change your religion

with "It is not the case that" yields

20)   It is not the case that I will marry you, if you change your religion.

This is at best ambiguous between the negation of (19) and something equivalent to "If you change your religion, I will not marry you".

Evidently we need some kind of bracketing device. We might write: "It is not the case that (I will marry you, if you change your religion)". In spoken English, a similar effect can be achieved by inflection, for example one which includes a small pause after "that".

The introduction of such special devices is typical of the formal logician's approach. One point of the devices is that they facilitate the characterization of relations which are of logical importance (like negation) purely in terms of the physical make-up of sentences.

The question remains open whether such a result could be achieved merely by tinkering with a natural language, or whether it requires starting from scratch. The idea of starting from scratch, constructing an artifical language constrained only by the demands of logic, has inspired a philosophical tradition. Russell, for example, coined the expression "philosophical logic" to represent his view that the workings of natural language, and of our thought, could be adequately represented only by an artificial language, the language of his *Principia Mathematica*.

With this approach comes a problem. How are whatever results are obtained for the artificial language to be applied to natural language and to our everyday thoughts? A project opens up, which I call the project of formalization. The idea is to pair each natural sentence with an artificial one. The latter is, or reveals, the logical form of the former. Thanks to the pairing, the results about validity which we have been able to obtain, with relative ease, for the artificial language can be transferred to the natural one. To put it in another idiom: the results about validity which we have obtained by expressing arguments in an artificial language become relevant only if these arguments are, or are specially related to, those we use in our everyday thought and talk. One demonstrates the relevance by showing how to pair natural language sentences with artificial language sentences in such a way that the propositions expressed by the former are the very same as, or specifically related to, the arguments expressed by the latter.

Within this tradition, the first question to ask about an argument expressed in a natural language is: what is its logical form? And the answer is to be given by translating the argument into some artificial language: by, as it is called, *formalizing* the argument. In the next four chapters, we examine in detail how the project of formalization proceeds.

Notes

1 A slightly fuller account is given in the glossary. The glossary glosses some terms which are not explained in the text.
2 The view that inductive "logic" is not a formal discipline has been given impetus by a famous discussion by Goodman [1955], Part II. For a general introduction to problems of induction, see Skyrms [1966], esp. chs **1–3**.

3 For a good discussion, see Plantinga [1974]. By logical possibility I mean what he calls "broadly logical possibility" (p. 3).

4 The contrast between whether a speaker is propounding an argument or not, and thus the contrast between whether he has said something appropriately assessed for validity or not, is not as clear as the text would suggest. This should be apparent from reflection on Exercise 13. Cf. also van Dijk [1977].

5 Square brackets [,], are used for greater legibility.

6 Cf. Anderson and Belnap [1975].

7 One could show this by applying the *propositional calculus*, if one were willing to make a certain assumption about the relationship between that language and English. See ch. 2.

8 Recall Exercise 27.

9 Logical constants are so-called because they receive a constant interpretation in formal logical languages. This answers the question which expressions a formal logician is treating as logical constants; but it does not answer the question whether he is right to treat them in that way. These matters are taken up again in ch. **6.5**.

10 The hope has a long history, going back at least to Leibniz. However, if it were made precise in terms of the technical notion of *decidability*, it is a hope which will be disappointed for any interesting logic. For a discussion of this issue, see e.g. Delong [1970], pp. 132ff., or Kirwan [1978], pp. 169ff.

11 The example is from Geach [1972], pp. 492–3.

2

Truth functionality

This chapter begins (in §§1–2) by introducing an artificial language: the language of propositional truth functional logic, here called **P**. Readers already familiar with this language should merely skim these sections, to check on the terminology and symbolism used here. (Of particular importance is a grasp of the precise notion of *interpretation*, in terms of which an appropriate notion of validity – here called **P**-validity – is defined.) In the later sections, the following question is discussed: what can the validity of arguments expressed in this artificial language tell us about the validity of arguments expressed in English? A crucial prior question will be whether the logical constants of **P** adequately translate the English expressions to which they are taken to correspond.

1   The classical propositional language

The two main features of **P** are these: (i) the only logical constants it recognizes are *sentence connectives*; (ii) all its sentence connectives are *truth functional*.

Sentences of **P** are composed of two kinds of symbol: the *letters* of **P**; and the *sentence connectives* of **P**. The letters of **P** are p, q, r, p' etc. (we here envisage an endless supply), and they are used to formalize sentences which express propositions.

*The **P**-logical constants* are the following sentence connectives:

¬   (corresponding to "it is not the case that"; the symbol is called "tilde")

&   (corresponding to "and"; called "ampersand")

∨   (corresponding to "or"; called "vel")

→   (corresponding to "if ... then ..."; called "arrow")

↔   (corresponding to "if and only if"; called "double arrow").[1]

*The **P**-sentences* are any of the following:

1) a letter, standing alone. (That is, the letters themselves also count as sentences.)

2) (a) any sentence preceded by "¬". We can write this more economically as follows: if X is a sentence, so is $\neg X$.

     (b) if X and Y are sentences, so are:
$(X \& Y)$
$(X \vee Y)$
$(X \to Y)$
$(X \leftrightarrow Y)$.

Examples: (i) "$\neg p$" is a sentence; for "p" is a sentence by (**1**), and "$\neg p$" results from it by preceding it by "¬"; so, by (**2a**), it is a sentence. (ii) "$(\neg p \ \& \ (r \leftrightarrow s))$" is a sentence, since "$(r \leftrightarrow s)$" is a sentence, by (**1**) and (**2b**), and so is "$\neg p$", which establishes, by a further application of (**2b**), that "$(\neg p \ \& \ (r \leftrightarrow s))$" is a sentence.

The above gives what is called the *syntax* of the language **P**: rules which determine what is to count as a sentence of **P**. (We will sometimes omit outer parentheses around **P**-sentences, provided that there is no danger of confusion.) We now turn to the *semantics* of **P**. These are rules which in some sense specify the meanings of sentences of **P**.

(1) Truth values

"Coal is white" is false, but "Snow is white" is true. We shall record this information by saying that "Coal is white" *has the truth value false* and "Snow is white" *has the truth value true*. We are thus thinking of the truth values, truth and falsity, as kinds of objects. True sentences stand in the special relation of *having* to the truth value *true*; false sentences stand in that very relation to *false*. This way of putting things has become standard, having been proposed, for subtle philosophical reasons, by Frege ([1892b], esp. p. 47). If you are reluctant to posit these abstract objects, true and false, you are in good company (see Dummett [1973] pp. 401–27). Everything in this book could be reformulated, so that "has the truth value true (false)" could be replaced by "is true (false)", and likewise for other idioms, given fairly minor reformulations. I have kept to the standard talk of truth values simply because it is standard.

(2) *Interpretations*

3) An interpretation of **P** involves the assignment of exactly one of the truth values, true or false, to each sentence-letter in **P**.

This definition does not allow one to speak of a sentence of **P** as, simply, true or false. Rather, we need to say that it is *true upon an interpretation* or *false upon an interpretation*. A sentence may be true upon one interpretation, false upon another.

As an account of meaning, this is incomplete in two ways. First, it is obvious that two sentences with the same truth value can differ in meaning, so an interpretation of **P** cannot attribute *meanings* to the **P**-sentences.[2] Secondly, (3) addresses only the **P**-letters, whereas there are many (infinitely many) **P**-sentences which are not **P**-letters.

The first incompleteness is not to be resolved here. The claim for **P** is that interpretations specify *enough* of the meaning for purposes of studying the validity of arguments, in so far as their validity turns on the presence of expressions corresponding to the logical constants of **P**. We will return to this claim in §11. The second incompleteness is resolved by introducing *rules of interpretation*. Their effect is to ensure that every interpretation of **P** determines a truth value for *every* **P**-sentence. The rules fix ways in which truth values are transmitted upwards, from the simplest sentences to more complex ones.

(3) *P-interpretation rules*

Henceforth "if and only if" will be abbreviated "iff".

4) For any interpretation of **P**, say i,
$\neg X$ is true upon i iff X is false upon i;
$(X \& Y)$ is true upon i iff X is true upon i and Y is true upon i;
$(X \lor Y)$ is true upon i iff X is true upon i or Y is true upon i;
$(X \rightarrow Y)$ is true upon i iff X is false upon i or Y is true upon i;
$(X \leftrightarrow Y)$ is true upon i iff either both X and Y are true upon i or both X and Y are false upon i.

One standard way to codify this information is by means of *truth tables*. In table 2.1 a "T" ("F") below an expression indicates that it is true (false) upon an interpretation which assigns the values to X and Y which are indicated at the left of the row. Thus, for example, the third row of the

Table 2.1

X	Y	$\neg X$	$X \vee Y$	$X \& Y$	$X \rightarrow Y$	$X \leftrightarrow Y$
T	T	F	T	T	T	T
T	F	F	T	F	F	F
F	T	T	T	F	T	F
F	F	T	F	F	T	T

table for the "→" column says that an interpretation upon which X is false and Y is true is one upon which $X \rightarrow Y$ is true.

Neither the interpretation rules nor the truth tables could be adequately represented by using just **P**-sentences. Suppose the rule for "¬" were written:

> "¬p" is true on an interpretation iff "p" is false on that interpretation.

The displayed condition tells us what the truth value of "p" preceded by a tilde is, relative to an assignment of a truth value to "p", but does not tell us what truth value will be accorded, relative to an assignment of a truth value to "p", to the result of prefixing a tilde to "q", nor what truth value will be accorded, relative to an assignment of a truth value to "p", to the result of prefixing a tilde to a complex sentence, not a sentence-letter, for example, to the result of prefixing "¬p" by a tilde. We thus use "X" and "Y" to stand for arbitrary **P**-sentences.[3] We could have attained the required generality without using "X" and "Y" by, for example, writing the rule for "¬":

> A **P**-sentence consisting of a tilde followed by a **P**-sentence is true upon an interpretation iff the latter sentence is false upon the interpretation.

In the light of the rules, an interpretation will determine a truth value for every **P**-sentence. There is a difference between the way an interpretation relates to **P**-letters and the way it relates to complex **P**-sentences. For **P**-letters, an interpretation is free to choose to assign either truth value. But once an interpretation has made its assignments to the **P**-letters, its freedom is over: the rules of interpretation determine how it must treat all the other **P**-sentences.

(4) Nomenclature

$\neg X$ is called the **P**-negation of X.
$(X \& Y)$ is called the **P**-*conjunction* of the **P**-*conjuncts* X and Y.
$(X \lor Y)$ is called the **P**-*disjunction* of the **P**-*disjuncts* X and Y.
$(X \rightarrow Y)$ is called the (*material*) **P**-*conditional* with **P**-*antecedent* X and **P**-*consequent* Y.
$(X \leftrightarrow Y)$ is called the (*material*) **P**-*biconditional* of X and Y.

(5) P-validity

5)   An argument in **P**, $X_1, \ldots X_n$; Y, is *P-valid* iff every interpreta-
     tion upon which all the premises are true is one upon which the
     conclusion is true.
     Abbreviation:

$$X_1, \ldots X_n \vDash_P Y.$$

The "**P**-" prefix and subscript will be dropped when there is no danger
of ambiguity. The point of it is to facilitate the formulation of various
comparisons, for example with English sentences. Thus we have already
defined *negation* for English. A question to be asked later is: is **P**-negation
essentially the same thing as negation? Equally, we have given a definition
of validity for English. A question to be asked later is: is **P**-validity
(represented by "$\vDash_P$") essentially the same thing as validity in English
(represented by plain, unsubscripted, "$\vDash$")?

The sentence connectives of **P** are so called because they *take* one or
more sentences to *make* a fresh sentence. Other languages than **P** contain
sentence connectives. For example, "It is not the case that" is a sentence
connective in English: it takes *one* sentence, say "John is happy", to form
a fresh sentence, "It is not the case that John is happy". The fresh
sentence is called the *negation*[4] of the original. "And" is a sentence
connective which takes *two* sentences to make a sentence. For example, it
can make "John is happy and Mary is sad" from the two sentences "John
is happy" and "Mary is sad". Let's call the sentence or sentences a
sentence connective takes to make a fresh sentence the *component(s)* of the
fresh sentence; and let's call the fresh sentence itself the *resultant*
sentence.

(6) Scope

6) The *scope* of an occurrence of a sentence connective is the shortest **P**-sentence in which it occurs.

Thus the scope of "&" in

    7) $\neg(p \mathbin{\&} q)$

is "$(p \mathbin{\&} q)$", whereas in

    8) $(\neg p \mathbin{\&} q)$

its scope is the whole of (**8**). An occurrence of a sentence connective is said to *dominate* the sentence which is its scope. Nomenclature of **P**-sentences is determined by the role of their *dominant* connective (see above).

2 Truth functional sentence connectives

This section gives an account of truth functionality. It can be skipped by those already familiar with the notion.

    Standard ("classical") propositional logic deals with sentence connectives having a special property: they are *truth functional*. The terminology derives from the mathematical notion of a function, and one can use this to give a mathematically precise definition of truth functionality.[5] Alternatively, we can define a truth functional sentence connective in more informal terms:

1)   *A sentence connective is truth functional* iff whether or not any resultant sentence it forms is true or false is determined completely by, and only by, whether its components are true or false.

    Take, for example, "it is not the case that ...". Suppose what fills the dots is true. Then the resultant is false. Suppose what fills the dots is false. Then the resultant is true. So "it is not the case that" is truth functional, according to the definition.[†]

    All the sentence connectives of **P** are truth functional, according to the definition, as can be seen from (**1.4**) above.

† See Exercise 32, page 344.

Not every expression which we are inclined to classify as a sentence connective is a truth functional one. For example, we might naturally think of "Napoleon knew that" as a sentence connective. It can take the sentence "St Helena is in the Atlantic Ocean" to form the sentence: "Napoleon knew that St Helena is in the Atlantic Ocean". However, it is not truth functional. The component is true, but this does not, in and of itself, determine whether or not the resultant sentence is true. The mere fact that "St Helena is in the Atlantic Ocean" is true does not settle whether Napoleon knew this or not.

A good test for truth functionality is the *substitution test*: if a sentence connective is truth functional, then the truth or falsehood of every resultant sentence which it forms depends only on the truth or falsehood of the components. So replacing a true component by another true one should make no difference to whether the resultant is true or false, and likewise for replacing a false component by another false one. Let us apply the substitution test to "Napoleon knew that St Helena is in the Atlantic Ocean". We assume that this resultant sentence is true. (Anyone who disagrees can choose their own example of a truth of the form "Napoleon knew that ...".) The component "St Helena is in the Atlantic Ocean" is also true. If "Napoleon knew that" were a truth functional sentence connective, it would form only sentences which pass the substitution test. But it does not. Consider any truth which Napoleon did not know, for example, "Quarks come in four colours". "Napoleon knew that quarks come in four colours" is false, whereas "Napoleon knew that St Helena is in the Atlantic Ocean" is true. So "Napoleon knew that" fails the substitution test. Substituting one true component for another *does* sometimes yield resultants which differ in truth value.[†]

Now suppose we are wondering whether "and" is truth functional. If it is, then it is clear that it will express the same truth function as "&", as specified by (**1.4**): that is, "*A* and *B*" will be true if and only if "*A*" is true and "*B*" is true. "Paris is west of Berlin" and "London is north of Paris" are both true. If "and" is truth functional, then the conjunction "Paris is west of Berlin and London is north of Paris" is true. Moreover *any* true conjuncts will yield a true conjunction. So we could replace, say, "Paris is west of Berlin", in the conjunction, by any other truth whatsoever, for example, "$5 + 7 = 12$", and the new conjunction ("$5 + 7 = 12$ and London is north of Paris") will be true. The results of this application of the substitution test are consistent with "and" being truth functional. The results do not establish the truth functionality of "and". For that, it

† See Exercise 33, page 344.

would be necessary to show that *whatever* sentences composed a conjunction, *any* replacement of either component by *any* sentence fails to affect the truth or falsity of the conjunction.

Instead of saying that a connective is truth functional, I shall sometimes say, equivalently, that it *expresses a truth function.*

One issue that will be of general concern when we come to compare **P** with English is this: **P** can represent only truth functional connections between sentences. English can, it seems, also represent non-truth functional connections between sentences. How extensive is this dissimilarity? And to what extent does it undermine the project of using **P**-validity to understand validity in English? The next section considers specific cases of formalizing English arguments in **P**, and using **P**-validity, where possible, to say something about validity. The section is more practical than theoretical. The theoretical questions raised will be discussed in subsequent sections.

3   Formalizing English in *P*

No one has ever supposed that **P**-validity exhausts the notion of validity, nor even that of formal validity. So what is at issue is whether **P**-validity gives a partial characterization of validity, or formal validity. The crucial questions are: can we be sure that if a rendering in **P** of an English argument is **P**-valid, the English argument is valid (or formally valid)? And can we be sure that if a rendering in **P** of an English argument is not **P**-valid, the English argument is not valid (or not formally valid)? The examples will suggest some detailed considerations which bear on these questions.

Following a standard terminological practice, we shall call rendering an English sentence or argument in **P** *formalizing* it (more fully, **P**-formalizing it).

Let us start with a straightforward argument to illustrate the method of formalizing:

1)   The battery is flat. If the battery is flat the car will not start. So the car will not start.

In formalizing, we stipulate a *correspondence scheme* between English sentences and **P**-letters. For (1) we might choose:

2) Let "*p*" correspond to "The battery is flat" and "*q*" to "The car will not start".

Using this correspondence scheme, (**1**) is formalized by

3) *p*, (*p*→*q*); *q*.

This is **P**-valid (in the sense of (**1.5**) above). Any interpretation, *i*, upon which all the premises are true is one upon which "*p*" is true. By the rule for "→" (specified in (**1.4**) above), if "*p*→*q*" is true upon *i*, then either "*p*" is false upon *i* or "*q*" is true upon *i*. So if both "*p*" and "*p*→*q*" are true upon *i*, then "*q*" is also true upon *i*. Hence every interpretation upon which all the premises are true is one upon which the conclusion is true. Standardly, one infers from the **P**-validity of (**3**), together with the correspondences of (**2**), to the validity of (**1**). The correctness of this inference is a main theme of this chapter. The examples of this section contribute to the answer, and an explicit statement is defended in §**10**.

A further straightforward example (compare (**1.10.1**)):

4) You can buy a ticket only if you have the exact fare. You haven't got the exact fare. So you cannot buy a ticket.

Let "*p*" correspond to "You can buy a ticket", "*q*" to "You have got the exact fare". Then the following formalizes (**4**) and is **P**-valid:

5) (*p*→*q*), ¬*q*; ¬*p*.

If the premises are to be true upon an arbitrary interpretation, *i*, then "*q*" must be false upon *i*, and so, by the "→" rule, "*p*" must also be false upon *i*, so "¬*p*" must be true upon *i*. So all interpretations upon which the premises are true are ones upon which the conclusion is also true.

We obviously cannot expect to make any inference from the quality of a **P**-argument to the quality of an English argument supposedly formalized by it unless the formalization meets some standard of adequacy. We stipulate that if a formalization is to be *adequate*, the associated correspondence scheme should be such that if we replace the **P**-letters by the corresponding English sentences, and then replace the **P**-connectives by the corresponding English connectives, the result is a sentence (argument) that says the same as the original English. (Note that what English expressions correspond to which **P**-connectives was stipulated in §**1**.) Let us call the result of applying the correspondences to an argument the *recovered* argument. The proposed standard of adequacy is that the argument to be formalized should say the same as the recovered argument.

A related test of a formalization's adequacy can be given in terms of the notion of an *intended interpretation*. An intended interpretation is one which assigns to the sentence-letters the same truth values as the ones the corresponding English sentences possess.[†] A necessary condition for adequacy is that *every* sentence in the formalization should be true (false) on an intended interpretation iff the corresponding English sentence is true (false). In other words, an intended interpretation, by assigning the "right" truth values to the sentence-*letters*, must thereby assign the right truth values to the complex sentences (those which are not letters). We do not always know what the truth values of the English sentences are, so we cannot always apply this test for adequacy.

On either formulation, the adequacy of a formalization is always relative to a correspondence scheme. Hence presenting a correspondence scheme is an essential part of presenting a formalization.

There may be more than one adequate formalization. Consider the argument:

6) If the figure is closed and has sides of equal length, then it is square or rhomboid. The figure is closed. The figure has sides of equal length. So it is square or rhomboid.

We could set up the following correspondence scheme:

7) "*p*" corresponds to "if the figure is closed and has sides of equal length, then it is square or rhomboid", "*q*" to "The figure is closed", "*r*" to "The figure has sides of equal length" and "*s*" to "The figure is square or rhomboid".[6]

Relative to the correspondences of (7), the **P**-formalization of (6) is:

8) *p*, *q*, *r*; *s*.

According to the definition (1.5) of **P**-validity, (8) is not **P**-valid, for the premises are all true and the conclusion false upon an interpretation which assigns truth to "*p*", "*q*" and "*r*" and falsehood to "*s*". Intuitively, (6) is valid, yet by our stipulations (8), given (7), is adequate.

† See Exercise 34, page 345.

An alternative correspondence scheme is as follows:

9) "*p*" corresponds to "the figure is closed", "*q*" to "the figure has sides of equal length", "*r*" to "the figure is square" and "*s*" to "the figure is rhomboid".

Relative to (**9**), an adequate formalization is:

10) $(p \,\&\, q) \rightarrow (r \vee s), p, q; (r \vee s)$.

(I omit parentheses where no confusion can result.) (**10**) is **P**-valid. For consider an interpretation, say *i*, upon which all the premises are true. Then "*p*" and "*q*" are true upon *i*. So, by the rule for "&", "$(p \,\&\, q)$" is true upon *i*. The rule for "→", together with the supposition that the premise "$(p \,\&\, q) \rightarrow (r \vee s)$" is true upon *i*, ensures that "$(r \vee s)$" is also true upon *i*. So any interpretation upon which all the premises are true is one upon which the conclusion is true. For short:

11) $(p \,\&\, q) \rightarrow (r \vee s), p, q \vDash_{\mathbf{P}} (r \vee s)$.

So (**10**), relative to (**9**), is a candidate for demonstrating the validity of the English.

We could have used the following correspondence scheme:

12) "*p*" corresponds to "the figure is closed", "*q*" to "the figure has sides of equal length" and "*r*" to "the figure is square or rhomboid".

Relative to (**12**), the formalization is:

13) $(p \,\&\, q) \rightarrow r, p, q; r$.

This is **P**-valid.[†] So (**13**) is also a candidate for demonstrating the validity of the English.

All of (**8**), (**10**) and (**13**), coupled with their correspondence schemes, count as adequate formalizations of (**6**): intuitively, they are faithful to the meaning of the English, or at least to those aspects of the meaning of the English that are relevant to propositional logic. But as only (**10**) and (**13**) are **P**-valid, only they could purport to demonstrate the validity of (**6**).

† See Exercise 35, page 345.

(8), (10) and (13) differ in how much of the structure of (6) they make manifest. (8) captures the least, (10) the most. (10) captures more than is necessary to demonstrate the validity of (6), (8) captures less than is necessary. How much of a given sentence's structure needs to be reflected in its **P**-formalization will vary, depending on specific facts about the argument in which the sentence occurs. I shall say that the more structure an adequate formalization captures, the *deeper* it is. When people speak of "the" *logical form* of a sentence, they have in mind a formalization which goes as deep as possible.

If a formalization is invalid, can we infer that the argument formalized is also invalid? Consider the following example:

14)   If you are good at mathematics, you will find logic easy. But you are not good at mathematics. So you'll find logic hard.

Let "*p*" correspond to "You are good at mathematics", and "q" to "You will find logic easy". Then a candidate for a formalization of (11) is:

15) $(p \rightarrow q), \neg p; \neg q.$

This is not **P**-valid. For consider an interpretation, *i*, which assigns truth to "*q*" and falsehood to "*p*". The conclusion is false upon *i*. But both premises are true upon *i*, by the rules for "¬" and "→". Is (15) an adequate formalization of (14)?

To obtain the *recovered argument* from (15) replace the letters by the corresponding English sentences as specified in the correspondence scheme, "→" by "if ... then ..." and "¬" by "it is not the case that ". The result is:

16)   If you are good at mathematics then you will find logic easy. It is not the case that you are good at mathematics. So it is not the case that you will find logic easy.

The sentence from (14) "You'll find logic hard" is replaced by "It is not the case that you will find logic easy". However, you can fail to find something easy without finding it hard. So (16) does not say the same as (14) and so (15) is an inadequate formalization of it. Does this show that there is no inference from the invalidity of (15), together with the associated correspondences, to the invalidity of (14)?

Consider a different correspondence scheme. Let "*p*" correspond to

"You are good at mathematics", "q" to "You will find logic easy", "r" to "You'll find logic hard". Then the **P**-formalization is:

17) $(p{\to}q), \neg p; r.$

This is, obviously, **P**-invalid.[†] Moreover it does not fail the condition for adequacy. Yet (17) is, intuitively, in some respect a less good formalization of (14) than is (15). Can we make sense of this intuition?

(14) is an argument which someone might actually propound in good faith, believing it to be valid. (17) gives no hint about how this mistake is possible, whereas (15) does. (17) is miles from anything that looks valid. But (15) might for a moment look valid, because of its passing resemblance to the **P**-valid

18) $(p{\to}q), \neg q; \neg p.$

One of the advantages of formalization is that it makes logical mistakes easier to spot. It is intelligible that one who is reasoning in English should think that the validity that attaches to an argument having the logical form of (18) attaches to (14). This explanation is abetted by (15), but not by (17).

If a formalization that counts as inadequate by the present standards can none the less be useful in explaining how someone might wrongly think an argument valid, would it not be better to revise the standards of adequacy? Consider again the first correspondence scheme of translation, and the relationship between (16) and (14). It is logically impossible for someone to find logic hard, yet not fail to find it easy. As we shall say, "You will find logic hard" entails "You won't find logic easy". In general, A entails B iff it is logically impossible for A to be true yet B not true. If A entails B but B does not entail A, we say that A is *stronger than* B and that B is *weaker than* A. The argument that one recovers from the formalization (15) by using the correspondences thus has a *weaker* conclusion than the conclusion in (14). However, if some premises are inadequate to establish even a weaker conclusion, they are obviously inadequate to establish a stronger one. So if we allow that the invalidity of (15) establishes the invalidity of (16), then it should also be allowed to establish the invalidity of (14).

This suggests the following emendation of the account of adequacy: we allow that an invalid formalization *may* be adequate to an invalid

† See Exercise 36, page 345.

argument even if, though the premises of the recovered argument say the same as the original premises, the recovered conclusion is weaker. A parallel relaxation, a case of which will be considered shortly (see (**21**) and (**22**)), would allow that a valid formalization may be adequate to a valid argument even if the recovered argument has weaker premises.

To be able to use these relaxations, however, we would need already to be in a position to recognize the validity of arguments in English, whereas we were investigating whether we could use **P** to test for this property. A test which required us to recognize the target property before we could apply the test would not be much help. So I shall not allow the relaxed standards of adequacy.

Is there any way, using **P**-methods, that we can establish the invalidity of (**14**)? One possibility we would need to consider is that (**14**) has a *deeper* formalization which is valid. Think how mistaken it would have been to have concluded that (**6**) was invalid on the basis of (**8**). A necessary condition for there being no deeper formalization is that every truth functional sentence connective in English be somehow reflected by sentence connectives in the formalization. This condition is clearly met by (**15**).

However, the fact that the deepest **P**-formalization is invalid is still not sufficient to establish the invalidity of the English. This is because there are valid arguments whose validity cannot be represented in **P**, for example:

19) All football supporters are interested in sport. Some football supporters are hooligans. So some hooligans are interested in sport.

The following correspondence scheme will ensure we go as deep as we can: let "*p*" correspond to "All football supporters are interested in sport", "*q*" to "Some football supporters are hooligans" and "*r*" to "Some hooligans are interested in sport". This correspondence goes as deep as possible, for there are no words in the sentences corresponding to **P**-letters which correspond to any of the **P**-connectives. Hence no further complexity in the formalization could be justified. Yet the resulting formalization is the **P**-invalid:

20) *p, q; r.*

The conclusion is that we cannot hope that all validity should be reflected as **P**-validity.

Sometimes an argument's premises are stronger than required for the conclusion. A putative example is:

21)   If the mare dies, the farmer will go bankrupt, and then he will not cultivate the ground. The result will be that the wheat will fail, and this, in turn, will lead to local food shortages. Then the revolutionary spirit of the people will become inflamed, and they will man the barricades. So if the mare dies, the local people will man the barricades.

An appropriate correspondence scheme is:

"p" for "The mare dies"
"q" for "The farmer will go bankrupt"
"r" for "The farmer will cultivate the ground"
"s" for "The wheat will fail"
"t" for "There will be local food shortages"
"u" for "The revolutionary spirit of the people will become inflamed"
"v" for "The local people will man the barricades".

A possible candidate formalization is:

22) $(p \rightarrow q)$ & $(q \rightarrow \neg r)$, $(\neg r \rightarrow s)$ & $(s \rightarrow t)$, $t \rightarrow (u$ & $v)$; $p \rightarrow v$.

The formalization corresponds only approximately to the original. There is a syntactic aspect: for example, the sentence "This will lead to local food shortages" is formalized as a conditional, whereas it is not one. In the project of formalization, this kind of reorganization is standard. There is a semantic aspect: explicitly causal idioms like "the result will be" and "this will lead to" have been formalized by weaker **P**-conditionals, e.g. "$\neg r \rightarrow s$". What the correspondences recover from the latter is entailed by "If the ground is not cultivated, then, as a causal consequence, the wheat will fail", but does not entail it.

However, (22) is **P**-valid. So, provided we have no qualms about the adequacy of "$p \rightarrow v$" as a formalization of the conclusion, and provided we have the general assurance that **P**-validity can establish validity, we can use (22) to establish the validity of (21), despite the fact that the former does less than full justice to the strength of the latter's premises. The idea is that (21) would still have been valid, even if the premises had been weaker, as weak as the premises of the argument recoverable from (22).

This approach is not available when the **P**-conclusion (more exactly, the conclusion recoverable from the **P**-formalization) is weaker than the English conclusion. For example, consider

> **23)**   Putting garlic in the salad will make Richard think that we care nothing for his preferences, and if he thinks that he will be upset. So putting garlic in the salad will make Richard upset.

A suitable correspondence scheme is:

"*p*" for "Garlic will be put in the salad"
"*q*" for "Richard will think we care nothing for his preferences"
"*r*" for "Richard will be upset".

The obvious formalization is:

> **24)** $(p \rightarrow q)$, $(q \rightarrow r)$; $(p \rightarrow r)$

and this, plainly, is **P**-valid. However, "If garlic is put in the salad then Richard will be upset" is weaker than "Putting garlic in the salad will make Richard upset". The falsehood of "Garlic is put in the salad" is enough for the truth of "If garlic is put in the salad, Richard will be upset" *on the assumption that "→" correctly translates the "if" that occurs here.* (To check this, consult the rule for "→" in §1.) However, the falsehood of "Garlic is put in the salad" is not enough for the truth of "Putting garlic in the salad will make Richard upset". So the **P**-validity of (**24**) does not show that the premises of (**23**) establish the stronger conclusion: "Putting garlic in the salad will make Richard upset".

I think that (**23**) *is* valid. The trouble is that, like (**19**), its validity cannot be shown by **P**-formalization. As we shall say: it is not *valid in virtue of its* **P**-*logical form.*

The next example shows how **P**-formalization may be used to resolve ambiguities:

> **25)**   John will choose the colour for his new bathroom and will paint it with his own hands only if his wife approves. But his wife doesn't approve. So he will not paint it with his own hands.

A suitable correspondence scheme is:

"*p*" for "John will choose the colour for his new bathroom"

"*q*" for "John will paint his new bathroom with his own hands"
"*r*" for "John's wife approves".

Two formalizations are possible, depending on how we understand the
organization of the first sentence:

    26) $((p \,\&\, q) \rightarrow r), \neg r; \neg q.$

    27) $(p \,\&\, (q \rightarrow r)), \neg r; \neg q.$

(**27**) is **P**-valid, but (**26**) is not.[†] One cannot speak of the validity or
invalidity of (**25**). Rather, one must speak of it as valid on one reading (the
one corresponding to (**27**)), and invalid on another (the one corresponding to (**26**)).[‡] The formalizations treat the ambiguity of the first premise of
(**25**) as a matter of scope. In (**26**), "→" dominates the first premise: it has
wide scope relative to "&". Analogously, we shall say that (**26**) treats "if",
in the first premise of (**25**), as having wide scope relative to "and". In (**27**),
"&" dominates the first premise: it has wide scope relative to "→".
Analogously, we shall say that (**27**) treats "and", in the first premise of
(**25**), as having wide scope relative to "if".
    The final example shows how finding the **P**-logical form of an
argument may lead one to the view that it is valid, even though one was
not clear whether it was valid or not when one looked at the English
version:

    28) If common sense is correct, then physics is true. If physics is
           true, then common sense is incorrect. Therefore common
           sense is incorrect.

Using obvious correspondences, this formalizes to the **P**-valid

    29) $p \rightarrow q, q \rightarrow \neg p \vDash_{\mathbf{P}} \neg p.$[¶]

    Two vital questions have been raised in this section: can the validity of
an argument be inferred from the **P**-validity of its **P**-logical form? And
can the invalidity of an argument be inferred from the **P**-invalidity of its

---

† See Exercise 37, page 345.   ‡ See Exercise 38, page 345.
¶ See Exercise 39, page 345.

P-logical form? The second question has been answered negatively. The first question has not been explicitly answered, though the formalizations of examples like (**1**), (**4**) and (**6**) may encourage optimism. I return to these questions more systematically in §**10** below.

4 Comparison of *P*-connectives and English

In the examples of §**3**, we have taken for granted that the connectives of **P**, though given their official definitions by the interpretation rules, correspond closely to their English counterparts. It is now time to examine that assumption.

First, we must try to state what sort of correspondence we require. The simplest idea is that when we recover an argument from a logical form by applying the correspondences, the result should be the original argument. But this standard is unduly restrictive. The tradition allows for a certain amount of reorganization, allows one, at the very least, to match "unless" with "∨".† A more relaxed standard is that the recovered argument should have all the validity-relevant features of the original.

If this standard is to be met, it is clear that the **P**-connectives must have all the validity-relevant features of the English expressions they are matched with in formalizing. Since validity is definable in terms of truth conditions, the requirement can be put: the **P**-connectives must make the same contribution to truth conditions as the English connectives with which they are matched. The contribution to truth conditions of a **P**-connective is given by the appropriate interpretation rule. The standard is thus that the English expressions should express the same truth function as the **P**-connectives with which they are matched. A precondition for an expression's expressing a truth function is that it be a sentence connective. This is because it is sentences that are true or false (possess truth values), and a truth function fixes a truth value from (a sequence of) truth values. (For example, the truth function expressed by "&" fixes the value false for a conjunction on an interpretation iff one of the conjuncts is false on the interpretation.) The discussions thus tend to fall into two parts: putative cases in which the English expression corresponding to a **P**-connective is not a sentence connective, and putative cases in which, though a sentence connective, it is not truth functional.

The important notion of *implicature* which is mentioned here is discussed in more detail in §**6**.

† See Exercise 40, page 346.

(1) "¬" and "not"

The **P**-connective "¬" corresponds closely to the English word "not", and similar phrases like "it is not the case that", *as these are used to form negations*. If "John is here" is true (false), then "John is not here" and "It is not the case that John is here" are false (respectively, true). So it appears that on at least some occurrences, "not" is a sentence connective and expresses the same truth function as "¬".

There are cases in which "not" does not form a negation. For example, if "Some cows eat grass" is true, it does not follow that "Some cows do not eat grass" is false. Here "not" does not even seem to function as a sentence connective,[7] let alone as a sentence connective expressing the same truth function as "not".

There are other cases in which "not" seems not to function as a sentence connective. For example:

1) Not Abraham, but George, chopped down the cherry tree.

Here, on the face of it, "not" attaches to a name rather than a sentence. However, it could be argued that (1) is merely a fanciful way of expressing

2) Abraham did not chop down the cherry tree but George did

and here "not" does, after all, form a sentence from a sentence (from "Abraham did chop down the cherry tree"), and in a way that expresses the truth function expressed by "¬".

There are no cases in which "not" appears to function as a sentence connective, but a non-truth functional one.[†]

(2) "&" and "and"

The **P**-connective "&" is fairly closely matched by the English "and". "Two is an even number and three is an odd number" is indeed true iff "Two is an even number" is true and so is "Three is an odd number". But there are, arguably, some discrepancies.

(a) Cases in which "and" appears not to function as a sentence connective, in which case it certainly could not be translated by "&", which *is* a sentence connective:

3) Tom and Mary came to dinner.

† See Exercise 41, page 346.

On the face of it, far from taking two sentences to make a fresh sentence, "and" in (**3**) takes two *names* to make a complex subject expression "Tom and Mary". In this sort of case you might argue that appearances are superficial, and that (**3**) abbreviates

    4)   Tom came to dinner and Mary came to dinner.

Even if the suggestion works for some cases in which "and" is, superficially, a name connective rather than a sentence connective, it may not work for all. Consider:

    5)   Tom and Mary got married.

Arguably, this is not reducible to

    6)   Tom got married and Mary got married.

For, it may be argued, (**5**) entails that Tom married Mary, but (**6**) does not.

This may be disputed. An objection is that (**5**) does not *entail*, but only implicates, that Tom married Mary (see §**6** for this distinction). The idea is that (**5**) would still be *true*, though perhaps misleading, even if Tom and Mary got married, but not to each other, so that there is nothing here to show that "and" does not express the truth function of "&". To lend this case some plausibility, it is worth reflecting that the following is not inconsistent:

    7)   Tom and Mary got married, but not to each other.

Yet one would expect that if (**5**) entailed that they married each other, (**7**) *would* be inconsistent.

It is harder to apply the idea used in (**6**) to:

    8)   Tom and Mary are compatriots.

For

    9)   Tom is a compatriot and Mary is one too

is nonsense, and it is not easy to think of an alternative reduction of (**8**) in which "and" is uncontroversially a sentence connective.[†]

† See Exercise 42, page 346.

Here is another case in which "and" may be argued not to be a sentence connective:

10)   Some girls are pretty and flirtatious.

Here "and", at least superficially, joins two adjectives, "pretty" and "flirtatious", to form a complex adjectival expression. Perhaps (10) is an abbreviation of a sentence in which "and" is genuinely a sentence connective. But what sentence? Not

11)   Some girls are pretty and some girls are flirtatious.

For (11) does not have the same truth conditions as (10): (11) would be true, yet not (10), if it were the case that ugly girls, and only ugly girls, flirt.

12)   John washed immediately and thoroughly.

Here "and", at least superficially, joins two adverbs, "immediately" and "thoroughly", to form a complex adverbial expression. Perhaps (12) is an abbreviation of a sentence in which "and" is genuinely a sentence connective. But what sentence? Not

13)   John washed immediately and John washed thoroughly.

For (13) does not have the same truth conditions as (12): (13) would be true yet not (12), if John washed twice, once immediately but not thoroughly, and once thoroughly but not immediately.

(b) Cases in which "and" is a sentence connective but is, allegedly, not truth functional, and so not equivalent to "&".[8]

A standard kind of example is:

14)   Mary got married and had a baby.

15)   Jane Austen died in 1817 and was buried at Winchester.

Here there is no question about "and"'s claim to be a sentence connective. (There is a slight element of abbreviation: the second component elides the name in each case.) But it is argued that "and" cannot be translated by "&". If it could be, then the truth or falsehood of (14) and (15) would depend on nothing more than the truth or falsehood of the components. The objection is that this is not so: (**14**) and (**15**) require for their truth that the event reported in the second component occur *after* that reported in the first.

From the interpretation rule for "&", we know that "X & Y" is true iff "X" is true and "Y" is true. So "X & Y" is true iff "Y & X" is true: as we shall say, "X & Y" and "Y & X" are *equivalent*. This fact is reflected in the general truth about **P**-validity:

16) $X \& Y \vDash_{\mathbf{P}} Y \& X.$

This is to be read as follows: if X and Y are any **P**-sentences, the result of forming their conjunction in that order serves as a premise to a **P**-valid argument if the conclusion is the conjunction of X and Y in the other order.

If "&" expressed the same truth function as "and" in (**14**) and (**15**), the following arguments would be valid:

17) Mary got married and had a baby. So Mary had a baby and got married.

18) Jane Austen died in 1817 and was buried at Winchester. So Jane Austen was buried at Winchester and died in 1817.

Those who think that the premises do not entail the conclusion will, obviously, hold that (**17**) and (**18**) are not valid, so that in these cases "and" does not express the truth function that "&" expresses.

The standard response is to say that (**17**) and (**18**) are valid, the contrary appearance being created by the fact that (**14**) *implicates*, but does not entail, that Mary got married before having the baby, and similarly for (**15**).†

(3) "v" and "or"

The nearest English equivalents to "v" are "or" and "either ... or ...". These are certainly sometimes sentence connectives, as in:

19) You're a fool or you're a rascal.

20) Either you're a fool or you're a rascal.

† See Exercise 43, page 346.

It certainly seems that (**19**) and (**20**) are true if and only if at least one of "You're a fool" and "You're a rascal" is true. So there is a case for saying that "or" expresses the same truth function as "v".

Another class of cases in which "v" seems close to "or" is provided by a certain kind of game. "I'll give you a clue: either William hid the silver or Tom hid the gold ... Which box contains the silver, which the gold, which the lead?"

As with "and", I divide the alleged discrepancies into those which allege that "or" (or "either ... or ...") is sometimes not a sentence connective, and those which allege that on some occurrences as a sentence connective it does not express the same truth function as "v".

(a) Cases in which "or" appears not to function as a sentence connective.

21)   Tom or Mary could help you.

Here "or", at least superficially, joins not two sentences to form a sentence, but two names to form a complex subject expression. (**21**) cannot be regarded as an abbreviation of

22)   Tom could help you or Mary could help you,

for most people hear (**21**), but not (**22**), as meaning that Tom could help you and so could Mary.[†]

23)   Every number is odd or even

should be compared with (**10**). At least superficially, "or" here joins two adjectives, "odd' and "even", to form a complex adjectival expression. Could (**23**) be an abbreviation of a sentence in which "or" is genuinely a sentence connective? What sentence? Not

24)   Every number is odd or every number is even

for this plainly means something different from (**23**). Indeed, (**23**) is true, (**24**) false.

A problematic case is:

25)   He asked whether John would win or not.

† See Exercise 44, page 346.

We cannot understand this as meaning

> **26)** He asked whether the following is true: John will win or John will not win.

Everyone knows that "John will win or John will not win" is true: *this* cannot have been what the questioner wanted to know. But it is not easy to see how "or" as a sentence connective could be used to express what is being asked.

(b) Cases in which "or" is a sentence connective but is, allegedly, not equivalent to "∨".

These cases are of two kinds: (1) those in which it is agreed that "or" expresses some truth function, and the disagreement is over whether it expresses the same one as "∨"; (2) those in which it is contended that "or" does not express a truth function.

(1) "∨" expresses what is standardly called *inclusive disjunction*. If "X" and "Y" are both true, so is "$X \vee Y$". It is sometimes claimed that "or", and, more especially "either ... or ...", express exclusive disjunction. The exclusive disjunction of X with Y is true just on condition that *exactly* one of "X" and "Y" is true. We can, of course, easily *define* a **P**-connective which expresses this function.[†] And if "or" does, sometimes or always, express exclusive disjunction, we need, sometimes or always, to avoid matching it simply with "∨".

> **27)** This number is odd or this number is even

might be offered as a candidate example of exclusive disjunction. However, the case is inconclusive. We must admit that the truth of both disjuncts is excluded, but it remains to be shown that the excluding is done by "or" rather than by the particular senses of the disjuncts, which already preclude their joint truth. This phenomenon is entirely consistent with "or" expressing inclusive disjunction.

A better example is:

> **28)** You are welcome to come to dinner on Monday or Tuesday.

If "or" is a sentence connective here, it presumably connects "you come to dinner on Monday" and "you come to dinner on Tuesday", so that (**28**) is an abbreviation of something like:

† See Exercise 45, page 347.

29)   I would welcome your making it true that: you come to dinner
on Monday or you come to dinner on Tuesday.

But many people hear (28), and so, presumably, (29), as constituting an
invitation for just one dinner, not two. If this is right,[9] there is a case for
thinking that "or" sometimes expresses exclusive disjunction, and so on
such occasions should not be matched simply with "∨".[†]

(2) The most important reason for thinking that "or" is not truth
functional issues from such cases as:

30)   Either the superpowers will abandon their arms race, or there
will be a third world war.

The suggestion is that (30) requires for its truth not merely the truth of at
least one (or exactly one) of its disjuncts, but in addition that there be
some special connection, presumably in this case causal connection,
between the falsehood of one disjunct and the truth of another. (30)
asserts, it may be said, that the arms race will *lead to* war.

If this is right, then we ought to be able to discover failures of the
substitution test (see §2). The test is rather hard to apply in such a case,
because there is likely to be disagreement about whether the disjuncts are
true or false. Suppose we think that (30) is true, but that the first disjunct
is false (i.e. it is false that the superpowers will abandon their arms race).
Then, if the "or" it contains is truth functional, we ought to find the
following true, despite the fact that no one could for a moment suppose
that there is any causal connection between the truth of the first disjunct
and that of the second:

31)   Either $2+2=22$ or there will be a third world war.

But, holding our suppositions firmly in mind (the truth of (30), the
falsehood of its first disjunct), it actually seems rather unlikely that one
should find (31) false.

(4) "→" and "if"

The nearest English equivalents of "→" are: "If ... then ..."; "... if ...";
and "... only if ...".[10] I will concentrate on the first of the idioms, calling
"if ... then ..." sentences *conditionals*, the sentence filling the first blank
the *antecedent*, the sentence filling the second blank the *consequent*. A

† See Exercise 46, page 347.

common view is that conditionals cannot be adequately formalized by "→".

One ground for this view is that a sentence "if A then B" requires for its truth some special connection between what would make "A" true and what would make "B" true, a causal connection, for example. No such connection is required for the truth of "$p \to q$". For example, consider a volume of water which in fact will not be heated to 90° at any time in the coming year, so that the sentence "this volume of water is heated to 90° at some time during the coming year" is false. Let this correspond to the **P**-letter "p". Let "q" correspond to "this volume of water will turn to ice". It follows that

32) $p \to q$

is true upon an intended interpretation,[†] whereas, so the claim runs,

33)   if this volume of water is heated to 90° at some time during the coming year, this volume of water will turn to ice

is false. (32) is an inadequate formalization, since "if" does not express the same truth function as "→".[‡]

The question of the relation between "→" and "if" has been very widely discussed in recent years. I shall divide up the issues as follows. I shall begin, in this section, by trying to demarcate the area of the controversy: it concerns "indicative" rather than "subjunctive" conditionals, but it is not easy to give adequate criteria for this distinction. In subsequent sections, I present the case against the truth functional interpretation (§5), the outline of an implicature defence against this case (§§6, 7), and finally (§8) some general arguments in favour of the truth functional interpretation. The issue is not resolved in these discussions, and the question is taken up again in chapter **3**, where two alternative accounts of "if" are presented.

Consider the following two "if" sentences:

34)   If Oswald didn't shoot Kennedy, someone else did.

35)   If Oswald hadn't shot Kennedy, someone else would have.

It would be perfectly reasonable to regard one of these sentences as true and the other false. For example, you might reasonably believe (**34**),

† See Exercise 47, page 347.   ‡ See Exercise 48, page 347.

simply on the ground that someone did shoot Kennedy (he wasn't poisoned, etc.). Yet you might not believe (**35**), and would not if you thought that Oswald was a maniac working alone, and that no one else would have wanted Kennedy dead. The fact that (**34**) and (**35**) may diverge in truth value may encourage the view that we cannot have a uniform account of the two kinds of sentence: we will need one account of the way in which "if" works in *indicative* conditionals, like (**34**), another of the way in which it works in *subjunctive* conditionals like (**35**).[11] The target of our discussion is the claim that indicative (rather than subjunctive) conditionals can be adequately formalized by material conditionals, those dominated by "→".

The attempt to formalize (**35**) as a material conditional leads to an instructive kind of nonsense:

36)   (Oswald hadn't shot Kennedy)→(someone else would have shot Kennedy).[12]

A truth functional connective like "→" can be used to connect two sentences which are capable of having truth values, that is, of being true or false. But the components of (**36**), as they occur there, do not have this capacity. Ask yourself whether "Oswald hadn't shot Kennedy", as it occurs in (**35**) or (**36**), is true or false. You will see that, for a sentence of this kind, in a context of this kind, the question doesn't make sense.[13]† The sort of component that "→" requires must be capable of being assessed for truth or falsity as a self-standing sentence. On the face of it, then, sentences like (**35**) are not examples of "if" as a non-truth functional sentence connective, but of "if" as an expression which is not a sentence connective at all.

There are other ways in which, even where the mood is not explicitly subjunctive, the English conditional does not link two sentences that are capable of being true or false when standing alone, so that "if" is not functioning as a sentence connective. One kind of example will be familiar from (**10**) and (**23**):

37)   If someone is in debt, he should curb his expenditure.

Here the consquent "he should curb his expenditure" is not a complete sentence, since the referent of "he" is not determined. Hence the consequent is not capable, on its own, of being true or false. Hence we cannot translate (**37**) as

† See Exercise 49, page 347.

38)   (Someone is in debt)→(he should curb his expenditure).†

There are also conditionals which it is not obviously correct to classify as indicative conditionals, even though their components are in the indicative mood, and appear capable of being used self-standingly. For example, someone contemplating the future might truly affirm:

39)   If John dies before Joan, she will inherit the lot.

There is no doubt that the components are capable of being evaluated for truth and falsehood as self-standing sentences. However, it also seems that (**39**) is equivalent to

40)   If John should die before Joan, she would inherit the lot.

This means that there is a case for saying that (**39**), despite being grammatically indicative, is best classified as a subjunctive rather than as an indicative conditional.

There are other uses of "if" which we would not wish even to try to formalize using "→", for example

41)   John wonders if his life is meaningful.

It is literally true that each of "John wonders" and "John's life is meaningful" are capable of being evaluated for truth and falsehood as self-standing sentences, but it should also be clear that their use as self-standing sentences differs from their use in (**41**). This could be brought out by sentences similar to (**41**) but in which the mood is not indicative, for example

42)   John wanted to know if he had been born in wedlock.

It is not that this is a subjunctive conditional. It is not a conditional at all. Rather, "if" in (**42**) is being used to form an indirect question, and could be replaced by "whether".

The examples show that there is no simple definition of the category of sentences – indicative conditionals – we are after. I hope it has none the less given an intuitive feel for what that category is.

† See Exercise 50, page 347.

5   The case against the material implication account of "if"

The view to be attacked in this section is that we can adequately formalize (indicative) conditionals by "→"-sentences; equivalently, that conditionals are **P**-material implications; equivalently, that "if" (as it occurs in the cases under discussion) expresses the same truth function as "→". As a preliminary, let us see if we can establish the following:

>   1)   Any truth functional occurrence of "if ... then ..." expresses the same truth function as "→".

(The phrasing allows for the possibility of non-truth functional occurrences of "if ... then ...".) If we can establish (1), then the question whether conditionals are material conditionals becomes the question whether "if", as it occurs in English conditionals, is truth functional.

If "if ... then ..." functioned in the same way on all its occurrences, there would be a simple way to establish (1). It would be enough to find "if ... then ..." sentences whose truth values, in point of their components and resultant, match the truth table for "→". For example, suppose we find a true "if ... then ..." sentence with false antecedent and true consequent, perhaps "if New York is south of Beirut, then New York is south of Paris". The supposition that "if ... then ..." is always truth functional would enable us to infer that *any* "if ... then ..." sentence with false antecedent and true consequent is true, which corresponds to the third line of the truth table for "→" (table 2.2). It is not hard to do the same for the remaining three lines of the table.[†] The result would be a correct table in which "if A then B" replaces $X \rightarrow Y$.

The trouble with this approach is its assumption that "if ... then ..."

Table 2.2

X	Y	$X \rightarrow Y$
T	T	T
T	F	F
F	T	T
F	F	T

† See Exercise 51, page 347.

makes the same contribution in all four of the sample sentences. If it were ambiguous, having one meaning in one of the sentences, another in another, we would not have shown that on any of its meanings it expresses the "→" truth function. Suppose one of our sample sentences were "If you need bandages, there are some in the first aid box" (cf. Exercise 48). It might be argued that on this occurrence it expresses a different truth function from the one it more typically expresses, on the grounds that this sentence is true iff its consequent is.

Here is an argument which goes some way towards meeting this difficulty. Suppose some arbitrary sentence "if A then B" is truth functional with respect to A and B. Suppose also that:

2)   Using "if ... then ..." as it occurs in "if A then B", every instance of "if A and B then A" is true.

3)   "and" expresses the same truth function as "&".

4)   Using "if ... then ..." as it occurs in "if A then B", there are falsehoods of this form.

We can argue as follows, starting from (2). Suppose A is false; that is, take an instance of "if A and B then A" in which A is false. Then "A & B" is also false. So by the assumption of truth functionality, every instance of "if ... then ..." with false antecedent and false consequent is true, establishing the fourth line of a truth table for "if ... then ...". Now suppose that A is true. There are two subcases. On one of them, B is true, so "if A and B then A" is a truth with true antecedent and true consequent, which, given truth functionality, establishes the first line of the table. On the other subcase, B is false, so "if A and B then A" is a truth with false antecedent and true consequent, which, given truth functionality, establishes the third line of the table. The second line of the table is then established by (4). So given (2)–(4), we can establish (1). The upshot is that the question whether (some occurrence of) "if ... then ..." expresses the same truth function as "→" reduces to the question whether (that occurrence of) "if ... then ..." is truth functional, that is, expresses some truth function or other. In this section, I present the case for a negative answer to this question.

(1) The falsehood of X upon an interpretation is enough for the truth of $X \rightarrow Y$ upon that interpretation. But the falsehood of the antecedent is not enough for the truth of an indicative conditional formed with "if" or "if ... then". Examples:

5)   If ice is denser than water, then ice floats on water

6)   If ice does not float on water, then ice floats on water

are usually held to be false. But, using "p" to correspond to "ice is denser than water", and "q" to correspond to "ice floats on water", "$p \to q$" and "$\neg q \to q$" are both true upon an intended interpretation, viz. one in which the truth values of the **P**-letters match those of their corresponding English sentences, viz. one which assigns false to "p" and true to "q".

This is connected with a fact about validity. In **P** we have:

7) $X \vDash_{\mathbf{P}} [\neg X \to Y]$, whatever Y may be.[†]

This says that any **P**-sentence constitutes the premise of a **P**-valid argument for any **P**-conditional having the negation of that sentence as antecedent. But, it is claimed, the following is false for English:

8) $A \vDash$ [if not-A then B], whatever B may be.

That is, the fact corresponding to (7) does not obtain for English. You cannot always validly infer from an English sentence to an arbitrary indicative conditional having the negation of that sentence as its antecedent. If this contrast is correct, then "if" does not make the same contribution to validity as "$\to$" does; hence it does not make the same contribution to truth conditions; in particular, it is not truth functional.

(2) An interpretation upon which Y is true is one upon which $X \to Y$ is true. But the truth of the consequent is not enough for the truth of an indicative conditional formed with "if" or "if ... then". An example is:

9)   If ice is as dense as lead, then ice floats on water.

A connected claim is that whereas

10) $Y \vDash_{\mathbf{P}} [X \to Y]$, whatever X may be,

it is not the case that

11) $B \vDash$ [if A then B], whatever A may be.[‡]

(3) Whereas

† See Exercise 52, page 347.   ‡ See Exercise 53, page 347.

12) $\neg(X\to Y)\models_\mathbf{P} X,$

it is not the case that

13)   [it is not the case that (if A then B)]$\models A.$

For example, it is claimed that the following is invalid:

14)   It is not the case that if the number three is even then it is prime. So the number three is even.[†]

(4) Whereas

15) $\neg(X\to Y)\models_\mathbf{P}\neg Y,$

it is not the case that

16)   [it is not the case that (if A then B)] $\models$ it is not the case that B.

For example, it is claimed that the following is invalid:

17)   It is not the case that, if I go to the party tonight, I shall get drunk tonight. So I shall not get drunk tonight.

(5) Whereas

18) $(X\to\neg Y),\ Y\models_\mathbf{P}\neg X,$

it is not the case that

19)   [if A then not-B, B] $\models$ not-A.

For example, it is claimed that the following is invalid:

20)   If it rains, then it will not rain heavily. It will rain heavily. So it will not rain.

(6) "If" is not transitive, whereas "→" is. By the transitivity of "→" is meant the following:

† See Exercise 54, page 348.

21) $X \to Y$, $Y \to Z \vDash_P X \to Z$.

The non-transitivity of "if" can be analogously represented by the claim

22)   if A then B, if B then $C \nvDash$ if A then C.

This is said to be established by the invalidity of arguments like:

23)   If Smith dies before the election, Jones will win.

If Jones wins, Smith will retire from public life after the election.

So, if Smith dies before the election, he will retire from public life after the election.

(7) Whereas

24) $(X \& Y) \to Z$, $X \vDash_P Y \to Z$,

the alleged invalidity of the following example purportedly shows that the analogous fact does not hold for English:

25)   If this room is getting warmer and the mean kinetic energy of the molecules of its contents remains the same, then the scientific world will be astonished.

This room is getting warmer.

So if the mean kinetic energy of the molecules of this room's contents remains the same, then the scientific world will be astonished.[†]

In other words, it is claimed that

26)   [if $(A$ and $B)$ then C, A] $\nvDash$ if B then C.

(8) Whereas

27) $(X \to Y) \vDash_P (X \& Z) \to Y$, whatever Z may be,

the alleged invalidity of the following example purportedly shows that the analogous fact does not hold of English:

† See Exercise 55, page 348.

28) If I put sugar in this cup of tea it will taste fine.

So, if I put sugar and also diesel oil in this cup of tea, it will taste fine.[14]

In other words, the claim is that

29) [if A then B] $\nvDash$ [if (A and C) then B], whatever C may be.

(9) Whereas

30) $\vDash_P (X \to Y) \vee (Y \to X)$,

it is not the case that

31) $\vDash$ Either (if A then B) or (if B then A).

Suppose that Peter says that there will be a third world war and Quentin denies this. Then it seems that both "If Peter is right, then so is Quentin" and also "if Quentin is right then so is Peter" are false, so that substituting these sentences in (31) leads to a falsehood.[15] †

One who would defend the view that "if" or "if ... then ..." are truth functional against these charges will need some impressive resources. They will include the notion of implicature, mentioned, but not yet discussed.

I offer no separate discussion of the correspondence between "↔" and "if and only if".

6 Implicature

How is one to tell whether, for example

1) Jane Austen was buried at Winchester and died in 1817

is true or not (cf. (4.15))? Clearly it would help to know where and when Jane Austen died. There can be no general answer to how we know things like this, but let's pretend we do know that (1) correctly specifies the place and date of Jane Austen's death. This still leaves unresolved the question of whether (1) is true.

† See Exercise 55, page 348.

One way to proceed is to imagine someone addressing this remark to us. The thought irresistibly presents itself that there is something wrong. But does what is wrong consist in the remark failing to be true?

It has to be agreed that there are all sorts of ways in which a remark can sound "wrong" even if it is true; and sound "right" even if it is false. On being introduced to an ugly person, it would be wrong – morally wrong – to say

2)   You are without question the ugliest person I have ever met,

even if (2), as uttered in the circumstances, would be a literal truth. Now suppose that someone enquires whether you know anything about birds, and you wish to convey, in as graphic a fashion as possible, that you know next to nothing. You say

3)   I can't tell a crane from a canary.

This, we shall suppose, is not true. Still, it can be a perfectly proper thing to say – a permissible, and readily understood, exaggeration. No one would take you quite literally.

The first point to note, established by these examples, is that the truth of a remark is neither necessary nor sufficient for it being a right and proper one.

The next important point is that it is possible to *convey* something without strictly and literally *saying* it. It is probably this distinction which accounts for the point of the preceding paragraph. By uttering (2) you convey that you care nothing for the other person's feelings, though you do not say this. What is right about (3) is that it correctly conveys that you know next to nothing about birds.

The distinction is made even more graphically in a famous example. *H* asks *S* his opinion about Jones's qualities as a philosopher. (Jones is a student of *S*'s, and *H* has every right to know *S*'s opinion.) *S* replies:

4)   Jones has beautiful handwriting

and says nothing further. *S* conveys that Jones is a bad philosopher, but he does not say this.†

I once asked the famous philosopher, Gilbert Ryle, whether he liked music. He replied

† See Exercise 56, page 348.

5)   I can tell the difference between loud and soft.

Ryle conveyed that he did not like music; but this is not what he said.
In *Can You Forgive Her?* Mr Bott says to Alice Vavasor:

6)   The frost was so uncommonly severe that any delicate person
     like Lady Glencowrer must have suffered in remaining out so
     long. (Trollope [1864] p. 366)

Alice knew that Mr Bott knew that Alice had been out with Lady
Glencora, yet Mr Bott had made no enquiry about Alice's health.
According to Trollope, Mr Bott thereby conveyed that Alice was not
delicate (and so not upper class). But he did not say this. Note that what is
conveyed may be false even when what is said is true.

The nurse, beaming brightly, announces to the newly delivered mother

7)   Congratulations! It's a baby.

The nurse conveys (presumably in jest) that there was some significant
possiblity of the mother having given birth to something other than a
baby. But she does not say this.[†]

H. P. Grice coined the phrase "implicature" to apply to what, in such
examples, is conveyed but not said.[16] A truth may implicate something
false. (7) would normally be an example of this, but any of the other cases
might be examples, in appropriately altered circumstances. This possibility could be exploited by someone wanting to defend a truth functional
interpretation of (1). An utterer of (1) may implicate that the burial
occurred before the death, and this may be false; but all that (1) strictly
and literally says is that the two events occurred, so it is true. Hence it is
not after all a counterexample to the truth functionality of "and".

Grice claimed that what he called "cancellability" is a mark of
implicature, and will help us differentiate what belongs to implicature
and what belongs to strict and literal saying.[17] If *A* entails *B*, then "*A*, but
not *B*" will be contradictory. Thus "Tom is a bachelor" entails "Tom is
unmarried", and "Tom is a bachelor, but he's not unmarried" is
contradictory. This shows that Tom has to be unmarried, if what is
strictly and literally said by "Tom is a bachelor" is to be true.

By contrast, the following, uttered in the situation we envisaged for (4),
is perfectly consistent:

† See Exercise 57, page 348.

8) Jones has beautiful handwriting – though I don't mean to suggest that he is other than an excellent philosopher.

This shows that Jones's being a bad philosopher, or being thought by S to be such, is not a necessary condition for the truth of (4).

A defender of the truth functional interpretation of "and" can allow that in some sentences in which "and" occurs, something more like "and then" is implicated. But cancellability will show that it belongs to implicature only, and not to what is strictly and literally said. For example, in the circumstances envisaged for (4.14), one could quite consistently have said:

9) Mary got married and had a baby, but I'm not willing to pronounce on the correct order of these occurrences.

And in place of (1) it would be quite consistent to say

10) Jane Austen died in 1817 and was buried at Winchester, but I'm not saying which happened first.

(A plausible conversational background for this case would be one in which the general topic of discussion is cases of people being buried alive. One wishes to make clear that one is carefully remaining neutral on the question of whether Jane Austen was an example.)

The phenomena we have discussed in connection with "and" can be reproduced merely by the use of separate sentences. There is little to choose between (4.14) and

11) Mary got married. She had a baby.

Putting things in this order might well implicate that the corresponding events occurred in that order. But it would be absurd to suggest that either of the sentences, or the utterance as a whole, would be false (as opposed to, say, misleading) if the birth preceded the wedding. That the same phenomenon can arise in the absence of "and", and in a context in which it cannot be attributed to the truth conditions of any expressions involved, suggests that it should not be attributed to the truth conditions of "and".

Let us now resume the essential features of an implicature defence of a truth functional interpretation, applying it to the interpretation of "and" as expressing the truth function expressed by "&".

The objector claims that there are cases in which a compound has a truth value inconsistent with the truth functional interpretation. In the present case, the allegation is that conjunctions can fail to be true, even when both conjuncts are.

The defence consists in saying that a sentence which is strictly and literally true may have false implicatures. The objector, it will be claimed, has mistaken the falsity of an implicature for the falsity of the sentence itself – that is, for the falsity of what the sentence strictly and literally says. So the examples allegedly of false conjunctions with true conjuncts are really examples of true conjunctions which have false implicatures.

In the case of "and" this defence seems to me correct. But can the same defence speak to other alleged mismatches between English and **P**?

7 "If": implicature in defence of the truth functional interpretation

I envisage two strategies at the disposal of the defender of the truth functional interpretation of "if". The first is to exploit the notion of implicature to try to defuse alleged counterexamples. The second strategy, deferred until the next section, is to provide direct arguments for the interpretation.

We have made use of the distinction between saying and conveying, in particular the distinction between saying and implicating, while giving almost nothing by way of an account of the principles which govern it. We have mostly appealed to examples. The only theoretical point made was the Gricean claim that implicatures are cancellable. To launch a defence of the truth functional interpretation of "if" requires a more systematic account of implicature.

Consider an utterance which, though true, would give rise to a false implicature. For example, suppose that Jones really does have beautiful handwriting, and that you utter (**6.4**) in the envisaged circumstances. It will be entirely reasonable for your hearer to infer that you think ill of Jones as a philosopher. Suppose that in fact you think highly of him. Then, in uttering (**6.4**) you have spoken truly, but misleadingly. You *should not* have asserted (**6.4**). In the context, (**6.4**), though true, had (as I shall put it) a *low degree of assertibility*. One goes some way to providing an account of implicature by giving principles which determine features of an utterance which raise or lower its degree of assertibility.

One relevant feature (suggested by Grice) is that an utterance is more assertible the more informative it is, relative to the needs of the conversation. For example, if you are asked where Tom is and you know he is in the library, you should not reply

1) He is either in the library or in a lecture

despite the fact that (1) is true. (1) is less assertible, because less informative, than the equally true

2) He is in the library.

The principle is:

3) The more informative a true utterance is (relative to the conversational needs in question), the more assertible it is.[†]

A consequence of this principle is that very uninformative true utterances will normally have very low assertibility.[‡]

Let us apply this to the problem for the truth functional account of "if" posed by (4.10) and (4.11). Consider the argument

4) Ice floats on water. So, if ice is denser than water, it floats on water,

and the allegation that this argument is invalid (having true premise but false conclusion), whereas formalizing "if" by "→" yields a P-valid argument. Then the defence of the truth functional interpretation using (3) claims that the conclusion is not really false, and the argument is not invalid. If we think the conclusion is false, it is because we imagine it being asserted in the context of the propounding of an argument like (4). In such a context, the premise is much more informative than the conclusion, which is very uninformative. Anyone in a position to use the argument must be in a position to assert its premise. So (3) has it that, as a conclusion of the argument, "if ice is denser than water, it floats on water" is highly unassertible. If anyone wrongly thinks it is false, it is because they confuse low assertibility with falsehood.

One problem with this account is that it is open to question whether (3) is true. But there is also a more internal problem. Consider

† See Exercise 58, page 348.   ‡ See Exercise 59, page 348.

5)   Ice floats on water. So, if ice does not float on water, it floats on water.

Here, again, we have an apparent counterexample to the truth functional interpretation of "if", for we have an argument which is intuitively invalid, having true premise but false conclusion, but which would be **P**-valid were "if" formalized by "→".

However, the same defence cannot be brought to bear on this case as on (**4**). To see this, one must appreciate that

6) $\vDash_P X \leftrightarrow (\neg X \rightarrow X)$.

In other words, any sentence is equivalent to the material conditional having it as consequent and its negation as antecedent. Hence in one good sense of what it is for two sentences to be equally informative, a sense which the *proponent* of the truth functional interpretation of "if" cannot very well despise, the premise and conclusion of (**5**) are equally informative. Hence the defence of the truth functional interpretation which relies on (**3**) cannot be applied here; and that casts doubt on whether its application to (**4**) was legitimate. (**4**) and (**5**) appear to pose a common problem, requiring a single solution.

An alternative defence is based on the following principle:

7)   The utterer of a conditional implicates that he has good grounds, of certain standard kinds, for his utterance.

Standardly, a conditional is used only when the truth or falsity of the components is not known. One standard kind of ground is that one has good evidence for a generalization of which the conditional is an instance. For example, one may well know, from general principles governing floating and density, that anything denser than water will sink in water. You may be ignorant of the density of some particular substance, say bakelite, relative to water. Still, you know from the general principle that if bakelite is denser than water, it will sink in water. The defence of the truth functional interpretation based on (**7**) has it that many of the apparent anomalies can be explained in terms of the falsity of implicatures of this kind.

For example, (**4.33**), viz.:

if this volume of water is heated to 90° at some time during the coming year, this volume of water will turn to ice will be held to implicate that there is some general connection between heating water and it turning to ice. Since there is no such connection, the implicature is false. When people allege that (**4.33**) is false, they are responding to the falsity of the implicature. On the assumption that this volume of water will not be heated to 90° at any time during the coming year, what (**4.33**) strictly and literally says is true, or so the defender of the truth functional interpretation will urge. The different approaches are marked by the fact that those who are confident of the falsity of (**4.33**), unlike the defender of the truth functional interpretation, feel no need to find out whether or not its antecedent is true.

The style of defence has some plausibility for a number of the alleged counterexamples to truth functionality given in §5. For example, it says something relevant to (**5.14**), viz.:

> It is not the case that if the number three is even then it is prime. So the number three is even.

To serve as a counterexample, it is necessary that the premise be true. For this, it is necessary that

8)   if the number three is even then it is prime

be false. But, the defence goes, what we are responding to when we think (**8**) is false is merely the false implicature that being even is in general a sufficient condition for being prime. This is only an implicature: what (**8**) strictly and literally says is true.

However, there are counterexamples about which (**7**) has little to say. For example, it is hard to see how it could even address the problem posed by (**5.23**), viz.:

If Smith dies before the election, Jones will win.
If Jones wins, Smith will retire from public life after the election.
So, if Smith dies before the election, he will retire from public life after the election.

There may well be appropriate general grounds for the premises: for the first, that it is a two-horse race; for the second, the honourable Smith's sincere declarations. It seems that only a determination to defend a truth functional interpretation come what may would lead one to accept that the conclusion is, despite appearances, true.[†] Likewise, (**7**) would appear unable to address (**5.20**), viz.:

† See Exercise 60, page 348.

If it rains, then it will not rain heavily.
It will rain heavily.
So it will not rain.

And it has nothing to say about the sentence discussed in connection with
(**5.30**):

> Either, if Peter is right then so is Quentin, or, if Quentin is right
> then so is Peter.

Before turning to other matters, I shall consider one further principle
that might serve the cause of defending a truth functional interpretation
of "if".

> **9)** It is conveyed (but not said) that a conditional is "robust with
> respect to its antecedent".

One principal use of conditionals is in modus ponens arguments: ones
having the form

> **10)**   If A, then B
>
> A
>
>   Therefore B.

For this use to be possible, it must be possible to hold both to the
conditional, "if A then B", and to its antecedent, "A". This means that it
has not to be the case that were one to come to have evidence for "A",
one's evidence for "if A then B" would thereby be undermined: in short,
"if A then B" must be robust [18] with respect to "A". If a conditional were
not robust with respect to its antecedent, then one could never use it in a
modus ponens argument, for one could not have a body of evidence which
would support both of the needed premises.

Here is an example of failure of robustness in a different connection.
Suppose in January I am convinced that:

> **11)**   John will finish his book by April, or at any rate by May.

If I learn in June that the book is still unfinished, and hence infer

12)   John did not finish his book by May

I do not combine (**11**) and (**12**) and go on to infer:

13)   John will finish his book by April.

Rather, I abandon (**11**). We can express the phenomenon by saying that (**11**) is not robust with respect to the falsity of its second disjunct.

Let us apply (**9**) to the allegedly invalid pattern of argument (**5.8**):

$A \vDash$ [if not-A then B], whatever B may be.

If I used this to establish a conditional, it would not be robust with respect to the antecedent. For to use the argument in this way depends upon having good evidence for "A". Hence, subsequently acquiring evidence for the antecedent of the conditional would lead me to abandon the conditional, rather than use it to infer B. We are invited to conclude that arguments of this kind are valid, though as the conclusions will be very fragile indeed with respect to their antecedents, by principle (**9**) they will have very low assertibility. Confusing low assertibility with falsehood, we wrongly think the conclusions are false, and so think that the pattern of argument is invalid.[†]

An interesting application of (**9**) is to (**5.20**).[‡] However, it is far from clear that (**9**) could deal with all the alleged counterexamples: (**5.23**) and (**5.30**) again appear to be resistant.

Perhaps what is needed is some combination of the principles we have discussed, supplemented with further principles as well. But you may be impatient: isn't all this rather *ad hoc*? And in any case, how convincing is the crucial claim that we mistake low assertibility for falsehood? Let us ask ourselves not whether we would be happy to *assert* the sentences having low assertibility but whether we would happily *believe* what they say. How could assertibility enter into belief? Yet we are inclined to suppose, for example, that even if we believe that A it does not follow that we are in error if we fail to believe that if not-A then B, with respect to every B. It is not obvious that the case against the truth functional interpretation essentially involves thinking of particular utterances of conditional sentences, as opposed to thinking of how things are, and ought to be, with conditional beliefs.

It is hard to find neutral ground from which to assess such issues. If

---

† See Exercise 61, page 349.   ‡ See Exercise 62, page 349.

they can be treated adequately, it will be on the basis of comparing the truth functional account of indicative conditionals with alternatives (see chapter 3). I turn now to the second strategy available to the defender of the truth functionality of "if".[†]

8   "If": direct arguments for the truth functional interpretation

I shall consider two direct arguments for the truth functionality of "if". They both proceed by making claims about the validity of argument patterns involving "if", differing only in which argument patterns they select.

(1) The first argument

The following argument appears valid:

1) Either the butler or the gardener did it. Therefore if the gardener didn't do it, the butler did.

Suppose, as is natural, that the first premise is equivalent to

2) (the butler did it) ∨ (the gardener did it).[19]

Then it is equivalent to

3) (the gardener didn't do it)→(the butler did it).

Then the validity of (1) guarantees the validity of

4) (the gardener didn't do it)→(the butler did it), therefore if the gardener didn't do it, the butler did.

Assuming that this holds generally as a matter of form, and is independent of the particular subject matter of gardeners and butlers, then we can conclude

† See Exercise 63, page 349.

5) Any sentence of the form "$A \to B$" entails the corresponding sentence "if A then B".

It is generally accepted, and seems in any case obvious, that

6) Any sentence of the form "if A then B" entails the corresponding sentence "$A \to B$".

(5) and (6) together ensure that "$\to$" and "if" are equivalent, and thus they ensure the truth functionality of "if".

The most promising way to avert the conclusion of this argument is to deny the validity of (1). This avenue is discussed in chapter **3.1**.[†]

(2) *The second argument*

First premise:

7) A, if A then $B \vDash B$.

This is the principle of modus ponens, and is generally uncontested. It ensures that a true antecedent and a false consequent is enough for a false conditional.[‡] If we imagine trying to construct a truth table for "if" on the lines of the truth table for "$\to$", this fact secures the correctness of the second line of the truth table (table 2.3).

Second premise:

8) If $[A_1, \ldots A_n, B \vDash C]$ then $[A_1, \ldots A_n \vDash$ if B then $C]$.

This is a more controversial premise. It says, in effect, that if we have a valid argument for a conclusion, any one of the premises can be dropped, provided that it is made the antecedent of the conditional whose consequent is the original conclusion; and then the reduced premises will entail the new, conditional, conclusion. Let us take this principle, which might be called that of "conditional proof", on trust for the moment, and show how it would establish the remaining properties needed for the truth functionality of "if". Later we shall see how one might argue for the principle of conditional proof itself.

Let us recall two results from chapter 1:

† See Exercise 64, page 349.   ‡ See Exercise 65, page 349.

Table 2.3

A	B	If A then B
T	T	
T	F	F
F	T	
F	F	

1.6.5)   If $[A_1, \ldots A_n \vDash C]$ then $[A_1, \ldots A_n, B \vDash C]$, whatever B may be.

1.6.7)   If C is among the $A_1, \ldots A_n$ then $[A_1, \ldots A_n \vDash C]$.

As a special case of the latter, we have

9) $B \vDash B$.

So, using (**1.6.5**), we have

10) $B, A \vDash B$.

So, by (**8**), we have

11) $B \vDash$ if A then B.

This shows that a true consequent is sufficient for a true conditional,[†]

Table 2.4

A	B	If A then B
T	T	T
T	F	F
F	T	T
F	F	

† See Exercise 66, page 349.

establishing the first and third lines of the table. Our table now becomes table 2.4.

Another result from chapter **1** is:

1.6.8)   If $[(A_1, \ldots A_n) \vDash]$ then $[A_1, \ldots A_n \vDash B]$, whatever B may be.

This says that if the premises of an argument are inconsistent, then the argument is valid, no matter what its conclusion is. Hence we have:

12) B, not-$B \vDash A$.

So by (**8**):

13) $B \vDash$ if not-B then A.

This establishes that the falsity of the antecedent is enough for the truth of a conditional, and so establishes the third (again!) and fourth lines of the table. We now have the complete table (2.5) for "if A then B". We can conclude that (**7**) and (**8**) between them entail that English conditionals expressed by "if ... then ..." are truth functional, and express the same truth function as "→".[20] The question now is whether (**7**) and (**8**) themselves can be justified.

(**7**), expressing modus ponens, is hardly ever questioned, so I shall assume that we can accept it without further ado. (**8**), the principle of conditional proof, has been found more controversial, so let us see if we can find an argument for it. It will be easier if we abbreviate "$A_1 \ldots A_n$" as "A", and so write (**8**) as follows:

14)   if $[A, B \vDash C]$, then $[A \vDash$ if B then $C]$.

Table 2.5

A	B	If A then B
T	T	T
T	F	F
F	T	T
F	F	T

This antecedent of this conditional means (using an equivalent of the definition of "⊨" in Chapter **1.3**):

> **15)** It is logically necessary that, if A and B are true, then so is C.

So:

> **16)** It is logically necessary that, if A is true, then, if B is true, so is C.

So:

> **17)** It is logically necessary that, if A is true, then if B then C is true.

But this means:

> **18)** $A \vDash$ if B then C.   QED.

The principles of reasoning involved seem to be modest. But they have been firmly resisted by theorists who contest the truth functional interpretation of "if". Until we examine alternative theories of "if", our conclusion has to be simply that such theories will have to provide good reasons for thinking that the reasoning (**14**) to (**17**) above is unsound.[†]

We can summarize the conclusions of §5 and §8 as follows: there are apparently compelling arguments for the conclusion that "if" is not truth functional, and apparently compelling arguments for the conclusion that it is not. Appearances must deceive. Some attempts to say how they deceive are considered in chapter **3**.

In considering the use of **P**-validity in the investigation of validity in English, we will explore the consequences both of "if" being truth functional and of it not being truth functional.

9   Non-truth functionality in English

To deepen understanding of claims to the effect that some English expression is a truth functional sentence connective, let us examine

† See Exercise 67, page 349.

claims to the effect that an expression is a sentence connective, but one which is not truth functional.

(1) John believes that

This expression seems to be syntactically like "not". That is, it seems to be a unary sentence connective, taking a sentence like "the earth is flat" to form a new sentence, "John believes that the earth is flat". So the expression appears to function as a sentence connective.[21] If it is a sentence connective, it is certainly not truth functional: John may have both true and false beliefs, so the truth value of "John believes that A" is not a function of the truth value of "A".

(2) Because

On at least some occurrences, "because" seems to function as a binary sentence connective (one which takes two sentences to make a sentence). For example

  1)   John shot Robert because Robert betrayed him.

If we take this appearance at face value, the straightforward response is to see "John shot Robert" (abbreviate to A) and "Robert betrayed John" (abbreviate to B) as the two components, in which case we are forced to regard "because", thus used, as non-truth functional. This view could be vindicated by attempting to construct a truth table in which we envisage various theoretically possible truth values for the components, and ask, with respect to each, whether (1) would then be true or false (table 2.6). A necessary condition for the truth of "because" sentences is the truth of both components. But this is not sufficient. If it were, the result of inserting "because" between arbitrary truths would always be a truth, and this is plainly not so.[22]

Table 2.6

A	B	A because B
T	T	?
T	F	F
F	T	F
F	F	F

Frege ([1892b] pp. 76–7) suggested that "because" sentences are truth functional, but are not a truth function of the components we have so far identified. Discussing the sentence

2)   Because ice is less dense than water, it floats on water

he suggested that there is a concealed component, in addition to "ice is less dense than water" and "ice floats on water". The concealed component, he suggested, is

3)   Whatever is less dense than water floats on water.

The truth function is simply that of conjunction, applied to the *three* components.

A difficulty with this suggestion (and I think the difficulty is insuperable) is that there is no systematic way of eliciting the concealed component. What stands to (1) as (3) does to (2)? It would be absurd to suggest that (1) entails that whoever Robert betrays shoots him, or that whoever betrays John is shot by him, or that whoever betrays anyone is shot by the betrayed.

A promising suggestion is that "because" is not really a sentence connective at all, but rather functions, somewhat analogously to "therefore", to mark an act of inference. We saw in chapter **1.5** that something like "*A* therefore *B*" cannot be evaluated as true or false. Rather, we have to ask whether the argument thereby presented is valid. "Therefore" is not a sentence connective, since what is produced by placing it between two sentences is not itself a *sentence*, something capable of having a truth value.[†] Similarly, perhaps "because" forms not a sentence, but something more like an argument. (Very often, as in (1), one would be expected to evaluate the argument by inductive rather than deductive standards.)

How could this suggestion be tested? One relevant point is that certain complexes containing "because" appear not to operate in the way one expects of genuine *sentences*. For example, if ¢ is a sentence connective, then "if *A*, then (*B* ¢ *C*)" is a conditional, and on every account of conditionals neither the truth of *B*, nor of *C*, nor of *B* ¢ *C* can be necessary for the truth of the conditional. But consider

4)   If John comes to the party, then (Mary will leave because she has quarrelled with him).

† See Exercise 68, page 349.

It would seem that (4) is true only if Mary has quarrelled with John, and this would be hard to explain if "because" were a sentence connective.[†]

(3) But

I hold that "but" differs in meaning from both "and" and "&". This does not show that "but" does not express the conjunction truth function, for there can be differences of meaning which do not impinge on truth. When "but" functions as a sentence connective (which it does not always do)[‡] it expresses precisely the conjunction truth function – that expressed by "&". The only doubt is whether the truth of both A and B is enough for that of "A but B". This doubt is assuaged by the reflection that it is correct to infer from the falsehood of "A but B" that at least one of A, B is false.

The additional component in the meaning of "but", as opposed to "and", consists in the fact that one who asserts that A but B represents himself as supposing that in the context there is some contrast between the truth of A and that of B, something surprising, poignant, or worthy of special note or emphasis in the fact that both are true. Sometimes the element of surprise or whatever derives from A itself, as in the stock example "She was poor but honest". Sometimes it derives from something else, as in "The best recent book on perception is Jackson's but unfortunately it's not available in paperback".[23]

(4) When

Consider

    5)   When beggars die, there are no comets seen.

The following argument might be used to show that "when", on such an occurrence, is a non-truth functional sentence connective: "Beggars die" and "There are no comets seen" are both declarative, indicative sentences, capable of being used self-standingly, and each evaluable for truth and falsehood. So "when" is a sentence connective. But it is not truth functional, since the truth of the two components leaves undetermined the truth value of the whole.[¶] Equally, it fails the substitution test.

---

† See Exercise 69, page 349.   ‡ See Exercise 70, page 349.
¶ See Exercise 71, page 349.

If (5) is true, it does not remain so when "Comets are seen" replaces "Beggars die".

The argument is fallacious, for, *as they occur in (5)*, "Beggars die" and "There are no comets seen", are not genuine self-standing sentences. (5) means something like

6) Any time at which (only?) beggars die is a time at which no comets are seen.

Here there is no plausibility to the idea that there are two sentential components. Used self-standingly, "Beggars die" means something like

7) Every beggar dies sometime;

and "There are no comets seen" means something like

8) No comets are visible now/ever,

context determining the choice of "now" or "ever". It is plain that this is not what the expressions mean as they occur in (5).

"When", as it occurs in (5), is thus not an example of a non-truth functional sentence connective, since in that occurrence it is not an example of a sentence connective.[24] This does not preclude there being other sentences in which it is a genuine sentence connective.[†]

10 From *P*-validity to validity

Logic is the study of reasoning, and of one vital feature that reasoning should possess, namely validity. Introducing the language **P** affords the prospect of serving in two ways our desire to understand validity in English: first, it might enable us to attain useful generalizations about validity in English, and offer ways of testing for validity in English in doubtful cases; secondly, the definition of **P**-validity may have virtues lacked by the original definition of validity in chapter **1**.

In §**3** we discussed some ways of formalizing English arguments, but we left two questions with at best incomplete answers. One was: what can be inferred from the fact that an argument's logical form is **P**-valid? The

† See Exercises 72–3, page 349.

other was: what can be inferred from the fact that an argument's logical form is not **P**-valid?

With some qualification, the answer to the first question is that one can infer the validity of the English. What we have to show is that if a **P**-argument ϕ is **P**-valid and is an *adequate* formalization of an English argument ψ, then ψ is valid.

The adequacy of the formalization ensures that, where ϕ' is the argument recovered from ϕ by applying the correspondences associated with the formalization, ϕ' says the same as ψ. The notion of "saying the same" that is required here is that the sentences of ϕ' should have the same truth conditions as the corresponding sentences in ψ. For validity can be defined in terms of truth conditions, and if two arguments are related as ψ and ϕ' then, necessarily, both or neither are valid. So it will be enough if we can show that if ϕ is **P**-valid then ϕ' is valid.

A necessary condition for there being any adequate formalizations is that the **P**-connectives make the same contribution to truth conditions as the corresponding English expressions. We saw that this was doubtful in the case of "$\rightarrow$" and "if". Let us suspend this doubt for the moment. Then the following argument establishes what is needed:

1)   (i)   Suppose ϕ is **P**-valid.
     (ii)   Then every interpretation upon which the premises of ϕ are true is one upon which the conclusion is true.
    (iii)   Hence whatever may be the truth values of the components of ϕ', if the premises of ϕ' are true, so is the conclusion.
     (iv)   Hence, of logical necessity, all the conditions under which the premises of ϕ are true are conditions under which its conclusion is true.
      (v)   Hence, the truth conditions of the premises are contained within those of the conclusion.

(v) is equivalent to the claim that ϕ' is valid, in the sense of $\vDash$.

One crucial step is from (ii) to (iii). This essentially requires that every English expression in ϕ' corresponding to a constant in ϕ expresses the same truth function as the **P**-constant. Those who dispute that "if" is truth functional will dispute this step.

Another crucial step is that from (iii) to (iv). The former makes no explicit mention of logical necessity, the latter does. How can the intrusion of this notion be justified?

Consider a very simple argument, like

2) Either John is happy or Mary is. But John is certainly not happy. Therefore Mary is happy.

No doubt there are all sorts of intrinsically different conditions under which the premises are true. For example, John's unhappiness might stem from frustrated ambition, misfortune in love, or whatever. (iii) appears not to allude to all these possibly different conditions, yet (iv) does: it speaks without qualification of *all* the conditions under which the premises are true.

Conditions, like anything else, can be classified either coarsely or finely. It is not that (iii), as applied to (2), fails to speak to some conditions under which the premises are true; rather, it classifies these conditions rather coarsely. In this example, they in effect fall into a single category: those in which "John is happy" is false (as required for the truth of the second premise), and in which "Mary is happy" is true (as required for the truth of the first premise, given what is required for the truth of the second). However the conditions in this category may vary, they all have the common property of verifying the conclusion. Hence all the conditions – all logically possible conditions – under which the premises are true are conditions under which the conclusion is true.

As I mentioned, if "if" is not truth functional, (iii) does not follow from (ii). We will continue to assume that, even so, "if" entails "→", and that it is only the converse entailment that fails. In other words, an "if" sentence is stronger than the corresponding "→" sentence.

There are two cases, which have to be treated separately.

(a) The English argument contains "if" at most in the premises. In this case, the inference from **P**-validity to validity goes through. The reason is that the **P**-premises will be weaker than the English premises, and strengthening an argument's premises cannot invalidate it (cf. **1.6.5**).

(b) The English argument contains "if" in the conclusion. This divides into a number of subcases, depending on whether "if" is or is not the dominant connective. Let's just consider the case in which it is. Then the **P**-conclusion is weaker than the English conclusion. That the **P**-premises establish the weaker leaves open that they might not be strong enough to establish the stronger conclusion. So there is no correct inference from **P**-validity to validity.

The second question requires a more complex answer. If a **P**-argument is invalid, there are three possibilities for the validity of the English argument:

(1)   It is valid in virtue of its **P**-logical form.
(2)   It is valid, but not valid in virtue of its **P**-logical form.
(3)   It is invalid.

I shall illustrate these possibilities in turn.

The first possibility is illustrated by (**3.8**): we saw that the formalization of an argument which is not only valid, but valid in virtue of its **P**-logical form, may be adequate yet invalid, through failing to be deep enough.

The second possibility is illustrated by (**3.19**). This shows that not all validity is validity in virtue of **P**-logical form.

The third possibility is illustrated by (**3.14**).

P-logic offers us, naturally enough, no way of differentiating between cases (**2**) and (**3**), so the strongest conclusion we can reasonably draw from the **P**-invalidity of the deepest **P**-formalization we can find is that validity in virtue of **P**-logical form is a property which the formalized English argument lacks.[†]

We said in chapter 1 that the definition of validity there offered used a notion which could do with elucidation: that of it being *logically impossible* for all the premises to be true, yet the conclusion false. The definition of **P**-validity dispenses with this notion, and employs instead the notion of interpretation.

Generalizing from the remarks adduced to support the move from (iii) to (iv) in (**2**), one way to classify the logical possibilities for the truth of a sentence dominated by a truth functional sentence connective is according to the truth values for the components which determine the sentence itself as true. *All* conditions for the truth of the sentence will be embraced by this classification, though, of course, there are finer classifications available which this one ignores. The notion of a **P**-interpretation thus gives a partial elucidation of the notion of logical impossibility (or logical necessity) used in the definition of validity in chapter 1. The range of logical possibilities can be divided into interpretations. A sentence whose formalization is true on no interpretation represents a logical impossibility. A sentence whose formalization is true on all interpretations represents a logical necessity.

We can draw a stronger conclusion: if an English argument's formalization is valid, the argument itself is not merely valid, but formally so. This is because its validity depends only upon the meanings of the logical constants (the expressions corresponding to the **P**-connectives) and the pattern of occurrence of the sentences.

† See Exercise 74, page 350.

A traditional aim of logic has been to render mechanical the determination of the validity of arguments. **P** satisfies this aim. One can construct truth tables to check mechanically for the **P**-validity of **P**-arguments. There are also many computer programs which will do the same job.[25]

Finally, we should note that ⊨**P** has the general properties ascribed to ⊨ in chapter **1.6**.†

† See Exercise 75, page 350.

Notes

The history of logic goes back at least to Aristotle (384–322 BC). A classic account is Kneale and Kneale [1962]. Propositional logic was known to the Stoics (third century BC) but their work was lost (see Kneale and Kneale, ch. 3). George Boole developed a system equivalent to propositional logic in the mid-nineteenth century. The first wholly modern presentation is Frege [1879]. Good introductory texts are Lemmon [1965] and Hodges [1977].

A good discussion of (among many other things) the relationship between the **P**-connectives and English connectives is in Strawson [1952] ch. 3, section II.

On conditionals, the contrast between (**4.34**) and (**4.35**) derives from Adams [1970], and his [1975] contains a number of arguments against the material conditional interpretation. I have drawn heavily on Jackson [1979]; see also his [1981], and Lewis [1976] pp. 142–5 and postscript. Jackson [1987] unfortunately appeared too late for me to take it into account. Further references on conditionals appear in the bibliographical notes to ch. 3.

For Grice's work on implicature, see Grice [1961] (the first presentation of the notion, rather deeply embedded in a discussion of perception, but from which the famous (**6.4**) is drawn), [1975] (where he explicitly addresses an implicature defence of the truth functional interpretation of "if") and [1978]. See also McCawley [1981] ch. 8, esp. section 3.

The butler and gardener example (**8.1**) comes from Stalnaker [1975], which contains a very sophisticated account of why we tend *wrongly* to suppose that it is valid.

For the relation between the classical conception of validity (here represented by ⊨) and the material implication interpretation of conditionals, see Read [1988], ch. 2.

Frege's [1892b] defence of the truth functionality of natural language is contained in one of the most famous and important articles in philosophical logic.

---

1 " − " and " ∼ " (the last being a tilde properly so-called) are sometimes used in place of "¬"; ".", " ∧ " or simple juxtaposition in place of "&", " ⊃ " instead of "→", and "≡" instead of "↔". For a fuller list of variants, see Kirwan [1978] p. 280.

2 You might reasonably have another worry: if **P** is really a language, then, setting aside ambiguity as a special case, only one interpretation can be correct. The only reply is: in this sense, **P** is not a language! For further discussion see below, ch. **6.6**, Smiley [1982], Kirwan [1978] pp. 3–8, 32–41.

3 "*X*" and "*Y*" are called *metalinguistic variables* relative to **P**.

4 Compare the definition in ch. **1.4**.

5 Where a sentence connective ¢ takes some number n of sentences to make a sentence, let's call ¢ an *n-ary* sentence connective. We can write an arbitrary sentence resulting from applying ¢ to the appropriate number of sentential components, $x_1 \ldots x_n$, as $¢(x_1, \ldots x_n)$. An n-ary truth function is a function from n-ary sequences of truth values to a truth value. An n-ary sentence connective, ¢, *expresses* an n-ary truth function f, iff f is an n-ary truth function and for every sentence $¢(x_1, \ldots x_n)$, where Σ is the sequence of truth-values possessed by $x_1, \ldots x_n$, $f(\Sigma)$ is the truth value of $¢(x_1, \ldots x_n)$.

6 Officially, "*s*" is not a **P**-letter, but let it be regarded as an abbreviation of "*p'''*"; similarly for "*t*", "*u*", etc.

7 The emphasis is on "seem", for in chapter **4.2** a semantics is offered which unifies these kinds of occurrences of sentence connectives with their occurrence as explicit sentence connectives.

8 It has also been suggested that there are cases in which "and" is a sentence connective but not a binary one: see McCawley [1981] pp. 78–81.

9 The best putative examples of exclusive disjunction tend to involve rather complex and poorly understood constructions, like the one in the example. It may well be that it is these constructions, rather than "or", which are responsible for any exclusivity. For a recent illuminating account see Higginbotham [1988], esp. pp. 226–7.

10 For a justified protest at the standard treatment of "*A* only if *B*" as an idiomatic variant of "if *A* then *B*" see McCawley [1981] pp. 49–54.

11 It is one thing to say that the two types of sentences require different accounts, and another thing to say that there is no uniform account of the meaning of "if". The word might have a single meaning in both sorts of sentence, the difference in their overall truth conditions being attributable to, say, their different moods, rather than to an ambiguity in "if". Arguably, it is a precondition for an adequate account of "if" that it discerns no ambiguity in the word as it occurs in both of (**34**) and (**35**). Such an account is mentioned in ch. **3.2**.

12 We extend the definitions of the **P**-connectives in an obvious way, so as to allow them to stand between English sentences. For example, if "A" and "B" are English sentences, then "$A \to B$" is true iff "A" is false or "B" is true.

13 There are other contexts in which this sentence certainly does have a truth value. But intuitively, in these other contexts the sentence requires a different interpretation from what is appropriate to (**35**) or (**36**).

14 Note, however, that, as in the case of (**4.39**), there is room to doubt whether this should be classified as indicative, rather than subjunctive. The firmly indicative

> If I put sugar in this cup of tea it tasted fine.
> So, if I put sugar and also diesel oil in this cup of tea, it tasted fine

does not yield any clear intuition of invalidity.

15 This powerful objection is adapted from Read [1988] pp. 23–6. The present work was in almost complete form when Read's book appeared; otherwise I would have been able to profit more from it.

16 Grice's phrase was "conversational implicature". He distinguished this from "conventional implicature". Arguably, some of the implicatures associated with the assertion of conditionals (though not, I think, conjunctions) should count as conventional rather than conversational. (See Jackson [1981].) To allow room for this alternative (which I do not explicitly discuss), I have mostly dropped the qualifier "conversational".

17 In Grice, cancellability is a sign of conversational, but not of conventional, implicature. See previous note. For example, Grice suggested that "and" and "but" agree in expressing the same truth function, but differ in that "but" has some kind of contrastive *conventional* implicature (see also below, §9), which is, accordingly, not cancellable.

18 The expression "robust", its use in a defence of a truth functional interpretation of "if", and example (**11**), all come from Jackson [1979]; these views can also be found, in a broader setting, in his [1987]. The most succinct account of the position can be found in the postscript to Lewis [1976] in Lewis [1986b].

19 Remember that we have extended the definitions of the **P**-connectives in an obvious way, so as to allow them to stand between English sentences. For example, if A and B are English sentences, then "$A \lor B$" is true iff at least one of "A", "B" is true.

20 For a more elegant derivation of the conclusion from these premises, see Hanson [1991].

21 For a different view, see ch. **4.16**.

22 Give an example of a false "because" sentence whose components are both true.

23 On "but", see Jackson [1981], §3. He does *not* use the example given above!

24 This point is made by Frege [1892b] p. 72. His overall aim was to show that *all*

sentence connectives are truth functional. Hence his account of "because", mentioned earlier.

25 **P** is decidable, that is, there exists a decision procedure for it (cf. e.g. Kirwan [1978] pp. 169ff.).

*3

Conditionals

The previous chapter presented a paradox: there are apparently compelling arguments both for and against the view that English conditionals are material implications. The aim of the present chapter is not to resolve the paradox, but to explore some alternative approaches to conditionals.

1 Conditionals and probabilities

Suppose there are two urns, A and B. Each contains a million marbles. All the marbles in A are white and so are all except one in B, where the exceptional one is black. Suppose you know all this. An urn is selected by some random method so that you don't know which is before you, and you take a marble at random without looking at it. You should certainly believe:

    1)   If this is urn A, I am not holding a black ball

and

    2)   If I am holding a black ball, this is urn B.

Should you believe:

    3)   If this is urn B, I am not holding a black ball?

It may be argued that you should, for (3) has a consequent which is made extremely probable by the antecedent: there's nearly a million to one chance against "I am not holding a black ball" being false, given that "This is urn B" is true.[1]

The example suggests the following general hypothesis: a conditional is highly probable (and so should be believed and asserted) iff the probability of the consequent is high, given the antecedent. In a more shorthand form, the hypothesis is that the probability of a conditional is equal to the corresponding "conditional probability".

Let us use "$Pr(A)$" to stand for "the probability of A", and represent probabilities by numbers in the interval (0,1). Instead of saying that the probability of A is fifty-fifty we will write "$Pr(A) = 0.5$". We will write the conditional probability of B, given A, that is, "$Pr(B)$, given A" as:

$$Pr(B|A)$$

which is standardly defined as:

$$Pr(B|A) = Pr(B \ \& \ A) \div Pr(A), \text{ if } Pr(A) > 0.\text{[2]}$$

(NB: if $Pr(A) = 0$, the conditional probability is not defined.)

The hypothesis suggested by (3) can now be written as follows:

4)   Pr (if A then B) = $Pr(B|A)$.

Strictly speaking, the probability we are talking about is relative to a person and a time. It is intended to represent the actual strength of the beliefs of a person, at a given time, if he satisfies some minimum standards of rationality. Two people, even if equally rational, may assign different probabilities to the same proposition, for one may have some information the other lacks. Hence the probability of A for one subject may differ from its probability for another. Similarly, a single person's probabilities may change, as he forgets what he knew or acquires fresh information. We will abstract for the moment from these relativities by pretending that we are dealing with the probability assignments of some one unspecified rational agent at some one unspecified time.

Suppose that (4) is the central fact about conditionals. What kind of logic would result? We saw that many of the objections to the material conditional interpretation of "if" pointed to alleged discrepancies between the logic of "→" and that of "if": cases of arguments containing "→" which are valid, even though the corresponding arguments in which "if" replaces "→" allegedly are not. Does (4) suggest a logic from which such discrepancies are absent?

Before we can address this question, we need a probabilistic conception of validity, that is, an account of what it is for an argument to be valid which draws on the notion of probability, rather than, as ⊨ does, on the notion of truth. A preliminary idea would be the following:

5) An argument, $A_1, \ldots, A_n; C$, is probabilistically valid (for short, $A_1, \ldots, A_n \vdash C$) iff it is logically impossible for the premises all to have high probability and the conclusion to have low probability.[3]

Relations among probabilities can be represented diagrammatically. We can associate a proposition with an area, and think of the points in the area as representing ways in which the proposition could be true: the larger the area, the more ways it could be true, and so the higher the probability. The areas we draw will lie within a square. If the probability of a proposition, A, is represented by the shaded area in figure 3.1, then the probability of its negation, not-A, is represented by the shaded area in figure 3.2. This corresponds to these facts: the ways in which A can be true exclude the ways in which not-A can be true; and the ways in which each can be true jointly exhaust all the possibilities.

The intersection of two areas measures the probability of the conjunction of the corresponding propositions: it marks out the possibilities which would verify both propositions. According to (4), the probability of "if A then B" is measured by the proportion of the A-area that falls within the B-area. If it all does, the probability is 1; if none does, it is 0.

Let us consider the probabilistic analogues of (**2.5.7**) and (**2.5.11**):

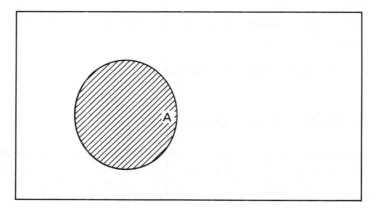

Figure 3.1

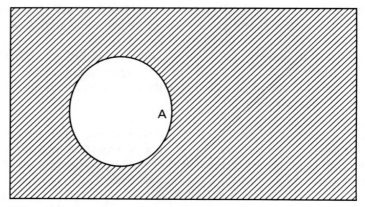

Figure 3.2

Figure 3.3   *Against* (7)

6) $A \vdash$ [if not-A then B], whatever B may be

and

7) $B \vdash$ [if A then B], whatever A may be.

We can use diagrams of the kind described to show that these are both
false. To show that (7) is false, note that in figure 3.3 "B" is highly
probable, occupying almost all the total area, whereas "if A then B" has
probability zero, since none of the A-area overlaps the B-area.[†]

† See Exercise 76, page 351.

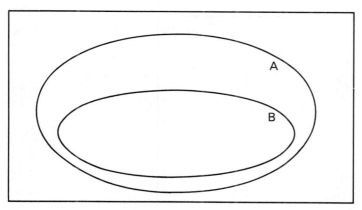

Figure 3.4   *Against* (**8**)

I now provide diagrams to show that not all the principles needed in the arguments for the truth functional interpretation given in chapter **2.8** are probabilistically valid.

The first argument for this conclusion depended essentially upon:

8)   *A* or *B* ⊨ if not-*A* then *B*.

The probability of a disjunction is represented by the sum of the areas corresponding to the disjuncts. In figure 3.4 "*A* or *B*" is highly probable (though admittedly *B* is not making a distinctive contribution!) but "if not-*A* then *B*" has probability zero. Hence the probability theorist has reason to reject the first argument given in chapter **2.8** for the truth functional interpretation of "if".

The second argument given for this conclusion depended upon the principle of conditional proof, (**2.8.8**):

If [A_1, ... A_n, *B* ⊨ *C*] then [A_1, ... A_n ⊨ if *B* then *C*].

In figure 3.5 each of *A*, *B*, and *C* have high probability (greater than 0.5). $\Pr(A)$ is represented by the long oblong, $\Pr(B)$ by the tall oblong on the right, and $\Pr(C)$ by the tall oblong on the left. This is consistent with the truth of *A*, *B* ⊢ *C*. However, only a small proportion of the *B*-area falls within the *C*-area, so "if *B* than *C*" has low probability, even though *A* has high probability. The probability theorist is thus in a position to reject the second of the arguments given in chapter **2.8** for the truth functional interpretation of "if".

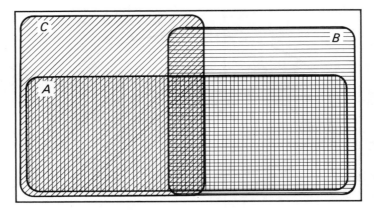

Figure 3.5   *Against conditional proof*

The probability theorist is not open to some of the objections directed against the truth functional interpretation of "if" (see chapter **2.5**). For example, the following, whose validity would represent a form of the logical principle called "contraposition",

    9)   If *A* then not-*B*; if *B* then not-*A*

is probabilistically invalid. This is shown by figure 3.6. "If *A* then not-*B*" has high probability, since almost all the *A*-area lies within the not-*B* area, whereas "if *B* then not-*A*" has probability zero. This has implications for (**2.5.20**).

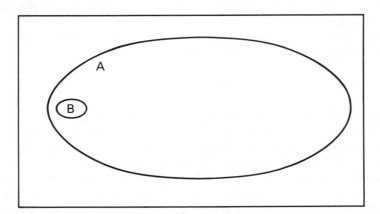

Figure 3.6   *Against contraposition*

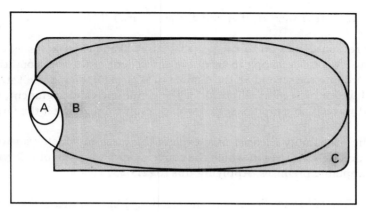

Figure 3.7   *Against transitivity*

    10)   If *A* then *B*, if *B* then *C*; if *A* then *C*,

which represents the alleged transitivity of "if", is also probabilistically invalid, as figure 3.7 shows. Here "If *A* then *B*" has probability 1 since the *A*-area (circle) is wholly included in the *B*-area (oval); "if *B* then C" has high probability, since most of the *B*-area is contained in the *C*-area (shaded); but "if *A* then *C*" has probability zero, since none of the *A*-area is contained within the *C*-area. This has implications for (**2.5.23**).[†]

    So far, I have considered only negative upshots of the probabilistic account of conditionals: cases of arguments which are probabilistically invalid, though valid in the standard sense, if "if" is truth functional. But are enough arguments probabilistically valid for the probability theory to have a chance of giving a correct account of our actual reasoning? Are there not plenty of arguments of the form of (**2.5.23**) which we unhesitatingly regard as valid, for example:

    11)   If the battery's dead, then the car won't start; if the car won't start, then I won't be able to get to work. So if the battery's dead, then I won't be able to get to work.

The fact that this argument is *not* valid, probabilistically, constitutes an objection to the probability interpretation.

    One response starts by pointing out that the following argument form is probabilistically valid:

† See Exercise 77, page 351.

12)   If A then B, if A and B then C; if A then C.

The response goes on to suggest that cases like (11) strike us as valid, because we are so happy to carry the antecedent of the first conditional forward into the second, in the manner of (12). (11), though itself strictly invalid, can be seen as elliptical for the genuinely valid argument, with equally acceptable premises, that would result from giving it the form of (12).

The probability theorist may explain the cases in which transitivity fails precisely as cases in which the conversion to the form of (12) yields unacceptable premises. Applying this to (2.5.23) yields:

13)   If Smith dies before the election, Jones will win.
      If Smith dies before the election and Jones wins it, Smith will retire from public life after the election.
      So, if Smith dies before the election, he will retire from public life after the election.

The second premise becomes unacceptable.

Let us summarize what has been achieved. The probability theorist offers theoretical reasons for denying the validity of arguments upon which the two cases for the material implication interpretation were based. But it remains an open question whether he is *right* to deny the validity of these arguments.

Let us consider first the claim that (2.8.1) is invalid:

Either the butler or the gardener did it. Therefore if the gardener didn't do it, the butler did.

This certainly strikes most people as valid. Yet, as we have seen, if its validity is a matter of its logical form one cannot escape the conclusion that "if" expresses the material conditional.

It must be stressed that the probability theorist need not deny the validity of (2.8.1). He need only deny that it is *formally* valid. When we affirm (8),

A or $B \vDash$ if not-A then B,

the generalization required for the argument for the truth functionality of "if", we affirm that it holds for every argument of that form, whatever sentences replace A and B. If the probability theorist can show that there is at least one invalid instance of (**8**), he can be indifferent to the validity of (**2.8.1**).[4] Here is a candidate for such an instance, modelled on the probabilistic counterexample to (**8**) given in the diagram:

    14)   Either his first card will be a court card or it will be an ace. Therefore, if his first card is not a court card, it will be an ace.[5]

Some people find this intuitively invalid. They claim that the premise could be true yet the conclusion false. This would be enough to refute (**8**). There is a straightforward probabilistic explanation. Suppose the pack has been denuded of deuces through tens (or alternatively suppose the pack has been rigged to bring the court cards to the top). Then the premise will have high probability, but the probability of the conclusion is zero. Notice, however, that opinions differ about the truth value of the conclusion. A defender of the material implication interpretation would claim that the conclusion is certainly true if his first card is an ace; and if the premise is true because the player's first card is a court card but not an ace, then the conclusion is also true (the antecedent of the conditional is false, so any consequent will yield a true conditional), though doubtless unassertible. We still seem to have found no neutral ground on the basis of which to decide the question of the validity of this argument.

In the case of some of the other argument patterns ruled invalid by the probability account, there are quite general considerations which are supposed to show that the invalidity is to be predicted. For example, the failure of transitivity follows from the following assumptions:

    15)   For some A, B, "if A then B" is *contingently* true: it is true, but the truth of A is logically consistent with the falsity of B. (For example, intuitively "if the wind blows, the cradle will rock" may be true even though it is logically possible for the cradle to remain still in the wind.)

    16)   Any conditional with a possibly true antecedent and a consequent incompatible with its antecedent is false.

Now suppose, for *reductio*, the principle of transitivity, (**10**). An instance of it is:

    17)   [if A and C then A, if A then B] ⊨ [if A and C then B]:

(putting "A and C" for "A", "A" for "B" and "B" for "C"). Given

18) ⊨ if A and C then A

we can infer

19) [if A then B] ⊨ [if A and C then B].[6]

If we take an instance of this in which the premise is contingently true (and (**15**) assures us that there is one), and in which we put "not-B" for "C", we have a counterinstance to (**19**), since the conclusion is false by (**16**). So seemingly modest assumptions about the validity-relevant features of conditionals ((**15**) and (**16**)) give a quite general reason for denying the transitivity of "if".[7]†

I shall shortly turn to some general objections to the probability theory that are independent of its pronouncements about validity. Before doing so, here are some detailed points.

First, even the examples supposedly best calculated to support the probabilistic theory, like (**3**), can be disputed. Many people feel that you shouldn't assert (**3**) if you know that there is *some* chance, however small, of the ball in your hand being black. "Probably, if this is urn B the ball in my hand is black" is held to be acceptable; but not the unqualified assertion of the conditional.[8]

Secondly, we have been riding uncritically along with a basic assumption of the probability theory: $Pr(B|A)$, and therefore, according to (**4**), Pr (if A then B), is undefined if $Pr(A) = 0$. However, there appear to be many uses of conditionals having antecedents with zero probability. If this appearance is genuine, then the probability theory is at best incomplete.

Consider:

20) If the longhorns lose, I'm a monkey's uncle.

This is clearly intended as an assertible conditional displaying great confidence in the longhorns. In other words, a typical asserter of (**20**) will assign Pr(the longhorns lose) = 0,[9] yet will be eager to affirm the conditional.

If that kind of example is thought to invoke a special idiom, consider instead a common kind of reasoning in mathematics. Aiming to disprove A, you find a B such that you can establish both "if A then B" and "if A then not-B". In classical mathematical reasoning, the two conditionals

† See Exercise 78, page 351.

license the derivation of not-A. Hence, if you know where your proof is going, you will affirm the conditionals while assigning zero probability to the antecedent.

If this is also thought too specialized a case, consider quite ordinary reasoning in which, perhaps to persuade someone of not-A, we follow through the consequences of A, expressing these by conditionals. Thus, wishing to persuade someone that no one broke into your house while you were away, you say:

> **21)** If someone broke in, they must have repaired the damage before they left,

intending the absurdity of the consequent to lead your hearer to recognize the certain falsity of the antecedent: i.e. to come to assign it, as you do, probability 0.

Thirdly, the probability theorist obviously has to explain away as incorrect the intuitive equivalence between conditionals and disjunctions.†

Finally, the probability theorist should explain why (4) holds. Two main kinds of explanation are reviewed in the next section.

2 Probabilities, propositions and assertion

One way to explain why (**1.4**) holds is to say that "if" is a non-truth functional sentence connective with a meaning which ensures the truth of (**1.4**). This view has been argued to be inadequate by David Lewis.[10] Partly under the influence of this argument, another line of explanation has been developed, which involves denying that conditionals are genuine propositions, possessors of truth value. On this view, what would normally be classified as asserting a conditional is really an act of conditional assertion. To utter "if A then B" assertively is not to assert that if A then B, but is rather to assert B, within the supposition that A. This explains (**1.4**) because the probability of an assertion of B, within the supposition that A, is just $\Pr(B|A)$.

I shall begin by giving a very rough account of Lewis's argument. We need first to define, more explicitly than we have so far done, a *probability*

† See Exercise 79, page 351.

function, Pr: this is an assignment of numbers to all the sentences of the language in conformity with the following rules:

1) $1 \geq \Pr(A) \geq 0$.
2) If $\vDash[A$ iff $B]$ then $\Pr(A) = \Pr(B)$.
3) If $\vDash[$if A then not $B]$ then $\Pr(A \lor B) = \Pr(A) + \Pr(B)$.
4) $\Pr(A$ and $B) = \Pr(A|B) \times \Pr(B)$, if $\Pr(B)$ is positive.
5) $\Pr(A) = 1 - \Pr(\text{not } A)$.
6) If $[\vDash A]$ then $\Pr(A) = 1$.

There are no constraints on how a probability function assigns numbers to atomic sentences.[†] Thus there is a probability function $\Pr_1$ such that $\Pr_1$ (the earth is flat) $= 0$ and also a probability function $\Pr_2$ such that $\Pr_2$ (the earth is flat) $= 1$, and indefinitely many other probability functions assigning intermediate values.

(**1.4**) has a hidden generality. It is intended to hold for all probability functions, or at least for all *reasonable* probability functions: all probability functions which could represent the way some rational agent assigns probabilities to his beliefs. One important feature which Lewis takes the class of reasonable probability functions to possess is being *closed under conditionalization*.[11] This means that if the class contains a function $\Pr_i$, and $\Pr_i(C)$ is positive, then it also contains a function $\Pr_j$, such that, for all A, $\Pr_j(A) = \Pr_i(A|C)$. The intuitive idea, of which this is a generalization, is that if a minimally rational agent could assign a specific probability to A conditionally upon C (which is possible only if the agent were to assign non-zero probability to C), then some minimally rational agent could assign the very same (absolute) probability to A.

The argument for Lewis's conclusion proceeds by *reductio ad absurdum*. Suppose that "if" is a non-truth functional sentence connective whose meaning ensures that *any* reasonable probability function meets the condition of (**1.4**). What has already been said ensures that reasonable probability functions also satisfy the following condition:

7) $\Pr((\text{if } A \text{ then } B)|C) = \Pr(B|(A \text{ and } C))$, if $\Pr(A \text{ and } C)$ is positive.

To establish (**7**), take an arbitrary member of the class of reasonable probability functions, say $\Pr_i$. Since we know that the class is closed under conditionalization, we know that there is another member, call it

† See Exercise 80, page 351.

Pr_k, such that, for some C where $\text{Pr}_i(C)$ is positive, for all A, $\text{Pr}_k(A) = \text{Pr}_i(A|C)$. We can then argue as follows:[12]

8) $\quad \text{Pr}_k(B \text{ and } A) = \text{Pr}_i((B \text{ and } A)|C) \quad$ [definition of Pr_k]

$$\qquad\qquad\qquad = \text{Pr}_i(A|C) \times \text{Pr}_i(B|(A \text{ and } C)) \quad [(4)]^\dagger$$
$$\qquad\qquad\qquad = \text{Pr}_k(A) \times \text{Pr}_i(B|(A \text{ and } C)) \quad [\text{df } \text{Pr}_k]$$
$$\text{Pr}_k(B \text{ and } A) = \text{Pr}_k(A) \times \text{Pr}_k(B|A) \quad [(4)]$$
$$\qquad\qquad\qquad = \text{Pr}_k(A) \times \text{Pr}_k \text{ (if } A \text{ then } B) \quad [(1.4)]$$

hence

$$\text{Pr}_k(A) \times \text{Pr}_k \text{ (if } A \text{ then } B) = \text{Pr}_k(A) \times \text{Pr}_i(B|(A \text{ and } C))$$

and so, given that $\text{Pr}_i(A \text{ and } C) > 0$:

$$\text{Pr}_k \text{ (if } A \text{ then } B) = \text{Pr}_i(B|(A \text{ and } C)),$$

which, given the definition of Pr_k, and the fact that Pr_i was arbitrary, amounts to (7).

We now take arbitrary sentences A, C, such that for some arbitrary member of the class of reasonable probability functions, $\text{Pr}(A \text{ and } C)$ and $\text{Pr}(A \text{ and not-}C)$ are both positive. It can be proved from (1) to (6) that, quite generally,

9) $\quad$ Pr (if A then B) = Pr((if A then B) and B) + Pr ((if A then B) and not-B).

Hence by (4):

10) $\quad$ Pr (if A then B) = (Pr ((if A then B)$|B$) × Pr (B)) + (Pr ((if A then B)$|$not B) × Pr (not B)).

Applying (7)

11) $\quad$ Pr (if A then B) = (Pr($B|(A$ and $B)$) × Pr(B)) + (Pr($B|(A$ and not $B)$) × Pr (not B)).

This simplifies to the absurd:

12) $\quad$ Pr (if A then B) = [(1 × Pr(B)) + (0 × Pr (not B))] = Pr(B).

† See Exercise 81, page 351.

(12) is absurd because there are plenty of examples in which the probability of a conditional differs from that of its consequent. Lewis gives the example of the probability function, Pr_j, of a rational agent about to throw a fair die. Pr_j(six will come up) = 0.166, whereas Pr_j (six will come up|an even number will come up) = 0.333, and, by the hypothesis (1.4), the latter value would be assigned to Pr_j(if an even number comes up, six will come up).

Lewis takes the absurdity as a refutation of the hypothesis that "if" is a non-truth functional sentence connective with a meaning which ensures the truth of (1.4). On the positive side, he argues that conditionals are material implications. True, they are not to be asserted just when they have high probability, for in general we do not have:

13) $Pr(A \rightarrow B) = Pr(A|B)$.

However, the assertibility conditions for conditionals implied by (1.4) are, he says, to be expected on the basis of the material implication interpretation, given certain principles about implicatures.

There are at least two alternative responses. One is to see (12) as a refutation not of the claim that (1.4) is to be explained by the very meaning of "if", but rather of the claim that reasonable probability functions sustain conditionalization, an assumption upon which the derivation of (12) given here essentially depends.[13] Another response is to say that there are no conditional propositions. An assertion of the form "If A, then B" should rather be regarded as an assertion of B conditional upon A than as an assertion of a conditional. This explains why (1.4) holds, without requiring "if" to be something which, if Lewis is right, it cannot be. I shall consider these two responses in turn.

What is it for an agent to assign a value to some expression $B|A$? One commitment appears to be that he would assign that value to B, were he to learn that A. So it seems as if the very notion of conditional probability ensures that a rational subject will "conditionalize": that is, will move from a probability function Pr to a conditionalized function Pr_C, on learning that C, that is, on moving to a function upon which C has probability 1. If this were right, it would justify Lewis's background assumption that the class of reasonable probability functions is closed under conditionalization. However, some doubts may arise.

One can reasonably doubt (7). More precisely, although (7) itself cannot be doubted, since it is a theorem of a calculus, one can doubt whether the calculus to which it belongs adequately reflects our intuitive concept of probability. Consider the following application of (7):

14) Pr((if the Democrats are elected then the Medicare program will improve)|the Medicare program will not improve) = Pr (the Medicare program will improve|(the Democrats will be elected and the Medicare program will not improve)),
if Pr (the Democrats will be elected and the Medicare program will not improve) is positive.

Here is an apparently reasonable distribution of subjective probabilities which is inconsistent with (**14**). I think it likely, though not certain, that the Democrats, if elected, will improve the Medicare program, but I think it unlikely that they will be elected, and thus unlikely that the program will, in fact, improve. The condition for (**14**) is met: I assign a low, but not zero, probability to the conjunction "The Democrats will be elected and the Medicare program will not improve". My confidence in the conditional "if the Democrats are elected then the Medicare program will improve" is undiminished by the relativization to "the Medicare program will not improve", so I assign a high probability to the expression on the left of the equality in (**14**). But I assign a low value, presumably zero, to the expression on the right side of the equality.[14]

The doubt connects with a doubt about Lewis's assumption that the conditionalization of a reasonable probability function is reasonable. Let Pr_i be the probability function described informally in the last paragraph. Is there a $Pr_{i'}$ meeting the condition

for all A: $Pr_{i'}A = Pr_i$ (A|the Medicare program will not improve)?

If I imagine what I ought to say about "if the Democrats are elected then the Medicare program will improve" by imagining myself to have come to know that, come what may, the program is not going to improve, then I can reasonably conclude that the probability should not change. I could take the information I imagine myself to have acquired as showing that the Democrats are not going to be elected. In other words, I could rationally hold that

$Pr_{i'}$ (if the Democrats are elected then the Medicare program will improve) = Pr_i ((if the Democrats are elected then the Medicare program will improve)|the Medicare program is not going to improve).

This reasoning, though natural, does not count as conditionalizing. The resulting $Pr_{i'}$ does not satisfy the calculus within which Lewis obtains his result. What would be needed for that result is an alternative style of reasoning, as follows: I should project myself into a world in which I know that the program will not improve, and imagine myself then evaluating the conditional "if the Democrats are elected, the program will improve". I should think to myself: "Now I am at a world at which the program will not improve, so probably what this means is that even if the Democrats are elected, their attempts at improvement will fail. So I should assign a low probability to the conditional." This method of conditionalizing conforms with (7), but it is unclear that the result is a reasonable probability function. It conflicts with the intuitively more plausible $Pr_{i'}$. Hence a probability function derived from a reasonable one by conditionalizing in the way Lewis demands may not be reasonable.

In summary, there is room for doubt whether the calculus within which Lewis's result is established, based upon a certain conception of conditionalization, accords with our intuitive conception of probability. Hence there is room for doubt whether his result establishes that there is no explaining (1.4) in terms of the meaning of "if ... then".

The other response I shall consider to Lewis's argument is that there are no conditional propositions: what appear to be assertions of conditionals are really only conditional assertions of their consequents. A conditional assertion can usefully be likened to a conditional bet. If you make the following bet

15)   If John runs, he'll be in Scotland by midnight

you neither win nor lose if John doesn't run. The question of winning or losing arises only if the condition for the bet is met, viz. that John runs.[15] This is analogous to the view that an assertive use of (15) introduces nothing to be assessed for truth or falsity unless John runs; and if John does run, truth (winning) is a matter of his getting to Scotland before midnight, and falsity (losing) is his failing to get there by midnight. The view that (15) is not a conditional proposition is just the view that (15) as a whole is not assessable as either true or false. If its antecedent is satisfied, then the question of truth or falsehood arises, but relates exclusively to the consequent of the conditional.

There are many reasons for which one might feel unhappy with the view that there are no conditional propositions. One is that we appear to debate the truth or falsehood of conditionals in just the ways we debate the truth and falsehood of what are indisputably propositions. Another is that it seems that conditionals can occur as, apparently, propositional components within larger propositions. The conditional assertion theorist can respond to the first of these charges, but the second is more resistant.

The conditional assertion theorist will be quick to point out that in what his opponent would describe as a debate concerning the *truth* of a conditional, the conflict is, not between one who affirms a sentence "if A then B" and one who affirms a sentence "not (if A then B)", but rather between one who affirms "if A then B" and one who affirms "if A then not-B". This is consistent with, and perhaps better explained by, the conditional assertion theory: one party asserts that B, within the supposition that A; the other party denies that B, within the same supposition.

There are two especially problematic ways in which, on the surface, a conditional can be embedded as a component of a larger sentence. One is the case in which the consequent is itself a conditional, so we have the form: "if A then (if B then C)". The immediate application of the theory yields the view that this is the assertion of "if B then C", within the supposition that A. However, the theory has it that there is really no such thing as the assertion of "if B then C". The natural thought, on behalf of the theorist, is that what is being asserted is C, within the supposition that both A and B. However, he will find it hard to explain why this should be so. For three reasons, he should not commit himself to

    16)   (if A then (if B then C)) is equivalent to ((if A and B) then C).

For, first, one might wonder how the conditional assertion theorist can apply the notion of equivalence to sentences which do not express propositions. To do so would seem to fly in the face of his own theory, since equivalence is defined in terms of truth values, and that is just what conditionals are supposed to lack. Secondly, (**16**) entails that conditionals are material conditionals. For from (**16**) we can infer

    17)   ((if $A \rightarrow B$) then (if (A then B)) is equivalent to (if (($A \rightarrow B$) and A) then B)

(putting "$A \rightarrow B$" for "A", "A" for "B" and "B" for "C"). Since the right-hand side of the equivalence is obviously true, we can infer the left-hand side. And that, together with the uncontentious

    18)   if (if A then B) then ($A \rightarrow B$)

gives the equivalence

19) (if A then B) iff $(A \rightarrow B)$.[16]

Thirdly, the discussion of the alternative response above suggested that there is an intuitive difference between "If the Medicare program will not improve then [I still hold that] if the Democrats win then it will improve" and "If the Medicare program will not improve and the Democrats win, then it will improve".

The other problematical embedding of conditionals has the form "if, if A then B, then C". If a conditional is not a proposition, it cannot be supposed, so this cannot be seen as the assertion of C, conditional upon if A then B. Nor can it be the assertion of C conditional upon the assertion of A and B, for this wrongly identifies this case with the previous one. Nor does it appear correct to say that there are no cases of this form in English. The following is perfectly idiomatic:

20) If the vase broke if dropped, then it was fragile.

The theory that there are no conditional propositions thus seems hard to defend.

We have been looking for an account of how **(1.4)** might come to be true. That it is made true by the meaning of "if ... then" came under attack from Lewis, though we saw that the attack might not be decisive. The theory according to which **(1.4)** is true because there are no conditional propositions, but only conditional assertions, faces the difficulties just mentioned about embeddings. What alternative accounts are there?

As mentioned earlier, David Lewis has suggested that something close to **(1.4)** can be explained on the assumption that conditionals are material implications. Our confidence in **(1.4)** largely reflects the fact that we regard a conditional as *assertible* iff the related conditional probability is high. Drawing on Gricean notions, Lewis suggests that *this* fact can be explained on the hypothesis that indicative conditionals are truth functional. It is misleading to assert "if A then B" if you think that A is false or that the probability of "A and not-B" is large relative to the probability of A. However, the probability of B given A is the probability of the material implication diminished by just these misleading factors. In other words:

21) $\Pr(B|A) = \Pr(A \rightarrow B) - (\Pr(\neg A) \times (\Pr(A \mathbin{\&} \neg B) \div \Pr(A)))$,
provided $\Pr(A) > 0$.

Lewis's explanation of (1.4) is that it measures (under the misleading title of probability) the degree of assertibility of conditionals, in a way that is explained by their being material conditionals. So, in summary, it would appear that a distinctively probabilistic theory has difficulty in providing a good explanation of (1.4) while, paradoxically, a material implication theory can explain, if not (1.4) itself, at least the basis upon which one might be tempted to hold (1.4).

3 Stalnaker's theory

How do we tell whether or not to accept or assert a conditional? Frank Ramsey[17] proposed the following test: believe "If A then B" just on condition that, were you to believe that A you would believe that B. In more detail, and in Robert Stalnaker's words, the idea is this:

> First, add the antecedent (hypothetically) to your stock of beliefs; second, make whatever adjustments are required to maintain consistency ... finally consider whether or not the consequent is then true. (Stalnaker [1968] p. 44)

This speaks only to the conditions under which a conditional should be believed. But we can use the basic idea to suggest a condition under which a conditional is true:

> Consider a possible world [possible situation] in which A is true, and which otherwise differs minimally from the actual world [the way things actually are]. *"If A, then B" is true (false) just in case B is true (false) in that possible world.* (p. 45)

This definition uses the notion of a "possible world", which has been widely employed in logic and semantics in recent years. Though the notion has a technical use, it is supposed to derive from quite non-technical ideas. Not all features of the actual world have to be as they are: some could have been otherwise. To consider a possible world is just to consider a way in which the world could have been different from the way it actually is.

Let us apply this to (2.4.39):

> If John dies before Joan, she will inherit the lot.

We will suppose that at the time of utterance both parties are alive. What is required for the truth of this conditional? Following Stalnaker's idea, it is true iff in the situation as close as possible to ours but in which John dies before Joan, Joan inherits the lot.[18] Thus if, for example, John's will actually leaves everything to Joan, a situation as close to ours as possible is one in which his will is unchanged, and hence a situation in which, given the normal flow of events, Joan will inherit the lot. (If John is going to die before Joan, then the situation to consider is the actual future course of events: nothing is more similar to the actual world than itself.) In this case, Stalnaker's theory seems to give precisely the right result.

It also delivers the right result in the case of (**1.21**),

> If someone broke in, they must have repaired the damage before they left,

which, arguably, has to be treated as undefined by the probability theory. Consider a world in which someone did break in, but which is otherwise as similar as possible to ours. In ours, there is no trace of violent entry. How could this be, unless the intruder had repaired the damage behind him? So the closest (that is, most similar) world in which the antecedent is true is one in which the consequent is true, so the conditional is true (and hence can be used to infer that no one broke in, given that no intruder would stop to repair the damage behind him).

Stalnaker explicitly considers, and rejects, the theory mentioned in chapter **2.4** that a conditional requires for its truth some connection between what would make its antecedent true and what would make its consequent true. He offers the following counterexample: suppose that at the time of the Vietnam conflict you are convinced that the United States will use nuclear weapons, come what may. You therefore also think that the Chinese entering the conflict would have no effect whatsoever on the US decision to use nuclear weapons. Yet you will confidently affirm the conditional:

> 1)   If the Chinese enter the Vietnam conflict, the United States will use nuclear weapons.

Notice that on Stalnaker's theory this conditional is, if your political views are correct, true, because any smallest revision of the actual world which verifies the antecedent will of course verify the consequent. But this is not to say that the truth of the consequent is *in general* sufficient for the truth of the conditional. On the contrary, this will not always hold: not, for example, in conditionals of the form "if not-A then A".

What impact does Stalnaker's account of the truth conditions of conditionals have upon validity? It turns out that his account of truth conditions validates (in the standard sense) just the arguments that are valid in the probabilistic sense.[†]

He agrees with the probabilistic theory in finding (**2.8.1**) invalid, and he offers a sophisticated explanation of the appearance of validity (Stalnaker [1980]). His idea is that (**2.8.1**) corresponds to a "reasonable" inference: one such that, in any context in which the premises are acceptable, so is the conclusion. But reasonable inferences can be invalid, and (**2.8.1**) is an example. If we think it is valid, it is because we mistake reasonableness for validity.

Everything hinges, of course, on the account of reasonableness. Without giving Stalnaker's whole theory, let us simply see how it is intended to apply to the example. A context in which the premise of the argument is acceptable is one which leaves open whether it was the butler or the gardener who did it. (If you knew who did it, you shouldn't assert the premise, but simply "the gardener did it", or "the butler did it", as the case may be.) When you have asserted or accepted the premise, the context includes the premise: it is now part of what is taken for granted. Hence when you evaluate the conclusion, you should consider a world in which the gardener didn't do it, but in which the premise holds true. Such a world must be one in which the butler did it. Reasonableness puts restrictions on what alternative worlds to consider in evaluating a conditional.

Even if reasonable, the argument is invalid, by Stalnaker's account. For there is no guarantee whatsoever that in the world most like ours, assuming that ours is one which verifies the premise, but in which the gardener did not do it, the butler did. One possibility is that our world is one in which the butler and the gardener were in league: they both did it. The most similar world to this verifying "The gardener did not do it" could well be one in which the victim was not murdered at all: a world in which, say, the victim fell (rather than being pushed). The relevant world cannot be exactly like our world, since in our world (we are supposing) the butler and gardener both did it. A world in which the butler–gardener conspiracy broke down might well be less similar to ours than one in which the conspiracy is in force, but the conspirators were thwarted at the last moment by their victim's accidental death. The truth of the premise

† See Exercise 82, page 351.

of the argument thus does not guarantee the truth of the conclusion, upon Stalnaker's semantics.

Stalnaker intends his account to apply to subjunctive as well as indicative conditionals. But how will it specify the difference between matched pairs of these different kinds of conditional? For example

> **2.4.34)** If Oswald didn't shoot Kennedy, someone else did.

> **2.4.35)** If Oswald hadn't shot Kennedy, someone else would have.

Let us apply Stalnaker's recipe to these successively, retaining the background assumptions we relied upon in chapter **2.4** to justify the reasonableness of accepting the first while rejecting the second.

For (**2.4.34**) we consider a world in which Oswald didn't shoot Kennedy. Remember that Oswald's guilt was a disputed question, but Kennedy's death from shooting was not. So a similar world will be one in which Kennedy died from a shooting, while not having been shot by Oswald. Such a world, obviously, verifies the consequent. Hence Stalnaker's account has it that (**2.4.34**) is true, and this is as it should be.

What should we consider for (**2.4.35**)? Although Stalnaker is not explicit, he clearly intends us to consider a world in which Oswald did not shoot Kennedy. (In other words, we revise the antecedent from "hadn't" to "didn't".) So it would appear that we have to consider just the same world as we did before. And just as we must revise the antecedent, to apply Stalnaker's recipe, so we have to revise the consequent. What we need to consider, then, is whether in a world in which Oswald did not shoot Kennedy, but which is otherwise as like as possible to this, someone else shot Kennedy. In short, we seem to have the same question before us as we did in the case of (**2.4.34**), in which case we should give the same answer. But this answer, that (**2.3.35**) is true, conflicts with our original intuition. (We imagined that we thought Oswald acted alone, there was no conspiracy, so if he hadn't shot Kennedy no one else would have.)

Stalnaker recognizes that there will be considerable context-sensitivity in point of what counts as the world most similar to ours in which the antecedent holds, and he suggests that this can account for the difference between (**2.4.34**) and (**2.4.35**). What remains unclear, however, is the relationship, which presumably ought to be systematic, between the contexts we envisage for (**2.4.34**) and (**2.4.35**), and the different dimensions of similarity relevant to the truth conditions. For (**2.4.35**) we want to count a world in which Oswald did not shoot Kennedy and

Kennedy was not shot as the most similar to ours; for (**2.4.34**) we want to count a world in which Oswald did not shoot Kennedy, and he was shot, as the most similar to ours. But what features of the context make for this difference? Clearly a vital difference is that in (**2.4.34**) it is taken for granted, as part of the background, that Kennedy was shot. In (**2.4.35**) this background assumption somehow gets suspended in selecting the relevant world. But what is the mechanism? It is certainly relevant that the envisaged use of (**2.4.35**), unlike (**2.4.34**), takes for granted that Oswald in fact did shoot Kennedy, yet any world which will satisfy Stalnaker's test will be one in which the presupposition is abandoned. Perhaps abandoning taken-for-granted facts induces much more relaxed standards of similarity. A world as similar as possible to this one but in which Oswald didn't shoot Kennedy is certainly not this world, in the context of (**2.4.35**), but is not certainly not this world in the context of (**2.4.34**).

The more obvious account of "most similar" in this case brings out both the indicative and the subjunctive as true. This is not invariable. Gibbard [1980] has suggested the following case, designed to show that Stalnaker's account works best for subjunctive conditionals, and the probability theory for indicative ones.

Jack and Stone are playing poker and Stone has just bet the limit in the final round. Zack has seen Stone's hand, seen that it is a good hand, and has signalled its contents to Jack. Zack remains unaware of the contents of Jack's hand. Stone, suspecting mischief, orders the room to be cleared. Not knowing the outcome of the game, Zack can confidently assert later:

2)   If Jack called, he won.

His grounds are that Jack is an experienced player, wants to win, and, knowing Stone's hand, knows whether or not he will win by calling. However, Zack can reasonably be doubtful about

3)   If Jack had called, he would have won.

After all, Stone's hand was good, and it's not likely that Jack had a better. Let us first confirm Gibbard's claim that the probability theory gives an acceptable account of (2), and then see how Stalnaker's account deals with the two sentences.

On the probability theory, (2) is assertible iff it has high probability, which, by (**1.4**), means that Pr (Jack won|Jack called) is high. The result is

that (2) is ruled highly assertible, as it seems to be, since, given that Jack
called, it is extremely likely that he won.

Turning now to (3), Gibbard argues that our intuitions harmonize with
Stalnaker's account. We in fact regard the truth or falsehood of (3) as
determined by whether or not Jack's hand was better than Stone's. On
Stalnaker's account, we must find a world as similar as possible to ours,
but in which Jack called. According to Gibbard, such a world is
determined by the following criteria:

> it is exactly like the actual world until it is time for Jack to call or
> fold; then it is like the actual world apart from whatever it is that
> constitutes Jack's decision to call or to fold, and from then on it
> develops in accordance with natural laws. (pp. 227–8: I have
> changed Gibbard's Pete to Jack)

However, it seems that a world in which Jack calls, and which is
maximally similar to ours, is a world in which Jack knows Stone's hand
and calls. Preserving similarity with our world, this can only be because
Jack knows that he has a winning hand. In this world, therefore, the
consequent of the conditional is true, rather than false, and so, contrary to
Gibbard's claim, Stalnaker's account seems not to deliver the desired
result for (3).

Gibbard envisages Zack's response to (3) as occurring when Zack is
still ignorant of what Stone did. The most natural setting for (3),
however, is one in which it is known that Jack folded.[19] Given that he
folded because he knew Stone had the better hand, Stalnaker's account
clearly works well. The alternative world is like our world up to the time
of Jack's decision. Before then, he knows he has a losing hand. If we now
add that he calls, then obviously the world will also contain the
proposition that he loses. So while there is room for doubt that the details
of Gibbard's defence help Stalnaker, it seems clear that there are cases for
which Stalnaker's account works perfectly.

It is worth comparing the ways in which Stalnaker's account treats
(2.4.35) and (3). In the latter, there is a feature of the actual world that
occurs before the time of the antecedent which makes the consequent true
(or very likely true): namely that Jack knew Stone's hand and knew that
he would lose if he called. This fact can be retained in the alternative
world. There is no corresponding feature in (2.4.35), and this is why
Stalnaker's account, on the most natural reading of the similarity
involved, fails to deliver the desired truth value.

Gibbard suggests that (2) is a totally different kind of conditional, for which the probability theory works well but – he implies – Stalnaker's account does not. But the discussion two paragraphs back shows that Stalnaker's account would evaluate (2) as true, which is the desired result. Gibbard has not brought a decisive case against that account.

One difficulty Stalnaker's account faces is that it stipulates that where the antecedent of a conditional is impossible – that is, true at no possible world – then the conditional is true.[20] This appears not to accord with the use of conditionals in mathematics. As we saw in connection with the probability theory, in an attempt to disprove A we search for a B such that we can establish both "if A then B" and "if A then not-B". On Stalnaker's theory, any old B will do, since, if A really is refutable, and the subject matter is mathematics, then the proposition is impossible. But we all take it to be a matter of the greatest importance which B to choose. The familiar proof that there is no greatest prime that begins:

If there is a greatest prime, it must be odd ...

has true premises; but a proof that began

If there is a greatest prime (p & $\neg p$)

(for some arbitrary p) should be faulted for having an unacceptable premise. Yet there is no doubt that, given $\neg(p$ & $\neg p)$, we can infer the desired conclusion.

Stalnaker could respond that the alternative "proof" is genuinely sound, but not persuasive. In general, when we look for a B such that "if A then B" and "if A then not-B", we are right to regard the choice as important. But it is not important to truth, only to persuasiveness: we have to find conditionals which even one not yet convinced of the falsehood of A will regard as true.

We saw earlier that Stalnaker's account does not definitely yield the intuitively correct truth value for (2.4.35). This fact should be set alongside a more general criticism of the account, which has been made by David Lewis. Consider the pair:

4)   If Bizet and Verdi had been compatriots, Bizet would have been Italian.

5)   If Bizet and Verdi had been compatriots, Verdi would have been French.[21]

Which, if either, is true on Stalnaker's account? The question resolves to this: which world is *the* world most similar to ours, but in which the composers are compatriots: one in which they are both French, or one in which they are both Italian? There appears to be no reasonable answer. But then Stalnaker's account is wrong to assume that there is any such thing as *the* most similar world at which the antecedent is true. This threatens the very structure of the account.[22]

Stalnaker has responded to this objection along the following lines. Conditionals are both vague and context-dependent. In this respect they resemble such sentences as

6)   He is tall.

This is context-dependent in at least two respects. First, one would rely on context to settle who was referred to by "he". Secondly, one would rely on context to determine the relevant comparison class for "tall". If the person referred to is an Eskimo child, then the appropriate comparison class is other Eskimo children of approximately the same age. If he is an American adult, he will have to be a good deal taller than were he an Eskimo child for (6) to be true of him, since in that case the relevant comparison class will be other American adults.

In addition to being context dependent, (6) is vague. Even when the comparison class is fixed, there can be borderline cases: people of whom it is neither definitely correct to affirm that they are tall, nor definitely correct to deny it.

Similarly, Stalnaker suggests that sentences like (4) and (5) are both context dependent and vague. The context can determine appropriate dimensions of similarity for the selection of an appropriate alternative world. But if a context fails to do this, then we have an indeterminacy: such sentences will be neither definitely true nor definitely false.

This response looks as if it could meet the objection. But can it meet a similar, yet more general, objection? Consider:

7)   If I have misremembered the date, the battle of Hastings was not fought in 1066.[23]

Intuitively, this is true. But on Stalnaker's account, we must evaluate the conditional by determining whether "the battle of Hastings was not fought in 1066" is true in the world most similar to ours in which I have misremembered the date of the battle. Since we are to make judgements of similarity which require similar worlds to retain as much as possible of what is taken for granted, and since it is taken for granted that the battle occurred in 1066, the most similar world in which the antecedent of (7) is true is one in which the consequent is false. Hence Stalnaker must rule (7) false rather than true. It is an open question whether this problem can be resolved by further refinement of the way in which context plus antecedent determine the relevant standard of similarity. One thing is certain: Stalnaker's account, as it stands, is at best incomplete.

Let us try to take stock. We have considered three theories of the indicative conditional: that it is material implication, that its probability is fixed in accordance with (**1.4**) – the probability theory – and, finally, Stalnaker's theory. One feature of all these theories is that they have to draw on the notion of implicature, or related phenomena, to accommodate some of the facts. The material implication theory makes heavy, and many think implausible, use of this notion with regard to the objections levelled against it in chapter **2.5**. The probability theory appeals to it in order to explain, for example, the apparent validity of some cases of transitivity. Stalnaker's theory appeals to it to explain the apparent validity of (**2.8.1**) (the butler or the gardener); and appeals quite widely to related phenomena, concerning the way in which the context of utterance affects interpretation, in his account of how the appropriate alternative world is selected.

The element of the probability theory which should be accepted is that indicative conditionals are generally highly assertible iff the related conditional probability is high. However, this much can be agreed by many different theories of the meaning of conditionals. If we interpret Lewis's result as entailing that there cannot be a connective whose meaning ensures the truth of (**1.4**), the most promising avenue for the probability theorist is to deny that there are conditional propositions, but this, as we saw towards the end of §2, faces difficulties. Moreover, all that (arguably) we really wanted from (**1.4**) is available to a material implication theory of conditionals.

Stalnaker's theory involves some highly intuitive ideas, but it leaves unexplained the way in which context effects differential determinations of the appropriate similarity relation. Although there is vagueness and indeterminacy in conditionals, we have many very firm and confident intuitions, which suggest some fairly fixed principles of selection are at work. Stalnaker's theory provides a framework within which such details could be elaborated.[24]

It may be useful to conclude by singling out, and presenting in table 3.1, some of the most pertinent logical considerations that have been at issue in these discussions. If you dispute a proposition in the row, you

Table 3.1

	→	Probability	Stalnaker
If A then B ⊨ if $(A$ and $C)$ then B	√	×	×
If A then B, if B then C ⊨ if A then C	√	×	×
If A then B ⊨ if not-B then not-A	√	×	×
B ⊨ if A then B	√	×	×
$A \vee B$ ⊨ if not-A then B	√	×	×
Conditionals have truth values	√	×	√
"If A then not-B" negates "if A then B"	√	×	?
"If A then B" may be true (probable) if A is impossible $(\Pr(A) = 0)$	√	×	√

should believe only a theory which places an " × " beside that proposition.[†]

[†] See Exercise 83, page 351.

Notes

A good introduction to recent work is the collection *Ifs* (Harper, Stalnaker and Pearce [1981]). Among many other excellent things, it contains a useful introductory survey by Harper, the famous Stalnaker [1968] and [1975], Gibbard [1981] and the classic Lewis [1976]. Jackson [1987] is excellent, but I saw it too late to make much use of it here.

The classic text for the probabilistic theory is Adams [1975], and the diagrams used here derive from him. A good discussion is Appiah [1985].

Stalnaker's theory (see Stalnaker [1968] and subsequent writings in the bibliography) is similar to one propounded by Lewis [1973a] and [1973b]. The main difference between them is that whereas Stalnaker's theory is intended to apply to all conditionals, Lewis's is intended to apply only to subjunctive (in his terminology "counterfactual") conditionals. I briefly present Lewis's theory in ch. 5.2. On Lewis's view the indicative conditional is material implication. Since his truth conditions for subjunctive conditionals depend on which worlds verify the related material implication, Lewis's theory shares with Stalnaker's the merit of providing an account which does justice to the apparent unequivocal meaning of "if" in both kinds of conditionals. For an account of Stalnaker's theory in a wider philosophical setting, and including replies to criticisms, see Stalnaker [1984], esp. ch. 7.

Among other articles from which I have profited, I will mention Dudman [1984a], [1984b] and [1987] and Bennett [1974].

1 The example comes from Appiah [1985] p. 172.
2 The definition will not resolve every perplexity, since the standard definition of probability for conjunctions is given by:

$$\Pr(B \ \& \ A) = \Pr(B|A) \times \Pr(A).$$

There is a case for saying that conditional probability should be taken as primitive. Certainly, a rational subject may justifiably assign conditional probabilities while simply refraining from assigning any absolute probabilities to the components. I can rationally assign a high value to "It is now raining in Moscow|the roofs there are wet" without having any reason to assign any absolute probabilities to the components. (Cf. Blackburn [1986a] p. 228.)
3 See Adams [1975] p. xi.
4 There would remain something to be explained: viz. to what (**2.8.1**) could owe its validity, if not to its logical form.
5 The court cards are ace, king, queen, jack.
6 Cf. Exercise 18a, p. 342. The principle is called the Suppression Principle.
7 The argument comes from Stalnaker [1984] pp. 123–4. He refers to what I call the transitivity of "if" as the validity of the hypothetical syllogism.
8 Cf. Dudman [1987].
9 The probability theorist may wish to insist that only necessary truths should be assigned probability 0 (by any agent meeting minimal standards of rationality). If this were accepted, then only the criticism of the following paragraph in the text is pertinent. However, the condition would separate the notion of probability from subjective notions of total certainty, and would represent those who have been persecuted for their beliefs as either irrational, or else less than totally confident in those beliefs. Neither representation is acceptable.
10 See Lewis [1976].
11 In his [1986c], pp. 587–8, Lewis shows that his conclusion can be obtained on the basis of a weaker assumption about the class of reasonable probability functions.
12 Cf. Blackburn [1986a], p. 220. Blackburn attributes this way of setting out the proof to Stalnaker.
13 I stress again that in Lewis [1986c] pp. 587–8, (**12**) is derived on the basis of a weaker assumption.
14 Cf. Blackburn [1986a] pp. 221ff. The phenomenon is analogous to Jackson's notion of robustness, mentioned in ch. **2.7**.
15 This is not to deny the possibility of a bet upon the proposition: (John runs)→(he'll be in Scotland by midnight). This bet is won if John does not run. This interpretation, however, is not the natural one in the case of (**15**).
16 This argument and the subsequent one come from Gibbard [1980]. Gibbard

thinks that the "no conditional propositions" view can be rescued from these objections by various case-by-case considerations, see esp. pp. 237–8.

17 Ramsey [1929] p. 143.

18 Should we assume that there is such a thing as *the* situation as close as possible to ours in which John dies before Joan? Might there not be many equally similar situations? The question was at issue in a famous debate between Stalnaker and Lewis, the former giving an affirmative, the latter a negative, answer. See (4) and (5) below.

19 It can reasonably be doubted whether Zack can correctly use the subjunctive conditional in the situation Gibbard envisages. The clearly assertible indicative "If Jack has called, he has won" does not appear to differ significantly from (2).

20 In Stalnaker [1984], he stresses that this is not an essential part of the theory, suggesting that one could, if one prefers, stipulate that all conditionals with impossible antecedents are indeterminate (p. 121 and n. 4). As he sees, this will not solve the problem of the *discriminating* use we appear to make of apparently impossible antecedents in mathematical reasoning. This problem, he says, requires thinking of mathematical falsehoods as not impossible: a heroic solution.

21 The examples come from Quine [1952] p. 15.

22 Note that exactly the same problem arises for the corresponding indicative conditionals.

23 Jackson [1981] p. 127.

24 Lewis [1976] pp. 145ff., suggests an important distinction bearing on this point.

4

Quantification

This chapter introduces a richer artificial language, **Q**, a language capable of representing *quantifiers* like "all" and "some". Readers already familiar with this language should merely skim the first two sections, to check on the terminology and symbolism used here. (Of particular importance is a grasp of the precise notions of interpretation and **Q**-validity.) Later sections consider problems of formalizing English in **Q**.

1 The classical quantificational language

In **P**, the smallest unit was the sentence-letter. It is clear, however, that English sentences are composed of parts that are not themselves sentences, and that such structure is sometimes relevant to validity. A sentence like "John runs" is composed of a *name*, "John", and a *verb*, "runs", yet the deepest **P**-formalization of this sentence is simply a **P**-letter, in which this structure is obscured. Thus the valid argument:

1) John runs; so someone runs

has no more revealing **P**-logical form than

2) $p; q$

which is plainly not **P**-valid.

The language **Q** is to include **P**, so that every **P**-sentence automatically counts as a **Q**-sentence, but is also to include further devices to reach structures **P** cannot reach.

Sentences of **Q** are composed of the following kinds of symbol:

Sentence-letters: p, q, r, p', ... etc.
These are exactly as in **P**.
Name-letters: α, β, γ, α', ... etc.
These will be used to correspond to ordinary English names like "Ronald Reagan".
Predicate-letters: F, G, H, F', ... etc.
These will be used to correspond to English verbs, like "runs" and "loves", some adjectives, like "hungry", and some nouns, like "man".

"John runs" will be **Q**-formalized as "$F\alpha$" (to be read "α is F" or "F of α"), and so will "John is hungry" and "John is a man".

Predicates: =
The only **Q**-predicate (as opposed to predicate-letter) is "$=$", the sign for identity (being the same as).

Hence we can formalize "Hesperus is Phosphorus" as "$\alpha = \beta$".

Variables: x, y, z, x', ... etc.
Their role will be explained shortly.
Operators: the sentence-connectives of **P**, together with "$\forall$" (the universal quantifier, corresponding to "all" or "every") and "$\exists$" (the existential quantifier, corresponding to "some" or "a").

Giving the meaning of the operators new to **Q** (as opposed to the sentence connectives carried forward from **P**) will be deferred until the basic ideas of how the language works have been introduced informally.

A *predicate* is an expression which takes one or more names to form a sentence. "Is a man" is a predicate which takes one name (e.g. "John") to form a sentence ("John is a man"). "Loves" is a predicate which takes two names (e.g. "John" and "Mary") to form a sentence ("John loves Mary"). "Is between ... and ..." is a predicate which takes three names (e.g. "Austin", "San Antonio" and "Waco") to form a sentence ("Austin is between San Antonio and Waco"). The number of names a predicate takes to form a sentence is called its *degree*. Thus "runs" is of degree 1, "loves" of degree 2, "is between ... and ..." of degree 3. Every predicate and predicate-letter of **Q** is stipulated to have some one fixed degree.[†] (In a fully explicit notation, we could superscript every predicate and predicate letter with its degree, e.g. writing "F^3" for a predicate-letter fit

† See Exercise 84, page 352.

to formalize a predicate of degree 3. We adopt the more flexible policy of letting predicate-letters take on whatever degree is appropriate to the task in hand.) Notice that the notion of a predicate embraces categories which traditional grammar either distinguishes (e.g. verbs, nouns, adjectives) or does not recognize (e.g. "is between ... and ...", which belongs to no single traditional grammatical category).

The *atoms* of **Q** are predicates or predicate letters, combined with an appropriate number of names (the number equals the degree of the predicate or predicate-letter). Examples of atoms which formalize the three sentences mentioned in the last paragraph are "$F\alpha$", "$F\alpha\beta$", "$F\alpha\beta\gamma$". Here, for the first example, "α" corresponds to "John", and "F" to the degree-1 predicate "is a man"; for the second, "α" corresponds to "John", "β" to "Mary" and "F" to the degree-2 predicate "loves"; for the third, "α" corresponds to "Austin", "β" to "San Antonio", "γ" to "Waco" and "F" to the degree-3 predicate "is between ... and ...".

The atoms can be compounded using the sentence connectives just as if they were **P**-letters. Thus all of the following are **Q**-sentences:

$F\alpha \rightarrow G\beta$.
$(H\alpha\beta \,\&\, F\alpha) \vee p$.
$F\gamma\alpha\beta \rightarrow \alpha = \beta$.

It remains to introduce the role of quantifiers and variables. Physicalists wish to affirm that everything is physical. They could use any of the following English, or more or less English, sentences:

Everything is physical.
Take anything you like, it is physical.
For all things, x, x is physical.

The standard **Q**-formalization of the English is more like the last of these variants:

3) $\forall x Fx$

read "for all x, F of x". Here the *quantifier* "$\forall$" corresponds to "for all things", and "F" corresponds to "physical". The variable, "x", plays something like the role of the pronoun "it" in "Take anything you like, *it* is physical". "Something is physical" would be formalized

4) ∃xFx.

(**3**) and (**4**) are called, respectively, a *universal quantification* and an *existential quantification*. Note, however, that

5) $(\forall xFx) \to (\exists xFx)$

is a conditional and neither a universal nor an existential quantification. This is because "→" has wider scope than either quantifier, and so is the *dominant operator*. (Compare the definition of scope in (**2.1.6.**))[†]

We can now give a fuller account of what it is to be **Q**-sentence:

6) X is a **Q**-sentence iff

    (i) X is a sentence-letter or

    (ii) X is an atom (i.e. a predicate-letter of degree n combined with n names) or

    (iii) there are **Q**-sentences Y and Z and a variable v not occurring in Y such that X is one of

        a) $\neg Y$
        b) $(Y \mathbin{\&} Z)$
        c) $(Y \lor Z)$
        d) $(Y \to Z)$
        e) $\forall v Y\star$
        f) $\exists v Y\star$ (where $Y\star$ results from Y by replacing at least one occurrence of a name by v).

For example, to show that (**3**) and (**4**) are **Q**-sentences, it is enough to point out that

  (i) "$F\alpha$" is a **Q**-sentence, being an atom,
 (ii) it does not contain "x".
(iii) "Fx" results from "$F\alpha$" by replacing at least one occurrence of a name by "x".

Given these facts, (**6**iiie) and (**6**iiif) respectively tell you that placing "$\forall x$" or "$\exists x$" before "Fx" yields a **Q**-sentence. Given this, (**6**iiid) tells you that (**5**) is a **Q**-sentence.

---

† See Exercise 85, page 352.

7) $\exists x \forall y Fxy$

is also a **Q**-sentence; we must see it as built up in stages. Starting with, say, "$F\alpha\beta$" we see that "$\forall y F\alpha y$" is a **Q**-sentence, and from this it follows that (7) is also.

8) $\exists y \forall y Fyy$

is not a **Q**-sentence, as we can see by trying the same pattern of construction. "$F\alpha\beta$" is a **Q**-sentence, so "$\forall y F\alpha y$" is also a **Q**-sentence, as before. But we cannot go on to infer that (8) is, since "$\forall y F\alpha y$" contains "y", and the condition at the beginning of (6iii) is not met.

2   Interpretations and validity

An interpretation of **P** is an assignment of truth values to all **P**-letters. The definition of the sentence connectives had the effect that every interpretation of **P** would determine a truth value for *every* sentence of **P**, not just the sentence-letters. We can see the process of interpretation as containing two elements: the assignment of entities (here, truth values) to the relatively simple expressions; and rules or definitions which bring it about that assignments to the simple expressions have determinate consequences for the complex expressions.

 P-validity, we saw, is defined in terms of interpretations as follows:

  2.1.5)   An argument in **P** is **P**-valid iff every interpretation upon which all the premises are true is one upon which the conclusion is true.

We now introduce a notion of *interpretation of* **Q** which will enable us to use the same definition of validity (with "**Q**" replacing "**P**" throughout). We first need to associate an interpretation with a *domain* of entities. These are, intuitively, the objects the language talks about. We may need to consider different domains in different contexts. Thus if, as a physicalist, I say, "Everything is physical", then I mean to speak of absolutely everything, or at least of all concrete things. But if, in a lecture, I say "There's no more chalk" the domain should be restricted to the objects in the room, or to hand: I say of all of these that none is chalk, and I do not say that no object in the whole wide world is chalk. The only restriction we place on domains is the assumption that every domain will contain at least one object. The stipulations with regard to the simple expressions are as follows:

1) In any interpretation of **Q**, with respect to any domain, D:
   all sentence-letters are assigned truth values (just as in **P**);
   all name-letters are assigned objects in D (e.g. "α" might be assigned Ronald Reagan, for a D which contains this man);
   all predicate-letters of degree n are assigned a set of n-tuples (n-membered sequences) of objects in D (e.g. "F" might be assigned a set of ordered pairs (2-membered sequences) such that the first member of the pair loves the second);
   " $=$ " is assigned the set of ordered pairs of members of D such that in each pair the first object is the same thing as the second;
   Variables are not assigned anything.

We can use expressons of the form "$i(\ldots)$" to denote what some **Q**-interpretation, i, assigns to some expression, ... So instead of saying that an interpretation, i, assigns Ronald Reagan to "α", we can say that $i(\alpha) =$ Ronald Reagan.

We now need to specify the rules whereby the interpretation of the simple expressions determines a truth value for every **Q**-sentence. For the sentence connectives, these are just the rules from **P**, but they are restated for completeness.

The expression "X^n_v" is to be read "the result of replacing every occurrence of the variable v in X by the first name-letter n not occurring in X". Talk of the *first* name-letter is to ensure that there is a unique result of the transformation. It presupposes some conventional (e.g. alphabetical) ordering of the name-letters.

2) For any **Q**-interpretation, i, with respect to a domain, D:

   (i) $\neg X$ is true upon i iff X is false upon i;
   (ii) $(X \,\&\, Y)$ is true upon i iff X is true upon i and Y is true upon i;
   (iii) $(X \lor Y)$ is true upon i iff X is true upon i or Y is true upon i;
   (iv) $(X \to Y)$ is true upon i iff X is false upon i or Y is true upon i;
   (v) an atom $\phi v_1 \ldots v_n$ is true upon i iff the sequence $\langle i(v_i) \ldots i(v_n)\rangle$ belongs to the set $i\,(\phi)$, and is false upon i iff the sequence does not belong to the set;
   (vi) $\forall v X$ is true upon i iff X^n_v is true upon i, and also upon every

other interpretation whose domain is D and which agrees with i except in point of what is assigned to n; otherwise it is false;

(vii) $\exists v X$ is true upon i iff X^n_v is either true upon i, or true upon some other interpretation whose domain is D and which agrees with i except in point of what is assigned to n; otherwise it is false.

The intuitive motivation for (v), (vi) and (vii) may not be apparent. I will try to explain how these clauses are arrived at.

For (v), consider an atom, say "$G\beta$". Consider an interpretation which assigns the set of 1-membered sequences whose members are presidents of the USA to "G" and Ronald Reagan to "β". Then "$G\beta$" is true upon this interpretation, since the one-membered sequence $\langle$Reagan$\rangle$ does belong to the set the interpretation assigns to "G".[1] Note that a given **Q**-sentence may be true upon one interpretation, false upon another. Consider an interpretation which assigns Reagan to "β" but the set of 1-membered sequences each of whose members is a Chinese emperor to "G". "$G\beta$" is false upon this interpretation.

For the point of the talk of *sequences* in (v), consider the **Q**-sentence "$F\alpha\beta$" and an interpretation i, which assigns to "F" the set of ordered pairs (i.e. 2-membered sequences) such that the first member of each pair loves the second. Suppose the interpretation assigns John to "α" and Mary to "β". Then (v) rules that "$F\alpha\beta$" is true upon this interpretation iff the ordered pair $\langle$John, Mary$\rangle$ belongs to $i(F)$, that is, iff John loves Mary. Notice that *order* is crucial here: while $\langle$John, Mary$\rangle$ belongs to $i(F)$, $\langle$Mary, John$\rangle$ may not.

Turning to (vi), intuitively we want to say that a universal quantification in English, say "Everything is physical", is true just on condition that "physical" is a word true of everything. Applying this to **Q**, we want to say that "$\forall x Fx$" is true upon an interpretation, with respect to a domain D, just on condition that it assigns to "F" the set of all the 1-membered sequences that can be formed out of members of D. To operate the rule in this case, we take some sentence, say "$F\alpha$", resulting from "$\forall x Fx$" by deleting the quantifier and the variable adjacent to it, and replacing the other variable-occurrence by a name-letter, and consider all the interpretations of the name-letter that agree on what set they assign to "F". For each object in the domain, D, one of these interpretations will assign it to "α". Just on condition that "$F\alpha$" comes out true on all these interpretations, "$\forall x Fx$" is true on the original interpretation. "$\forall x Fx$" is true on just those interpretations, i, such that $i(F)$ contains, for every member x of D, the sequence $\langle x\rangle$. Thus "$\forall x Fx$" is false, alas, on an interpretation with respect to the domain of all persons which assigns to "*F*" the set of all 1-membered sequences whose member is happy.

(vii) works just like (vi), except that instead of *every* alternative interpretation with respect to the name-letter we now talk of *some* alternative interpretation.

Those unfamiliar with the ideas used here are advised to continue with the following sections, in which they are applied, and return to this more abstract presentation of them later.

We can now define **Q**-validity:

> **3)** An argument is **Q**-valid iff for every non-empty domain, *D*, every interpretation with respect *D* upon which all the premises are true is one upon which the conclusion is true.

There are a number of different ways in which semantics for **Q** can be given. One motivation for my choice is that I wanted to bring out the unity of functioning of the sentence connectives as they occur in **P** and as they occur in **Q**. Consider a **Q**-sentence like

> **4)** $\exists x\ (Fx\ \&\ Gx).$

By our present criteria, "&" is not a sentence connective in (4) since "*Fx*" and "*Gx*" are not sentences. Some people like just to stipulate that they *are* sentences, but this loses the connection between something's being a sentence and its being usable in a complete act of communication, for example, to make an assertion. (Thus you cannot use the expression "*x* is a man" to make an assertion. For who is *x*?) On the semantics given here, "&", even as it occurs in (4), makes its contribution to truth conditions through functioning as a genuine sentence connective connecting the **Q**-sentences "*Fα*" and "*Gα*". The fact that it can be treated by exactly the same interpretation rule for both languages shows that its semantic functioning is the same in both.[2]

One way in which the syntactic and semantic classifications for **Q** could be made to coincide is to specify syntactic categories in terms of the "constructional history" of a sentence. Thus a sentence like (4) can be seen as having been constructed in stages, starting by joining, say, "*Fα*" and "*Ga*" by "&". At this point "&" is straightforwardly a sentence connective, since it joins two sentences to form a sentence. Thus if we specify an expression's syntactic category as that to which it belongs at the first point in which it enters a sentence's constructional history, we shall be able to count "&" (and the others) as syntactically, as well as semantically, sentence connectives.

3   Universal quantification

What is the test of the adequacy of a **Q**-formalization of an English sentence? As in the case of **P**-formalization there are two tests. One is that, relative to the intended domain, the **Q**-sentence should be true upon an intended interpretation iff the English sentence is true. The other test is that the recovered sentence or argument – the result of replacing the name- and predicate-letters by the English names and predicates specified in the correspondence scheme, and replacing the **Q**-operators by their stipulated English correlates – should say the same as the original.

The formalization of "Everything is physical" by "$\forall x Fx$" meets the first test. With respect to any domain, an intended interpretation assigns the set of all 1-membered sequences whose member belongs to the domain and is a physical thing to "F". The **Q**-sentence is true on this interpretation, with respect to this domain, iff everything in the domain is physical. Just as the English sentence does not make explicit the domain of the English quantifier (should it, perhaps, include only concrete things?), so the truth of the formalization is relative not only to an interpretation but also to a domain. We have already seen that the definition of validity quantifies universally over domains, so the relativization to domains does not hinder the central business.

Applying the second test of adequacy, we recover the original English sentence, give or take the insertion of "thing", via the correspondence of the **Q**-operators with English expressions stipulated in §**1**, and the correspondence scheme relating "F" to "is physical".

How should sentences like

1)   All men are happy

be formalized? The standard answer is:

2) $\forall x \ (Fx \rightarrow Gx)$,

with "F" corresponding to "is a man" and "G" to "is happy". Despite the fact that (2) contains no visible occurrence of "if", it is treated as if it were the same as

3)   For any object, x, if x is a man then x is happy.

Is this adequate?

To test its adequacy, let us take an unrestricted domain, D, and an intended interpretation, i: it will assign to "F" the set of all 1-membered sequences whose members are in D and are men, and to "G" the set of all 1-membered sequences whose members are in D and are happy. From (**2.2**vi) we know that "$\forall x(Fx \rightarrow Gx)$" is true upon i iff "$F\alpha \rightarrow G\alpha$" is true upon every interpretation, i', which agrees with i on the assignment to "F" and "G". Any such alternative i' falls into one of three classes:

(i) $i'(\alpha)$ is not a man. So, since $i'(F) = (F)$ and $i'(G) = i(G)$, $\langle i'(\alpha) \rangle$ does not belong to $i'(F)$. So "$F\alpha \rightarrow G\alpha$" is true upon i', by (**2.2**iv).

(ii) $\langle i'(\alpha) \rangle$ belongs both to $i'(F)$ and to $i'(G)$. In short, $i'(\alpha)$ is a happy man. "$F\alpha \rightarrow G\alpha$" is true upon i', by (**2.2**iv)

(iii) $\langle i'(\alpha) \rangle$ belongs to $i'(F)$ but not to $i'(G)$. In short, $i'(\alpha)$ is an unhappy man. "$F\alpha \rightarrow G\alpha$" is false upon i', by (**2.2**iv).

The upshot is that "$\forall x(Fx \rightarrow Gx)$" is true upon i iff there are no unhappy men in the – by hypothesis unrestricted – domain of i. So all we have to decide is whether (**1**) is true iff there are no unhappy men.

There is no doubt that (**1**) is false if there are unhappy men. The doubt concerns whether there being no unhappy man is sufficient for its truth. For suppose there were no men at all. Then there would be no unhappy ones. But would this make (**1**) true?

It has been argued[3] that if John has no children then

4)   All John's children are asleep

is not true, whereas a **Q**-formalization of (**4**) as (**2**) would be true, relative to a domain containing no child of John's, upon an interpretation assigning to "F" the set of all 1-membered sequences whose member is a child of John's, and to "G" the set of all 1-membered sequences whose member is asleep. The first set would be empty, and so all the relevant interpretations will make "$F\alpha$" false, and thus verify "$F\alpha \rightarrow G\alpha$".

An obstacle in the way of accepting this view is that it would lead us to hold that "all" is ambiguous. For the following is true, despite the fact that there are no bodies acted on by no forces:

5)   All bodies acted on by no forces continue in a uniform state of rest or motion.

If (**4**) is not true but (**5**) is, then it would seem that "all' must have a different meaning in the two cases, and this is hard to swallow. "All" is learnt once and for all. It is not like "bank", for which one has first to master one sense, and then, as a separate issue, master another.

This is a good opening for a conversational implicature response. (4) implicates that John has children, but does not entail it. To support this, consider that even a community whose natural language had **Q**-quantifiers would tend to utter sentences of the form "$\forall x\,(Fx \rightarrow Gx)$" only if they believed that something is F, because they would otherwise be more informative by saying that nothing is F.

The response as it stands cannot do justice to all the issues. It is intuitively incorrect to say that arbitrary quantifications with antecedent predicates true of nothing are true. For example,

6) All bodies acted on by no forces undergo random changes of velocity

is intuitively false, yet its standard **Q**-formalization as (2) would be true upon an intended interpretation (one assigning to "F" the set of all 1-membered sequences whose member is a body acted on by no forces; that is, the empty set).

The defence of (2) as **Q**-formalization of both (5) and (6) might draw on the idea that in some contexts the assertion of a quantified conditional (as (5) and (6) are taken to be) implicates that what is asserted is a law of nature. Since this is a false implicature in the case of (6), we think of (6) as unassertible and then, confusedly, come to think of (6) as false. A hearer, in order to understand a speaker's full intentions in uttering (5) and (6), would have to recognize him as intending to state a natural law. As one sentence expresses a natural law and the other does not there is room for a contrast between them, a contrast which need not be reflected in truth values. It is worth noting that any serious discussion using these sentences would naturally slip into counterfactual constructions:

If any bodies were acted on by no forces, they would continue in a uniform state of rest or motion

and

If any bodies were acted on by no forces, they would undergo random changes of velocity.

I now turn to more straightforward problems of formalizing universal quantifications in English.

7) Not all men are happy

is standardly formalized as

8) $\neg \forall x(Fx \rightarrow Gx)$.

On the other hand

9) All men are not happy

is probably ambiguous. One reading, encouraged by heavy stress on "all", makes it equivalent to (7), and so to be formalized as (8). The other reading, in my view more correct, treats it as appropriately formalizable as

10) $\forall x(Fx \rightarrow \neg Gx)$

and thus as equivalent to

11) No men are happy.

"Every", "any" and "whatever" are often to be formalized by "$\forall$". Thus (2) is standardly held to formalize all of:

12)   Every person in the room was happy.

13)   Any person who interferes will be shot.

14)   Whatever you buy you will charge to me.

Note, however, that this formalization is not always the "deepest". For example, it does not discern a possible conjunctive structure in "person in the room". A candidate for a deeper formalization of (12) is

15) $\forall x((Fx \ \& \ Gx) \rightarrow Hx)$

where "F" corresponds to "is a person", "G" to "is in this room" and "H" to "was happy".[†]

† See Exercise 86, page 352.

4 Existential quantification

The English sentence

    1)   Someone is happy

is standardly formalized

    2) $\exists x F x$.

An appropriate domain of interpretation is the set of persons. Given the correspondence of "F" to "is happy", for an intended interpretation, i, $i(F)$ is the set of all 1-membered sequences whose member is in the domain and is happy. Then (2) is true upon i iff for some interpretation i', where $i'(F) = i(F)$, $i'(\alpha)$ is happy, and this is intuitively the correct truth condition.

How should sentences like

    3)   Some elephants are greedy

be formalized? A standard answer is:

    4) $\exists x (F x \, \& \, G x)$,

pronounced: "there is an x such that F of x and G of x". Despite the fact that (3) contains no visible occurrence of "and", it is treated as if it were the same as

    5)   Something is both an elephant and greedy.

Is this correct?

Suppose an unrestricted domain. On an interpretation i which assigns to "F" the set of all 1-membered sequences whose members are elephants, and to "G" the set of all 1-membered sequences whose members are greedy, (4) is true iff at least one interpretation agreeing with i in the assignments to "F" and "G" assigns to "α" an object which is a greedy elephant. One doubt one might have about the formalization is that (3) appears to imply that there is more than one greedy elephant. Moreover, some have said that "some" implies "some but not all", so that (3) would imply that some elephants are not greedy, and (1) would imply that someone is not happy. These alleged implications appear stronger in some cases than others. For example,

6)   She ate some of the cakes

seems to imply quite strongly that some cakes were left and that more than one was eaten.

The "some but not all" implication could be approached in at least two ways. One could explicitly add to the formalization of, say, (3) that not all elephants are greedy. This is simply the denial of "All elephants are greedy", and so the formalization proposed is

7) $\exists x(Fx \& Gx) \& \neg \forall x(Fx \rightarrow Gx)$.

Alternatively, one could say that the implication is a matter of implicature only, and so need not be registered in the formalization. To support this view, one could say it would be misleading to use "some" if one knew something expressible by "all": misleading, but not false. The cancellability of the implicature would support this view.

8)   She (certainly) ate *some* of the cakes – *all* of them in fact

is consistent.

The implicature story does not work well for the implication from "some"-sentences to "more than one"-sentences. Since, for example,

9)   She ate a cake

does not entail

10)   She ate some cakes

they cannot both correctly be given the same formalization, viz. (4). Similarly, we do not have the cancellability one would expect from implicature, since the following is inconsistent:

11)   She ate some of the cakes – in fact, just one of them.

Compare also

12)   A hungry man is at the door

and

13) Some hungry men are at the door.

While the first could perhaps be adequately formalized by (4), the second could not be.

We should not infer that the meaning of "some" differs from that of "∃". The entailment in question need not be seen as due to "some", because where it obtains it is adequately explained by the presence of the plural nouns ("cakes" rather than "cake", "men" rather than "man").[†] In §9 we will see how the effect of plurals can be captured in **Q**. Probably the safest way to read "∃x" is "There is at least one thing, x, such that"; and we can be confident that, in using "∃" in the manner of (4), we are adequately formalizing English sentences containing quantifiers like "some", "something", "a" – as in "A man is here to see you" – just on condition that we can rephrase without distortion in terms of "There is at least one thing such that ...".

There is no need for **Q** to contain both "∀" and "∃", since they are interdefinable. This corresponds to facts about English: "everything is ..." means "it is not the case that something fails to be ..."; and "something is ..." means "it is not the case that everything fails to be ...".[‡]

5 Adjectives

A sentence like

1) Tom is a greedy man

is usually formalized as

2) $F\alpha$ & $G\alpha$,

with "F" corresponding to "is greedy", "G" to "is a man" and "α" to "Tom". The adjectival construction of the English, in which the adjective "greedy" qualifies the noun 'man', is rendered as a conjunction

† See Exercise 87, page 353.   ‡ See Exercise 88, page 353.

in **Q**. However, the truth conditions appear to come out correctly. (**1**) does indeed appear equivalent to

   3)   Tom is a man and is greedy.

Moreover, for an intended interpretation, i, one which assigns to "F" the set of all 1-membered sequences whose members are men, to "G" the set of all 1-membered sequences whose members are greedy, and to "α", Tom, (**2**) is true upon i iff Tom is a man and is greedy, which seems correct with respect to (**1**).

   In **Q**, adjectives like "greedy", noun constructions like "is a man" and verbs like "runs" are treated as all of essentially the same kind: they are alike considered predicates, and hence matched with predicate-letters in **Q**-logical forms. However, this policy ensures that some familiar English adjectival constructions cannot be adequately formalized in **Q** to a reasonable depth. Consider

   4)   Tom is a large man

and suppose we formalize this as (**2**): "$F\alpha \mathbin{\&} G\alpha$", with "$F$" corresponding to "is a man" and "G" to "large". Continuing this policy, how should we formalize

   5)   Tom is a businessman but not a large businessman?

The policy dictates:

   6) $H\alpha \mathbin{\&} \neg(H\alpha \mathbin{\&} G\alpha)$

with "H" corresponding to "is a businessman" and "G" to "large". However, (**4**) and (**5**) are consistent, whereas (**2**) and (**6**) are not.† Hence the policy of formalizing (**4**) and (**5**) by the conjunctive treatment of adjectives has failed. It has wrongly represented consistent sentences as inconsistent.

   There is a class of adjectives like "large" (and "heavy", "expensive" etc.) which resist **Q**-formalization. And there are other resistant adjectives. For example, it is obvious that

   7)   Tom is an alleged murderer

† See Exercise 89, page 353.

is not to be formalized as (**2**) (with "*F*" corresponding to "is a murderer" and "*G*" to "alleged"). As a further example, it would be reasonable to formalize

8)　Tom loves only happy women

as

9)　$\forall x(H\alpha x \rightarrow (Fx\ \&\ Gx))$

(with "*F*" corresponding to "is a woman", "*G*" to "is happy" and "*Hxy*" to "*x* loves *y*"). By contrast, it would not be reasonable, but nonsense, to adopt the same formalization of

10)　Tom loves only three women.

An intended interpretation would assign to "*G*" the set of all 1-membered sequences whose members are three. But it is nonsense to speak of an object (one single object) *being three*. There is no such property. I discuss in §9 how adjectives like "three" are **Q**-formalized.

In general, we can reflect adjectival modification by conjunction only if, where "*n*" is a name, "*A*" an adjective, and "*C*" the common noun it qualifies

11)　*n* is a(n) *AC* $\vDash$ *n* is *A* and *n* is a(n) *C*

is true (and has an intelligible conclusion).[4]

Adjectives (or adjectival phrases) satisfying (**11**) (like "happy", "greedy", "red", "weighs 12 pounds") are called "predicative". Among non-predicative adjectives, there are important distinctions to be made. For example, there is a category of adjectives which qualify other adjectives, for example "dark" (qualifying colour adjectives), and these, though ruled non-predicative by (**11**), behave rather differently from "large". Again, whereas the distinctive contribution of "alleged" cannot be captured in **Q**, even by devious means, the distinctive contributions of numeral adjectives like "three", though non-predicative, can be adequately represented in **Q**.

Of course, sentences containing non-predicative adjectives, like (**4**), can always be adequately **Q**-formalized. For example, we could let "*F*" correspond to "is a large man" and formalize (**4**) as "*Fα*". We can make sense of the idea of an interpretation which assigns to "*F*" the set of

1-membered sequences each of whose members is a large man. More radically, we can adequately formalize any English sentence by a sentence-letter. However, such **Q**-formalizations are not deep enough to bring out the distinctive contribution of non-predicative adjectives like "large". "Is a large man" is complex, composed of "large" and "man". This complexity of structure is not revealed in the envisaged **Q**-formalizations.

Adequacy does not set high standards upon formalizations. What is needed in addition is a reasonable depth, in particular enough depth to display the validity of formally valid arguments. Thus the deepest **Q**-formalization of

12)   Tom is a small businessman, therefore Tom is a businessman

is

13) $F\alpha$; $G\alpha$

(with "F" corresponding to "is a small businessman" and "G" to "is a businessman"), which is plainly **Q**-invalid. Does this matter? The idea of artificial languages was to capture generalizations about *formal* validity, but, by the standards set in chapter **1.10** and **1.11**, (12) is not formally valid.[†] While this may give the **Q**-formalizer relief from the burden of finding a formalization of (12) which reflects its validity, it also carries with it an objection concerning the cases in which there are validity-reflecting **Q**-formalizations. For example, the validity of

14)   Tom is a happy man, therefore Tom is a man

is reflected by the validity of the obvious **Q**-formalization. Hence the conjunctive method of **Q**-formalization has captured more validity than was intended: it has reflected as *formally* valid an argument that, though valid, is not *formally* valid, according to our original definition. There is a tension here. The original definition of formal validity could be revised, or else the project of formalization recharacterized.

I have given examples of supposedly predicative adjectives – "happy", "greedy", "red", "weighs 12 pounds" – but a doubt remains. Are all these adjectives really predicative? One sign of non-predicativity comes from the pattern discerned in the case of "large": you can be a mouse and

† See Exercise 90, page 353.

an animal and a large mouse without being a large animal. Suppose that humans are on the whole rather less greedy than most animals. Then you could be a man and an animal and a greedy man without being a greedy animal. Like "large", there is a relativization to a comparison class, which is introduced by the noun, and which could be expressed more explicitly by saying "He's large *for* a mouse, but not large *for* an animal", "He's greedy *for* a man, but not greedy *for* an animal". So perhaps "greedy", and by similar considerations, "happy", are, after all, non-predicative.[†]

The difference between "large" and "greedy" is this: "large", associating with the noun it qualifies, thereby determines, at least in part, the appropriate standards of size to be applied. This does not hold for "greedy" which in this resembles adjectives like "tall". Even when "tall" occurs modifying a noun as in "Tom is a tall man", it may not be this noun which determines the relevant comparison class. That determination is left to context. Being tall for a Swede involves being taller than does being tall for an Eskimo.

Does the growing shadow of non-predicativity extend even to an adjective like "red"? It's one thing for a house to be red (quite in order for the windows, doors, mortar and interior not to be red), another for a colour sample to be red (should be uniform red all over), another for a tomato to be red (*painted* red doesn't count) and yet another thing for hair to be red. Correct as these observations are, they do not establish non-predicativity. There is no general term, T, not itself having colour entailments, such that we have the pattern: "This is a tomato and a T and is a red tomato but not a red T". Moreover, (11) holds for "red" combined with any appropriate noun. So the phenomenon is not non-predicativity. Although the different nouns it qualifies impose different standards for what is to count as a red thing of that kind, it appears impossible for a thing to belong to two different kinds and count as red by the standards appropriate to one and as non-red by the standards appropriate to the other. By contrast, a non-predicative adjective, like "large", can introduce comparison classes which give different verdicts with respect to the applicability of the adjective.[‡]

6 Adverbs

In traditional grammar, an adverb or adverbial phrase is seen as

---

† See Exercise 91, page 353.   ‡ See Exercise 92, page 353.

modifying a verb. No such construction is available in **Q**. Hence there is no straightforward validity-reflecting **Q**-logical form of, for example:

1)   John walked quickly. Therefore John walked.

We could merge "walked" with "quickly", and formalize "walked quickly" by a single predicate letter; but we would have to choose a different letter to correspond to plain "walked", as it occurs in the conclusion, so the formalization would patently be **Q**-invalid.

However, we can reach a deeper formalization by a roundabout route, which has been proposed by Donald Davidson. It seems that

2)   John walked

is true iff *there is* something – a walk – which John walked. So, though there is no visible existential quantifier in (2), we could offer the following formalization of it:

3) $\exists x F \alpha x$

with "Fxy" corresponding to "x walked y", and "α" to "John".

Let us think what sort of properties John's walk might have. It could be quick or slow, uphill or downhill, with a stick or with a dog. This suggests that (1) should be formalized:

4) $\exists x (F \alpha x \ \& \ Gx); \exists x F \alpha x$

with "F" and "α" as before, and "G" corresponding to "is quick". Notice that it is perfectly in order to suppose an interpretation of "G" which assigns to it the set of 1-membered sequences each of which has a quick member: quick walks and quick swims and quick runs, but not slow walks, etc., and not trees or lamp-posts, since these things cannot be quick.

(4) is **Q**-valid. It is easy to see this informally. For any given domain, an interpretation, i, which brings the premise out true will assign a non-empty set to "F", and a sequence in this set will have $i(\alpha)$ as its first member. This is sufficient to ensure that the conclusion is true upon i. More informally still, the premise says that there is something with two properties – being F to α and being G – and it is clearly correct to infer from this that there is something with just the first of these properties.

In general the sort of thing that can be walked or can be quick is an *event*. We can sum up Davidson's proposal as involving two theses: (i) many verbs (those like "walk", "run" and "swim") introduce events, and many sentences containing these verbs, like (2), introduce (implicit)[5] existential quantification over events; (ii) adverbs are adjectives which qualify events. Thus Davidson's proposal sees the premise of (1) as containing an implicit existential quantifier, and sees "quickly" as an adjective whose intended interpretation associates it with a subset of events.

How widely can this account be applied? One class of adverbs can be exempted from the account at the outset: those which, like "allegedly", "necessarily", "probably" and "rarely", modify sentences rather than verbs: they are *sentence adverbs*.[6] It is plain that in

5)   John allegedly shot Jane

we are not saying that there is (really is!) a shooting with the property of being alleged. Rather we are saying something like: it is alleged that John shot Jane. We do not commit ourselves to John's having shot Jane or even to Jane's having been shot, and that is why we are not asserting that a shooting has a certain property. Similarly for the other cases.

To bring out the difference between sentence adverbs and the kind with which we are concerned, consider the ambiguity of

6)   John carelessly trod on a snail.

If we hear "carelessly" as a sentence adverb, we will regard this as equivalent to

7)   It was careless of John to tread on a snail.

If we hear "carelessly" as genuinely modifying "trod", we will reject this equivalence: John may have set out with full deliberation and care to tread on a snail, so (7) is false, but in the end executed his plan carelessly, so one reading of (6) is true.†

Does Davidson's account of adverbs hold of all those which are not sentence adverbs? There is another resistant category, exemplified by "intentionally". We cannot paraphrase "John intentionally trod on a snail" as "It was intentional that John trod on a snail", since the latter is ungrammatical. So it is not a sentence adverb. Yet we cannot see the

† See Exercise 93, page 354.

adverb "intentionally" as associated with a subset of events. The very same event may be an intentional switching on of the light and an unintentional scaring off of an intruder. Events as such can neither be intentional or otherwise.[7] Hence the adverb "intentionally" cannot be treated as an adjective applying to events.

Does Davidson's account hold of all adverbs which are neither sentence adverbs nor ones like "intentionally"? No: for in some cases – including "quickly" – the adjective that would be discerned on Davidson's account is non-predicative, and hence its distinctive contribution cannot be captured in **Q**. Suppose John swam the Channel, and thereby broke the record for cross-Channel swims. On Davidson's account

8)   John swam the Channel quickly

should be true in virtue of whatever swim verifies (**8**) being a quick one. On the other hand, compared to most Channel-crossings, John's self-propelled crossing was slow. So we also have as true

9)   John crossed the Channel slowly.

But the very same event is both the swim and the crossing, and verifies both (**8**) and (**9**). If we applied Davidson's account, we would be committed to the view that some particular event (John's swim, that is, his crossing) is both quick and slow, which is unacceptable.

Davidson's account should therefore be applied only to those adverbs not in any of the three resistant categories we have mentioned, viz.: sentence adverbs, those like "intentionally", and those corresponding to non-predicative adjectives. Whether there is a positive characterization of the remaining adverbs, and whether they genuinely form a special category of adverb, are questions we will not pursue. However, it is worth bearing in mind that failure of predicativity, or at least some closely related phenomenon, may be more prevalent than one might at first suppose.[8] For example, a walk can be uphill, and "uphill" appears predicative. A walk can also be a warning: walking might be the agreed danger signal. But it is open to question whether it is intelligible to qualify a warning as an uphill one. So we have the pattern: something is a walk and a warning and an uphill walk but not an uphill warning.[†]

† See Exercise 94, page 354.

7 Names

Names in English, like "Tom" and "London", are supposed to be **Q**-formalized by the use of name-letters. What is the common feature of names that justifies this common treatment? When we have answered this question, we will be in a position to say whether expressions like "the present Prime Minister of Great Britain", "water", "Pegasus", "Vulcan", which in many ways resemble names, should also be formalized by name-letters.

We need to begin by seeing how name-letters are interpreted in **Q**. It is simple: they are assigned an object (in the domain of interpretation). So one would expect the category of English names to be a category of expressions which stand for objects.

It seems clear that "Tom" and "London" fall into this category. But may it not turn out to embrace many other expressions as well? For example, why does not an expression like "happy", in the sentence "Tom is happy", count as standing for an object: the property of being happy, or, perhaps, the set of happy persons? Indeed, predicate-letters are interpreted in **Q** by being assigned objects (viz. sets), so should we not also construe English predicates as standing for objects, and thus, by the proposed criterion, counting as names?

Recall the definition of a predicate: it takes a name to form a sentence. What would happen if predicates were a species of name? Then a sentence like "Tom is happy" would be construed as two names juxtaposed. But two names juxtaposed cannot form a sentence, only a list. Hence predicates cannot count as names.

This argument does not prevent one seeing common nouns as names, and analysing, say,

1) Tom is a man

as

2) $F\alpha\beta$

with "α" corresponding to "Tom", "β" to "the property of being a man", and "F" to "having" (as when we say an object *has* a property). I know of no decisive objection to the policy, though we will not follow it.[9] Notice, however, that pursuing it cannot construe all predicates as names: there remains the predicate "having", matched to the predicate-letter "F".

Expressions like "the present Prime Minister of Britain" are called "definite descriptions". Many of them stand for objects, like the one just mentioned, so they are *prima facie* cases of names. Some, like "the golden mountain", do not. Hence if they are to be treated uniformly, they should not be formalized by name-letters. Other reasons for not formalizing them by name-letters are given in §10.

Does "water" stand for an object? One is inclined to answer affirmatively: of course, "water" stands for water! But what is this object, water? Not, presumably this puddle or glassful, but the scattered totality of water throughout the universe. There is no objection to allowing this totality to count as an object. Most objects are spatially cohesive in the sense that between any two spatially separated parts of the object lies a route (not necessarily a straight line) at each point of which is another part of the object. But there is no need to insist that all objects must be like this. So far, then, "water" passes the test for being a name.

If we do count it as a name, we need to be careful with our formalizations. For example, we will have to cope with sentences like:

3)   There is some water here.

Here a natural response is to abandon the name policy, and treat "water" as a predicate, formalizing (3) as

4) $\exists x(Fx \ \& \ Gx\alpha)$

with "Fx" corresponding to "x is water", "Gxy" corresponding to "x is present at y" and "α" corresponding to "here". The name policy can be adhered, too, however, if we adopt the more complex formalization:

5) $\exists x(Fx\beta \ \& \ Gx\alpha)$

with the correspondences for "G" and "α" as before, "β" corresponding to "water" (now treated as a name) and "Fxy" corresponding to "x is part of y".

More ingenuity is required to deal with sentences like

6)   Oil is lighter than water.

This sentence is true. But if we formalize as

7) $F\alpha\beta$

with "*F*" corresponding to "lighter than", and "oil" and "water" to "α" and "β" respectively, we say something which is true iff the weight of the total amount of oil in the universe is less than that of the total amount of water. This may still be true, but it is clearly not what (**6**) says. Rather, (**6**) says something like: any volume of oil weighs less than the same volume of water, and this would have to be the basis for an appropriate formalization, using the notion of "part of" as in (**5**).

In the case of words that are in many respects similar, for example "gold", a similar strategy would be called for in formalizing such sentences as

    8)   This ring is made of gold.

Made of gold it may be, but not composed of the scattered totality of all the gold in the world.†

Does "Pegasus" stand for an object? If it does, the object is a mythological one. Are there mythological objects, or is applying "mythological" a way of saying that there is no such object? Opinions differ. If, like me, you think that there are not *really* any such objects as Pegasus, you cannot formalize names like "Pegasus" by name-letters. One alternative route is given in §**10**.

Even if you think that there is such a thing as Pegasus, you will not, I presume, think there is any such thing as Vulcan. "Vulcan" was introduced in the nineteenth century by astronomers. They posited an additional planet, lying between the planet Mercury and the Sun, whose presence they felt was required to explain the course of Mercury. They were wrong: there is no planet between Mercury and the Sun. So you cannot think that "Vulcan" stands for an object, so you cannot formalize it by a name-letter. Again, an alternative is given in §**10**.

8 Identity

"=" is a genuine predicate (or, in some terminologies, a predicate constant). Unlike predicate-letters, which are assigned different sets in different interpretations, "=" is assigned a set specifiable in the same way in every interpretation: the set of all ordered pairs whose members belong to the domain and whose first member is the same as the second. This stipulation about interpretations reflects the intention that "=" should mean identity.

† See Exercise 95, page 354.

To conform to tradition, we will abbreviate expressions like "$= \alpha\beta$" and "$\neg = \alpha\beta$" to "$\alpha = \beta$" and "$\alpha \neq \beta$".

It follows from what was said in the first paragraph that

    1) $\alpha = \alpha$

is true on every interpretation, that is, $\vDash_Q \alpha = \alpha$. This is sometimes known as the "law of identity". It is enough to establish that

    2) $\vDash_Q \forall x\ x = x.$

By contrast, note that

    3) $\vDash_Q \forall x\ Fxx$

is false. For though (3) is true upon some interpretations (for example, those which assign to "F" what is assigned to "$=$"), it is false on others (for example, those which assign to "F" the set of ordered pairs such that the first is larger than the second).

The following true argument-claim illustrates the so-called principle of *substitutivity of identicals*:

    4) $F\alpha,\ \alpha = \beta \vDash_Q F\beta.$

This reflects the intuitive truth that if α is the same thing as β, then anything true of α is also true of β. The point can be put more generally as follows:

    5) An interpretation upon which "$\alpha = \beta$" is true is one upon which:

        "... α ..." is true iff "... β ..." is true.

The last line is to be read as follows: any sentence containing "α" is true (upon the interpretation) iff the corresponding sentence, but with "α" replaced by "β", on any number of occurrences, is true (upon the interpretation). A rough summing up: identical objects have identical properties.[†]

† See Exercise 96, page 354.

9 Numeral adjectives

Consider

1) There are two men at the door.

Though "two" is an adjective attached to "man", it does not modify "man" in the predicative way. We cannot formalize (1) as

2) $\exists x(Fx \, \& \, Gx)$

with "F" corresponding to "is a man at the door" and "G" corresponding to "two". An intended interpretation would have to assign to "G" the class of all one-membered sequences each of whose members is two, and this is nonsense: no (one) object is two.

However, we can provide adequate formalizations of sentences like (1) by using quantifiers and "$=$". (1) is probably ambiguous between

3) There are exactly two men at the door

and

4) There are at least two men at the door.

(To appreciate the difference, consider which of these is true if there are three men at the door.) We start by formalizing (4).
Consider

5) $\exists x \exists y(Fx \, \& \, Fy \, \& \, x \neq y)$.

This is true upon an interpretation iff at least two things form 1-membered sequences belonging to the set the interpretation assigns to "F"; for short, iff at least two things are F. Applying the interpretation rules of (2.2) in more detail, (5) is true upon an interpretation, i, iff "$\exists y(F\alpha \, \& \, Fy \, \& \, \alpha \neq y)$" is true on some interpretation, say i', agreeing with i on the assignment to F. This in turn is true upon i' iff "$F\alpha \, \& \, F\beta \, \& \, \alpha \neq \beta$" is true on some interpretation, say i'', agreeing with i' on the assignment to F. This holds iff i'' assigns different objects to "α" and "β", and assigns to "F" a set having at least two members. But i' and i must agree with i'' in the assignment to "F". So (5) is true upon an interpretation iff it assigns at least two things to "F". Hence (5) is an adequate formalization of (4).

We have already seen (§4) that "$\exists x Fx$" adequately formalizes "At least one thing is F". It is easy to see how to modify (5) so that it formalizes "At least three things are F". More generally, the following schema can be seen as giving instructions for formalizing "At least n things are F", for arbitrary numeral n:

6) $\exists x_1 \ldots \exists x_n (Fx_1 \ \& \ \ldots \& \ Fx_n \ \& \ x_1 \neq x_2 \ \& \ \ldots \& \ x_1 \neq x_n \ \& \ x_2 \neq x_3$
    $\& \ \ldots \& \ x_2 \neq x_n \ \& \ \ldots \ldots \& \ x_{n-1} \neq x_n).$

To apply this recipe, we imagine the variables to be ordered in some way, with "x_1" corresponding to the first variable in the ordering, and so on. The recipe enjoins you to write n instances of the existential quantifier followed by a variable (a different variable each time), followed by n instances of "F" followed by each of the variables, followed by instances of $v \neq v'$ for every pair of distinct variables. Applying it to the case $n = 4$, so that the aim is to formalize "At least four things are F", yields:

7) $\exists x \exists y \exists z \exists x' (Fx \ \& \ Fy \ \& \ Fz \ \& \ Fx' \ \& \ x \neq y \ \& \ x \neq z \ \& \ x \neq x' \ \& \ y \neq z$
    $\& \ y \neq x' \ \& \ z \neq x').$

"Exactly n things are F" is true iff (at least n things are F and at most n things are F). So if we can formalize "at most n", it will be easy to formalize "exactly n".

Suppose that

8) At most two things are F

is true. Then, if ever we find objects, o_1, o_2, o_3, which are all F, we know that we must have counted one object twice, which means that either $o_1 = o_2$ or $o_1 = o_3$ or $o_2 = o_3$. This justifies the formalization of (8) as

9) $\forall x \forall y \forall z ((Fx \ \& \ Fy \ \& \ Fz) \rightarrow (x = y \lor x = z \lor y = z)).$

Similarly

10) At most one thing is F

is formalized

11) $\forall x \forall y ((Fx \ \& \ Fy) \rightarrow x = y).$

In general, "at most n things are F" is formalized by the recipe:

12) $\forall x_1 \ldots \forall x_{n+1} ((Fx_1 \ldots \& \ldots Fx_{n+1}) \rightarrow (x_1 = x_2 \vee \ldots \vee x_1 = x_{n+1} \vee x_2 = x_3 \vee \ldots \vee x_2 = x_{n+1} \vee \ldots \ldots \vee x_n = x_{n+1})).$

To formalize (3) – "there are exactly two men at the door" – we can simply conjoin (5) and (9). However, the result is equivalent to the neater:

13) $\exists x\, \exists y (Fx \,\&\, Fy \,\&\, x \neq y \,\&\, \forall z (Fz \rightarrow (z = x \vee z = y))).$

This says that some x and y are F and are distinct, and that any F thing is one of these. Hence it adequately captures the idea that exactly two things are F. The general recipe on the lines of (13) is

14) $\exists x_1 \ldots \exists x_n (Fx_1 \,\&\, \ldots \,\&\, Fx_n \,\&\, x_1 \neq x_2 \,\&\, \ldots \,\&\, x_1 \neq x_n \,\&\, x_2 \neq x_3 \,\&\, \ldots \,\&\, x_2 \neq x_n \,\&\, \ldots \ldots \,\&\, x_{n-1} \neq x_n \,\&\, \forall x_{n+1}(Fx_{n+1} \rightarrow (x_{n+1} = x_1 \vee \ldots \vee x_{n+1} = x_n))).$

Although the recipies of (6) and (14) may be of some practical value in providing formalizations, they are rather cumbersome, and the ideas can be presented more neatly by a recursive definition. One can define numeral quantifiers of any of the forms "at least", "at most", "exactly". This is one way of defining the last, written in the form "$\exists^n$", where the superscript indicates the exact number:

15) $\exists^1 Fx =_{df} \exists x (Fx \,\&\, \forall y (Fy \rightarrow x = y))$

16) $\exists^n Fx =_{df} \exists x (Fx \,\&\, \exists^{n-1} y (Fy \,\&\, x \neq y)).$

The basis clause (15) captures the thought that something is F and that anything F is that thing, that is, that exactly one thing is F. The inductive clause (16) equates there being exactly n things which are F with there being an F such that exactly $n-1$ *other* things are also F.†

We can use these ideas to capture the plurality involved in sentences in which an existential quantifier attaches to a plural noun. Thus (4.3)

> Some elephants are greedy

may mean something like

† See Exercise 97, page 355.

17)   More than one elephant is greedy.

If so, it can be formalized as

18) $\exists x \, \exists y (Fx \, \& \, Fy \, \& \, Gx \, \& \, Gy \, \& \, x \neq y)$

with the intended interpretation assigning to "F" the set of 1-membered sequences whose member is an elephant, and to "G" the set of 1-membered sequences whose member is something greedy. If (17) seems more precise than (4.3), it may be that we need to turn to implicature for an explanation. If, looking at a large bag of nails, I exclaim,

19)   Some of the nails have gone rusty,

my audience may well expect there to be more than two rusty ones, but it seems wrong to say that this is actually entailed, rather than implicated, by (19).

Note that although we can always use one of the recipes (6), (12) and (14), or the quantifiers defined in (15) and (16), to formalize numeral *adjectives*, the numerals themselves, nouns in the series "one", "two", "three", cannot be formalized in the same way. In a sentence like

20)   Five plus seven equals twelve

there are, on the surface, no appropriate predicates to which a quantifier might attach to form a sentence. It would be forgivable to suppose that the deepest **Q**-formalization of (20) would use name-letters to correspond to the numerals.[10†]

10   Descriptions

We saw in §7 that there is a case for saying that definite descriptions like "the present Prime Minister of Great Britain" and "the moon" should not be formalized by name-letters. This section explores that case in more detail. First I present part of the only serious alternative style of **Q**-formalization, Bertrand Russell's Theory of Descriptions.

Russell's idea was that a sentence like

1)   The moon is cold

† See Exercise 98, page 355.

should be **Q**-formalized as containing an instance of (**9.14**) for $n = 1$:

2) $\exists x(Fx$ & $\forall y(Fy \rightarrow x = y)$ & $Gx)$

with an intended interpretation assigning to "F" the set of all 1-membered sequences each having a moon as member, and to "G" the set of all 1-membered sequences each having a cold member. (**2**) in effect says that exactly one thing is F and that thing is G. It thus treats (**1**) as false if there is more than one moon or less than one or if there is exactly one moon which is not cold, and otherwise true.

We can note at once that we need some relativization to a domain, or else (**2**) will be false on the intended interpretation whereas (**1**) is true. Many planets other than the earth have moons. Hence there is more than one moon. Hence (**1**), if formalized as (**2**) but with an intended interpretation having an unrestricted domain, is false. We should therefore stipulate that the intended interpretation's domain is restricted to, perhaps, Earth moons.

How does the name-letter treatment compare with Russell's?[11] One advantage of the latter is that it captures the validity of some arguments that the name-letter approach cannot capture, for example

3)   The moon is cold; therefore a moon is cold

and

4)   The author of *Mein Kampf* is a maniac, Hitler wrote *Mein Kampf*; therefore Hitler is a maniac.

The name-letter method yields the following:

5) $G\alpha; \exists x(Fx$ & $Gx)$

(the correspondences being "G" for "is cold", "α" for "the moon", "F" for "is a moon") and

6) $F\alpha, G\beta; F\beta$

(the correspondences being "F" for "is a maniac", "α" for "the author of *Mein Kampf*", "G" for "wrote *Mein Kampf*" and "β" for "Hitler"). Both of these are patently invalid, whereas (**3**) and (**4**) are valid.

On Russell's theory, by contrast, we get the following **Q**-valid formalizations:

> 7) $\exists x(Fx \ \& \ \forall y(Fy \to x = y) \ \& \ Gx); \ \exists x(Fx \ \& \ Gx)$

(the correspondences being "*F*" with "is a moon" and "*G*" with "is cold") and

> 8) $\exists x(Fx \ \& \ \forall y(Fy \to x = y) \ \& \ Gx), \ F\alpha; \ G\alpha$

(the correspondences being "*F*" with "wrote *Mein Kampf*", "*G*" with "is a maniac" and "*α*" with "Hitler"). The **Q**-validity of (7) should be evident. The **Q**-validity of (8) can be established as follows: an interpretation upon which the premises are true will assign a set containing just one 1-membered sequence to "*F*", that sequence's member is (by the second premise) assigned to "*α*", and that sequence (by the last part of the first premise) belongs to what is assigned to "*G*". That ensures that any interpretation upon which the premises are true is one upon which the conclusion is true also.

The fact that Russell's theory of descriptions makes it possible to reflect the validity of some arguments whose validity cannot be captured by the method of name-letters would decisively establish its superiority only if the arguments in question were *formally* valid. Here, however, we find a circle: given our present definitions, and given that the validity of the arguments depends upon the meaning of "the", the arguments are formally valid iff "the" is a logical constant, and, by the present standards, "the" is a logical constant iff Russell's theory of descriptions is correct. For then, and only then, is it definable in terms of the stipulated constants "all", "some", "if" and "is the same as".[12] We will assume, however, that the capturability of the validity of arguments like (3) and (4) gives us a reason for preferring it.[†]

We mentioned in §7 that another basis for favouring Russell's theory of descriptions depends upon the claims that (i) some definite descriptions – we will call them "empty" ones – stand for no object, for example, "the golden mountain"; and (ii) all definite descriptions should be treated in the same way. Since the empty descriptions cannot be adequately formalized by means of a name-latter, (ii) tells us that none should be. Since there are only these two candidates for methods of **Q**-formalizing, Russell's triumphs.

---

† See Exercise 99, page 356.

To reiterate why an empty description cannot be adequately formalized by a name-letter: our standard of adequacy is that upon an intended interpretation (in part determined by the correspondences), the formalization should be true iff the original English is true. But what could be an intended interpretation with respect to a sentence containing an empty description? We have to choose some object, since every interpretation assigns objects to all name-letters. Perhaps we should designate an arbitrary object, say the number 0 or the null set, as what an intended interpretation should assign to a name-letter which corresponds to an empty description.[13] But it seems that whatever object we choose we will get the wrong result. This is simply because some sentences containing empty descriptions are true and some are not. Thus

9) Someone was the unique author of *Principia Mathematica*

is not true (it was jointly authored by Russell and A. N. Whitehead) whereas

10) No one was the unique author of *Principia Mathematica*

is true. Formalizing by the method of name-letters would yield respectively:

11) $\exists x \; x = \alpha$

12) $\neg \exists x \; x = \alpha.$

Whatever object we choose as what an intended interpretation assigns to "α", (11) will be true upon the interpretation and (12) will be false. This gets the truth values just the wrong way about, so there is no adequate formalization to be had by this method.

One noteworthy consequence of Russell's theory of descriptions is that sentences which would naively be classified as identity sentences are formalized not as identity sentences but as existential quantifications. For example, we might naively classify

13) Scott was the author of Waverley

as an identity sentence. However, an appropriate **Q**-formalization, using Russell's theory of descriptions, is:

14) $\exists x (Fx \; \& \; \forall y (Fy \rightarrow x = y) \; \& \; x = \alpha)$

where "F" corresponds to "wrote Waverley" and "α" to "Scott". This **Q**-sentence, being dominated by an existential quantifier, must be classified as an existential quantification and not as an identity sentence. In particular, something of the form of (**14**) does not yield a premise fit for an application of (**8.5**), the substitutivity of identicals.

The principle that all definite descriptions should be treated alike is open to question. If we apply the same sort of principle to names, saying that all names must be treated alike, we find that we should treat "Vulcan" and "Reagan" alike. Since, for reasons already seen in §7, we cannot formalize "Vulcan" by a name-letter we also should not formalize "Reagan" by a name-letter. Russell himself accepted this consequence. Indeed, he held that the only expressions that could be adequately formalized by name-letters were expressions that we do not ordinarily classify as names (but rather as demonstrative pronouns): "this" and "that", as these are used to refer to momentary subjective experiences. These he thought deserved to be called names, from the logical point of view: as he put it, they, and they alone, are "logically proper names".[14] His reason was that he thought that they, and they alone, owed their meaning to their bearer. Just as a name-letter is assigned an object on any interpretation, so a logically proper name, to be meaningful, must have a bearer. Russell believed, for reasons we shall not explore, that this condition was not met by the expressions like "King Arthur" and "Aristotle" which we ordinarily call proper names.[†]

11   Existence

(**10.10**) raises the question of how we should formalize sentences which affirm or deny existence. That sentence could have been rephrased:

1) The unique author of *Principia Mathematica* does not exist.

This is not very idiomatic as compared with (**10.10**), but it seems to mean the same. We are perhaps more familiar with the phrasing of (**1**) in mathematical contexts, for example:

2) The greatest prime number does not exist.

We must also remember such truths as

† See Exercise 100, page 356.

3)   Vulcan does not exist,

where a name is used rather than a description.

There are two issues. One continues the question of §10: can we use name-letters in these contexts? The other issue is new: how shall we formalize "exists" and "does not exist"?

"∃" is the "existential" quantifier so one might expect it to have a role to play in formalizing assertions and denials of existence. On the other hand, "exists" is grammatically a predicate, so one might think to formalize it by means of a predicate-letter.

An example of a use of "exists" which cries out for treatment by "∃" is

4)   Mad dogs exist.

Though "exists" is grammatically a predicate, its function here is not to predicate existence of each and every mad dog. Rather, its role should be compared with "rare" in

5)   Mad dogs are rare.

"Rare" does not predicate rareness of each mad dog. Such predicates, then, are not appropriately formalized by predicate-letters. The interpretation of a predicate letter (of degree 1) will assign it a set of sequences of objects, of each one of which the predicate-letter is, on the interpretation, true. But "rare" and "exists" (at least as it occurs in (4)) are not expressions which intelligibly apply to any one object. This is even plainer if one considers denials of existence like:

6)   Mad dogs do not exist.

It is clear that there is no object in the universe of which it is being said that *that* object does not exist.

Let us explore in more detail the consequences of trying to use a predicate-letter to correspond to "exists" in these cases. We might try formalizing (4) by

7) $\forall x((Fx \ \& \ Gx) \rightarrow Hx)$

and (6) by

8) $\forall x((Fx \ \& \ Gx)) \rightarrow \neg Hx)$

with "*F*" corresponding to "is a dog", "*G*" to "is mad" and "*H*" to "exists". One trouble with this suggestion is that if there are no mad dogs both (**7**) and (**8**) come out true upon an intended interpretation, whereas (**4**), which (**7**) is supposed to formalize, should certainly be false in this case. If we try instead

    9) $\exists x(Fx \,\&\, Gx \,\&\, Hx)$

for (**4**), and

    10) $\exists x(Fx \,\&\, Gx \,\&\, \neg Hx)$

for (**6**) (with correspondences as before) we find that both require for their truth upon an intended interpretation that the domain should contain mad dogs which, on the face of it, is just what (**6**) denies.

One attempt to meet this difficulty involves distinguishing between *being* and *existence*. The category of being is the wider, embracing plenty of non-existent things, like Pegasus, the golden mountain, round squares, as well as existing things like Ronald Reagan and Italy. The existential quantifier, expressed in English by "there is", relates to the category of being, "exists" to the narrower category of existence. Thus

    11)   There is something which does not exist

expresses a truth, on this theory. It could be formalized, with correspondences as before, as

    12) $\exists x \neg Hx.$

Let us call the view that there are things which do not exist "Meinong's Theory of Existence" after a famous proponent of it. On this theory, there is no logical problem about using a predicate-letter to formalize "exists". For example, (**10**) would adequately formalize (**6**), and the envisaged objection would fail, since *there are* mad dogs, although according to (**6**) not one of them exists.

Although I do not know how to refute Meinong's Theory of Existence, I do not believe it.[†] I shall describe one non-Meinongian approach to the phenomena it addresses.[15] The essence of the alternative I shall now present is that Meinong's theory is false: by contrast, everything exists. It is possible, though not mandatory, to suppose that the **Q**-valid sentence

† See Exercise 101, page 356.

13) $\forall x \, \exists y (x = y)$

expresses just that thought.[†]

In the case of (**4**) and (**6**), non-Meinongian formalizations are attained by using the existential quantifier to correspond to "exists", rather than a predicate-letter:

14) $\exists x (Fx \,\&\, Gx)$

15) $\neg \exists x (Fx \,\&\, Gx)$

with "*F*" corresponding to "is a dog" and "*G*" to "is mad". So some sentences containing "exists" can be adequately formalized using "∃", and could only have been adequately formalized using a predicate-letter to correspond to "exists" at the expense of accepting Meinong's theory.

The cases so far discussed have had plural subjects (for example, "mad dogs"). I now turn to singular existential sentences, for example (**3**):

Vulcan does not exist.

This is a negative sentence, and we should also have an example of a positive one:

16)   The third man exists.

Imagine (**16**) to be uttered in connection with some intrigue. There is a debate about how many people were involved. You believe there were three, and you start using the phrase "the third man" with the intention of referring to that third person, whose name you do not know. You might well utter (**16**) to state your position on the controversy.

We cannot in any straightforward way apply the policy adopted in the plural case to the singular case. In (**4**),

Mad dogs exist,

the plural subject contained predicates for the quantifier to attach to. (**3**) does not on the face of it contain such a predicate. If we try to formalize it by using the existential quantifier, we might put:

$\neg \exists x Hx.$

† See Exercise 102, page 357.

But we need to ask what "H" corresponds to. It seems it must correspond to "Vulcan", but it cannot do that since it has to correspond to a predicate and "Vulcan" is a name. We will return to this question shortly.

In the case of (**16**), we would have exactly the same problem, if we were to think of "the third man" as formalizable by a name-letter. On the other hand, the fact that there are predicates contained in the phrase should encourage the view that they could be made accessible to existential quantification. The theory of descriptions given in §10 has the merit of taking seriously the fact that definite descriptions contain predicates. If we mechanically apply the theory to (**16**), the result is:

17) $\exists x(Fx \ \& \ \forall y(Fy \rightarrow x = y) \ \& \ Gx)$

where "F" corresponds to "is a third man" and "G" to "exists". However, if we deny the Meinongian view, we will hold that everything exists. In other words, the intended interpretation of "G" will be the set of all 1-membered sequences whose member is in the domain of interpretation. This means that "$\& \ Gx$" adds nothing to the truth-upon-an-interpretation conditions of (**17**). So we might as well formalize (**16**) as

18) $\exists x(Fx \ \& \ \forall y(Fy \rightarrow x = y))$

(with "F" as before). This is precisely how Russell's Theory of Descriptions claims that sentences like (**16**), existential sentences with a definite description in the grammatical subject position, should be formalized. The denial of (**16**) will be formalized by prefacing (**18**) with "$\neg$". These formalizations seem adequate, yet avoid Meinong's theory.

We saw in §7 that "Vulcan" cannot be adequately formalized by a name-letter. This blocks at once the straightforward idea that (**3**) should be formalized as

19) $\neg F\alpha$

with "F" corresponding to "exists" and "α" to "Vulcan". The previous paragraph, however, suggests how one might find existential quantifier formalizations of "exists" as it occurs in (**3**): by discerning a hidden predicate beneath the occurrence of the name "Vulcan". Two versions of this approach are available.

First, it is worth noting that

20) $\exists x \ x = \alpha$

asserts the existence of whatever the interpretation assigns to "α".[16] It does so by saying that something is identical to α, which, for the non-Meinongian, is just a way of saying that α exists. (This corresponds to the English equivalence between "Napoleon exists" and "There is something which is (identical to) Napoleon".) Here the name-letter "α" is converted into a predicate " = α" – an expression which when coupled with a name or name-letter forms a sentence. This is one way to discern a predicate hidden in the occurrence of a name, and of course it can be applied in just the same way to any name whatsoever.[17] Thus we could see (3) as saying that it is not the case that something is identical to Vulcan, and formalize it as:

21) $\neg \exists x \; x = \alpha$

with "α" corresponding to "Vulcan".

The trouble with this suggestion is that (21) is not true upon any interpretation, whereas (3) is true, so the formalization is inadequate.

The crucial fact here is that (20) is **Q**-valid. Let us see how. By (2.2), (20) is true upon an interpretation iff "$\beta = a$" is true upon some interpretation. But "$\beta = \alpha$" certainly is true on at least one interpretation: one which assigns the same object to both "α" and "β". Notice that we assume here that every interpretation is relative to a domain containing at least one object (else there would be no assigning one and the same object to the two name-letters). This is ensured by the stipulations of §2. (For a relaxation of these stipulations, see §20.)

I said that there were two versions of the search for a predicate hidden in a name. One is the method just described, in which a name, n, is held to conceal a predicate " $= n$". We have seen that this will not always yield the desired results. The other proposal is more radical. It is that names – or some names, to include at least "Vulcan" – are "really" definite descriptions. Thus "Vulcan" might be held to abbreviate the definite description "The intra-mercurial planet". Then (3) is equivalent to

22)   The intra-mercurial planet does not exist,

and we have already seen, in connection with (18), how one could give an adequate **Q**-formalization of a sentence like this.[†] If we hold to the view that everything classified as a name in English should be treated alike, and take it that the truth of (3) gives us a compelling reason to regard "Vulcan" as really a definite description, then we are committed, as

† See Exercise 103, page 357.

Russell was, to the view that what we ordinarily call names in English are really abbreviated or truncated descriptions.[18]

To summarize the main points: in formalizing plural affirmations and denials of existence, like "Mad dogs exist", the quantifier approach gives the right result and the predicate-letter approach the wrong result. Singular assertions and denials of existence superficially divide into two categories: (i) those involving a name, like "Vulcan exists" and (ii) those involving a description, like "The third man exists". Assuming that we adopt Russell's general recipe for formalizing descriptions, the quantifier does all that is needed in regard to (ii). Use of a predicate-letter corresponding to "exists" would be otiose, unless one accepts Meinong's theory. Existential sentences in (i) can only be adequately Q-formalized if they are seen as "really" in category (ii): that is, if the names in question are treated as "really" descriptions. Whether this seeming distortion of English can be justified is discussed in the next section, which, as it departs from the question of Q-formalization, can be omitted without loss of continuity.

Our policy will be to formalize English names by name-letters, unless, as in (3), this leads to an inadequate formalization.

*12 Are names "really" descriptions?

One of Russell's motivations for holding that English names are definite descriptions is that he was engaged in a project very similar to that of finding Q-formalizations of all English sentences. The only difference was that he was using a richer language than Q, but its richness has no bearing on the issue we are now discussing, so I shall from now on abstract from that difference. Russell certainly believed that some interpretation of Q would express anything that could be said or thought in *any* language, and Q would have the advantages of clarity, and accessibility to logical manipulations. This belief, coupled with a denial of Meinong's theory, entails that at least some names (e.g. "Vulcan") are descriptions, and yields a more general conclusion if we assume that all names should be treated alike. However, it would be wrong to see this as Russell's only reason. On the contrary, names, especially in the context of existential sentences, raise problems quite independently of the project of Q-formalization. This section briefly introduces some of those problems.

(i) Consider again (11.3):

Vulcan does not exist.

This is true (or so we all ordinarily believe). But how does the sentence work? It appears to introduce an entity, viz. Vulcan, and then go on to say of this entity that it does not exist. Unless we are Meinongians, a sentence which does that would be contradictory, and not true. The description theory dissolves this mystery. "Vulcan" does not introduce an entity, but rather a claim to the effect that exactly one thing has a certain property (being a planet between Mercury and the sun), and the rest of the sentence serves to negate that claim.

(ii) Consider how names are learned. It is as often as not by means of a definite description. "Who was Gödel?" you enquire. When I tell you that he was the German logician, best known for his proof of the incompleteness of arithmetic, you come to be in a position to use the name. A natural hypothesis is that the mechanism here is *definition*: the description defines the name, and thus gives you its meaning.

(iii) Consider those names which have bearers (like "Reagan" and unlike "Vulcan"). How is the knowledge of who the bearer is represented in your mind? A natural answer is that it is represented by a definite description. Thus if someone uses a name, N, he must be able to answer the question: "who or what do you mean by N?" and it seems the only appropriate answer he could make would consist either in pointing to the bearer of "N" or in citing a description true of it.

(iv) Consider an identity statement like

1) Lewis Carroll is Charles Dodgson.

This is no trivial truth, but a discovery that became known to a wider and wider circle in the late nineteenth century. Suppose that, instead of the view that names are really descriptions, we hold the view that a name simply stands for an object. Then it seems hard to explain how you could understand both of the names "Lewis Carroll" and "Charles Dodgson" without knowing that (1) is true. In understanding each, you must know who or what the name stands for, so how could you fail to know that they stand for the same object? Yet clearly one can understand (1) without knowing whether it is true, and this is incontrovertibly allowed for by the description theory: you may associate the names with different descriptions.

I now turn to some criticisms of the description theory, and comment on the above four arguments for it.

(a) It would seem that two people could both understand a name perfectly well, yet associate different descriptions with it. If you are a

White House janitor you may associate quite different descriptions with "Reagan" from those associated with the name by an El Salvadorian, perhaps living beyond the reach of television. So do we have to say that the name is ambiguous? This would seem implausible.

Russell was quite well aware of this issue, and explained how it was no objection to his theory (however objectionable it may be to theories sometimes attributed to him).[19] He allowed that two people could communicate satisfactorily using a name, even though they associated different descriptions with it, provided that the different descriptions were true of the same thing. The definite description which a name "really" is should not be thought of as something that has to be common to speaker and hearer when they communicate. Rather, the definite description which a name "really" is, is the one in the mind of the speaker or hearer on some specific occasion. There is no one description steadily and universally associated with the name.[†] Hence in one good sense of "meaning", that in which meaning is what is common to speaker and hearer in communication, Russell's theory is not intended as an account of the meaning of names.

This very fact, however, may make the logician regard the theory as unsuitable for his purposes. If someone asserts something, the logician will want to know the consequences of what is asserted. These consequences need to derive from something publicly shareable, to derive from something common to speaker and hearer in communication. If someone idiosyncratically thinks of Reagan via the description "The person who always misses the ashtray when stubbing out his cigarettes", we do not want to count as a consequence of this man's assertion that Reagan will give a press release today, that someone who misses the ashtray when stubbing out his cigarettes will give a press release today. Descriptions deriving from these subjective sources are not appropriate for the study of logic.[20]

(b) The argument of (i) is probably the strongest. Note that the strongest part relates only to names without bearers. Only very recently has a plausible alternative to Russell's theory for these names been offered.[21]

It may not be clear why one ought to treat names with and without bearers in the same way. Russell thought one should on the grounds that one might not know, of some group of names, which have bearers and which do not, even though one understood all the names in the group, so that the use of a name does not make the appropriate division. For

---

† See Exercise 104, page 357.

example, we can sensibly discuss whether or not Homer existed. In doing so, we use the name "Homer", presumably correctly, without knowing whether it is a name with a bearer or a name without. Logic is supposed to proceed without empirical knowledge, drawing just on the understanding of sentences. So logic ought not to discriminate between names with bearers and names without. This certainly puts the onus on the person wanting to discriminate to find a distinction between the two classes that shows up in the use of language.

(c) The argument of (ii) is not decisive, since the phenomena adduced are consistent with the view that the use of the description in learning is to get the learner to know *who or what the bearer of the name is*. This task can be achieved without giving an expression which means the same as the name.[22]

(d) It is not clear that, as (iii) alleges, we can always give a definite, uniquely identifying, description corresponding to every name we use with understanding. Suppose we notice someone on our way to work each day whom we inwardly name "Fred". We can recognize him when we see him, but any attempt to describe his features will probably (as the police know only too well) yield a description which fits hundreds of people.

(e) The argument of (iv) is fallacious. There is no valid argument from

$a = b$
He knows what the name "a" stands for
He knows what the name "b" stands for

to

He knows that "a" and "b" stand for the same thing.

Imagine Dodgson's colleagues, who obviously understand the name "Dodgson". In addition, they are aware of Lewis Carroll's success, and so understand the name "Carroll". They may have no basis for supposing that that Dodgson is Carroll.

(f) On Russell's theory of descriptions, "the" is a kind of quantifier, meaning "there is exactly one ...".[23] It is plain that we can understand a sentence like

2)   The inventor of the jet engine died in poverty

without (by one natural standard) knowing who the inventor was or even if there was one. (Perhaps the jet was the product of team research.) In particular, if there is a unique inventor, we need have no link with him in order to understand the sentence. (2), on Russell's theory, is a general statement: no particular object enters into its interpretation.

We tend to think of names otherwise. Understanding

>   **3)**   Frank Whittle died in poverty

does require knowing who Whittle was. We have to have met him or been told about him or seen traces of him in order to use his name.

Thus it seems that the conditions required for using a name are different from those required for using a definite description, and this counts against Russell's theory that names are descriptions.

These remarks scratch the surface of a large debate: see the bibliographical notes at the end of the chapter.

13 Structural ambiguity in English

Q has no structural ambiguity. One way to clarify structural ambiguities in English (see chapter **1.12**) is to provide distinct **Q**-formalizations, one for each distinct reading of the English. Thus structural ambiguity is treated as ambiguity with respect to logical form.

For example, it is often said that

>   **1)**   Everyone loves someone

is ambiguous between a (weak) reading upon which everyone is such that there is someone he loves, and a (strong) reading according to which some lucky person is loved by everybody. With "Fxy" corresponding to "x loves y" the unambiguous formalizations are, respectively:

>   **2)** $\forall x \exists y Fxy$

and

>   **3)** $\exists y \forall x Fxy$.

Note that

>   **4)** $\exists y \forall x Fxy \vDash_{\mathbf{Q}} \forall x \exists y Fxy$

but the converse does not hold. Roughly, an interpretation of the premise of (4) needs to find an object that everyone loves, so guaranteeing that everyone loves someone. An interpretation of the conclusion of (4) upon which, say, each person loves himself and only himself makes that sentence true but the premise of (4) false, which is why the converse argument is not **Q**-valid.

The inference from $\forall\exists$ to $\exists\forall$ is known as the "quantifier shift fallacy" and it is commonly attributed to philosophers and others. For example, there is an argument for a foundationalist view of knowledge which, denuded of some of its protective covering, runs as follows:

> 5) Every justification of a proposition has to end somewhere. Therefore some propositions cannot be justified, but have to be taken for granted.†

Some of the ambiguity of (**1.12.9**) was of this kind:

> Logic, epistemology and metaphysics are all the philosophical subjects there are. Nicholas has written a book about logic. Nicholas has written a book about epistemology. Nicholas has written a book about metaphysics. Therefore, Nicholas has written a book about every philosophical subject.

First, the correspondences:

"α" for "logic"
"β" for "epistemology"
"γ" for "metaphysics"
"δ" for "Nicholas"
"F" for "is a philosophical subject"
"Gxy" for "x has written y"
"Hxy" for "x is about y"
"$\mathcal{J}$" for "is a book".

The premises have no structural ambiguity, and can be formalized as follows:

> **6)** $\forall x(Fx \rightarrow (x = \alpha \lor x = \beta \lor x = y))$
> $\exists x(\mathcal{J}x \,\&\, G\delta x \,\&\, Hx\alpha)$
> $\exists x(\mathcal{J}x \,\&\, G\delta x \,\&\, Hx\beta)$
> $\exists x(\mathcal{J}x \,\&\, G\delta x \,\&\, Hx\gamma)$

† See Exercise 105, page 357.

The weak version of the conclusion upon which the argument is valid is:

7) $\forall x(Fx \rightarrow \exists y(\mathcal{J}y \,\&\, G\delta y \,\&\, Hyx))$.

The strong version of the conclusion upon which the argument is invalid is:

8) $\exists y(\mathcal{J}y \,\&\, G\delta y \,\&\, \forall x(Fx \rightarrow Hyx))$.

(7) is consistent with Nicholas writing various books, perhaps one on each of the three subjects. (8) requires him to have written a compendious book, treating all the subjects at once. In (7) the universal quantifier has wide scope relative to the existential quantifier. In (8) the scopes are reversed, the universal falling in the scope of the existential.

The use of variables in **Q** helps keep track of the application of quantifiers. Thus in (8) it is vitally important to know which quantifier is applying to the first position in "Hxy" and which to the second. This is shown by the attachment of "x" to both the universal quantifier and the first position, and "y" to both the existential quantifier and the second position.

Pronouns sometimes play a similar role on English, as in:

9)   Someone called today and *he* brought his wife.

They also play another role, as shorthand for the reapplication of a name, as in

10)   Oscar kissed Joan and *he* made *her* cry.

Here "he" and "her" are stylistic variants of the reuse of "Oscar" and "Joan".[†]

Sometimes it is unclear in English which role a pronoun is playing, for example "he" in

11)   If Oscar kisses anyone, he will be pleased.

The formalization upon which "he" is equivalent to a reuse of "Oscar" is

12) $\forall x(F\alpha x \rightarrow G\alpha)$

† See Exercise 106, page 357.

with "Fxy" corresponding to "x kissed y", "G" to "is pleased" and "α" to "Oscar". The formalization upon which "he" marks the application of the quantifier is

    13) $\forall x(F\alpha x \rightarrow Gx)$.

The word "only" often gives rise to ambiguity in English.

    14)   John only eats organically grown vegetables

would normally be interpreted in a way consistent with John eating meat, the claim being that, as far as vegetables go, all the ones he eats are organically grown. This reading is formalized

    15) $\forall x((F\alpha x \;\&\; Gx) \rightarrow Hx)$

with "Fxy" corresponding to "x eats y", "G" to "is a vegetable", "H" to "is organically grown" and "α" to "John".

In my view, the more correct reading of (**14**), the reading that would be favoured by teachers of English, entails that John eats nothing but vegetables. This reading is formalized

    16) $\forall x(F\alpha x \rightarrow (Gx \;\&\; Hx))$.

This is not literally a scope difference in **Q**: it is not that (**15**) and (**16**) differ only in point of the relative scopes of some pair of operators. The phenomenon is more like the ambiguity that can arise concerning the multiple qualification of noun phrases. For example, in

    17)   John is a dirty window cleaner

it is unclear whether "dirty" is meant to qualify the complex "window cleaner" or whether just the word "window". The alternatives are brought out respectively by the very approximate formalizations:

    18) $\exists x(Fx \;\&\; Gx \;\&\; H\alpha x)$

    19) $\exists x(Fx \;\&\; G\alpha \;\&\; H\alpha x)$

with "F" corresponding to "is a window", "G" to "dirty", "H" to "x cleans y" and "α" to "John". The formalizations are only very approximate, for it takes more (and also perhaps less) than cleaning one or a dozen windows to be a window cleaner.

We can use representations based on **Q** to clarify structural ambiguities in English sentences, even when those sentences resist **Q**-formalization in any reasonably revealing way. The technique involves mixing English and **Q**. For example,

20)   I am trying to buy a house

is ambiguous between the claim that there is a house I have set my eye on and towards which my buying efforts are directed, and a claim which can be true even if there is no such house – even if all I have done is ask the estate agents to send details. We could represent these claims as follows:

21) $\exists x(x$ is a house & I am trying to buy $x)$.

22)   I am trying to bring it about that: $\exists x(x$ is a house & I buy $x)$.

This is, or is analogous to, a scope distinction: in (21), "$\exists$" has wide scope relative to "trying", in (22) narrow scope. We sometimes express the first reading in English by saying "I am trying to buy a particular house". Since all houses are particular houses, "particular" here is serving not to qualify "house" but to effect a scope distinction.

In discussions of the theory of knowledge, it is often claimed that

23)   If you know you can't be wrong

is ambiguous in a way which can be represented as follows:

24)   necessarily $\forall x\ \forall y(x$ knows that $y \rightarrow y$ is true)

25) $\forall x\ \forall y(x$ knows that $y \rightarrow$ necessarily y is true).

This is also a difference of relative scope. A familiar view in epistemology is that (24) is true, but not terribly interesting, and (25), which indeed entails wholesale scepticism about the contingent, is false.

Russell gave a pleasing example of a scope distinction. He argued that

26)   I thought your yacht was longer that it is

could be heard as an absurd claim, deserving the reply: "Everything is just as long as itself". With "F" corresponding to "x uniquely numbers in meters the length of y", "Gxy" for "x is greater than y", and "α" for "your yacht" (ignoring, for present purposes, our earlier resolution to treat definite descriptions, like "the yacht which you own", by Russell's theory), the absurd claim could be represented as:

27) I thought that: $\exists x(Fx\alpha \ \& \ Gxx)$.

What the speaker of (26) meant, of course, is something more like:

28) $\exists x(Fx\alpha \ \& \ \exists y((\text{I thought that } Fy\alpha) \ \& \ Gyx))$.

Note that "Gyx" is not part of the thought I explicitly attribute to myself in uttering (26). A more long-winded clarification is: the length I thought your yacht was is greater than the length it actually is.

An important question in ethics is whether there can be genuinely incompatible obligations. One way in which the issue might be made more precise is by asking whether it is ever possible for the following both to be true:

29) You ought to do A

30) You ought not to do A

where "A" stands for some type of action. However, this is still not precise enough, for it is unclear whether (30) is really the negation of (29). Arguably, it is ambiguous between:

31) It is not the case that you ought to do A

and

32) You ought to do not-A.

In the first, "not" dominates the sentence, in the second it has narrow scope relative to "ought". Only if (30) is to be read dominated by "not" would the joint truth of it and (29) establish that there are incompatible obligations, in the logical sense of "incompatible". Logic alone finds nothing problematic in the joint truth of (29) and (32).[†]

† See Exercise 107, page 358.

14   Q-validity and decision

There is no method like that of truth tables for determining whether or not a **Q**-argument is **Q**-valid. There are, however, systematic methods, ones which can be applied quite mechanically and just on the basis of the physical make-up of the relevant **Q**-sentences, for determining, for any **Q**-valid argument, that it is **Q**-valid. The trouble is that if such a method has still not pronounced an argument valid (say after a hundred or a million steps), we do not know whether the right thing to believe is that the argument is not valid, or whether the right thing to believe is that the argument is valid but the method has not yet managed to show it. Hence there is no "decision procedure" for **Q**: no mechanical method for determining, with respect to an arbitrary **Q**-argument, and in a finite number of steps, whether or not it is **Q**-valid.

We can get a feel for **Q**-validity, without anything so grand as a systematic method, simply by working on some examples. It will probably be at first unobvious whether the following is **Q**-valid or not:

> 1) $\forall x \exists y (Fx \rightarrow Gy)$; $\exists y \forall x (Fx \rightarrow Gy)$.

A natural reaction would be to hold that it is invalid, by analogy with the invalid

> 2) $\forall x \exists y Fxy$; $\exists y \forall x Fxy$.

But this reaction would be incorrect. Let us consider how an interpretation, i, must be if it is to be one upon which the premise of (1) is true. It must either assign to "F" the empty set,[24] or else assign to "G" some non-empty set. The conclusion is true upon i iff "$\forall x (Fx \rightarrow G\alpha)$" is true upon some interpretation agreeing with i on "F" and "G". If $i(F)$ is empty then "$\forall x (Fx \rightarrow G\alpha)$" is true upon i (for reasons spelled out in §3); if $i(G)$ is non-empty then for some i' agreeing with i on "F" and "G", i'(α) is a sequence whose member belongs to $i(G)$, so again "$\forall x (Fx \rightarrow G\alpha)$" will be true on i. So however i makes the premise true, it will make the conclusion true also.

If an argument is **Q**-invalid, we can establish this if we can find a *counterexample*: an interpretation upon which the premise(s) are true and the conclusion false. For example, an interpretation which assigns to "F" the set of ordered pairs whose first member is smaller than the second,

and whose domain is the integers, is a counterexample to the **Q**-validity of
(**2**).

The equivalences between universal and existential quantifiers

    3) $\vdash_Q \exists x Fx \leftrightarrow \neg \forall x \neg Fx$
    4) $\vdash_Q \forall x Fx \leftrightarrow \neg \exists x \neg Fx$

can be confirmed by reasoning that uses the same equivalences in the
English which we use to describe the interpretations. Thus for (**3**) a
crucial consideration is that any interpretation, i, upon which "$\exists x\ Fx$" is
true is one such that there is an interpretation, agreeing with i on "F",
upon which "$F\alpha$" is true; that is, it is not the case that every
interpretation agreeing with i on "F" fails to bring out "$F\alpha$" as true; that
is, it is not the case that every interpretation agreeing with i on "F" brings
out "$\neg F\alpha$" as true; that is, "$\forall x \neg Fx$" is false upon i; that is, "$\neg \forall x \neg Fx$"
is true upon i.

Universal quantifications are true only upon interpretations which
make the corresponding existential quantification true, for example:

    5) $\forall x Fx \vdash_Q \exists x Fx.$

However

    6) $\exists x Fx \nvdash_Q \forall x Fx.$

Any interpretation which assigns a set other than one whose sequences
have as members everything in the domain serves as a counterexample to
the validity of the argument in (**6**).

The following general truth reflects the fact that we require an
interpretation to assign an object to every name-letter:

    7) $X \vdash_Q \exists v X^\star.$

This is to be read: for every sentence X for which there is an appropriate
$X^\star$ (i.e. one which results from X by replacing one or more occurrences
of a name-letter in X by a variable, v), every interpretation upon which X
is true is one upon which the result of prefixing $X^\star$ by "$\exists$" followed by v
is also true.

    8) $\forall x (Fx \rightarrow Gx), \exists x Fx \vdash_Q \exists x Gx.$

An interpretation, i, upon which the second premise is true must assign a set of non-empty sequences to "F"; but then for the first premise to be true, the conditional "$F\alpha \rightarrow G\alpha$" must be true upon i, whatever is assigned to "α", even if $i(\alpha)$ is in $i(F)$, and this means that $i(G)$ must be a set of non-empty sequences. Such an interpretation is one upon which the conclusion is true.

9) $\forall x(Fx \rightarrow Gx), \exists xGx \nvDash_Q \exists xFx.$

A counterexample is an interpretation over an unrestricted domain of existing things which assigns to "F" the set of 1-membered sequences each having a unicorn as member (that is, the set whose only member is the null sequence), and to "G" the set of 1-membered sequences whose member is a quadruped.

We need to distinguish between:

10) $\exists x(Fx \ \& \ Gx) \vDash_Q \exists xFx \ \& \ \exists xGx$

and

11) $\exists xFx \ \& \ \exists xGx \nvDash_Q \exists x(Fx \ \& \ Gx).$[†]

Working through examples like these should give a good feel for **Q**-validity, but what has become of an ideal mentioned earlier: that there be an entirely mechanical test for validity in an artificial language fit for logical purposes?

A decision procedure would indeed satisfy the hankering for mechanical tests. However, it can be proved that there is no decision procedure for **Q**. So that is a hankering which one must simply abandon.

There are systematic procedures which, for every **Q**-valid **Q**-sentence, will determine in a finite number of steps that it is valid, for example axiomatic, natural deduction and semantic tableaux methods of proof. As I mentioned earlier, the trouble is that as one is putting these procedures through their paces, passing from step to step in accordance with the instructions, there is no point at which one can say: we haven't proved the sentence valid, therefore it is invalid. True, there will be a proof of validity in a finite number of steps, if the sentence is valid, but one does not know what that number is, and any number of steps one has taken may fall just short of the number required for a proof of validity.

† See Exercise 108, page 358.

Working with these procedures leads to a sharpened perception of **Q**-validity, and the fact that these procedures exist is, of course, of great importance.[†]

15 Formalizing arguments

We gave an example of a valid argument which is not **P**-valid:

  1) John runs; therefore someone runs.

It is easy to see that there is a **Q**-valid formalization of it:

  2) $F\alpha \vDash_{\mathbf{Q}} \exists x Fx,$

with the obvious correspondences. We will for the moment take it for granted that the truth of (2) establishes the validity of (1).

Now for an old favourite:

  3) All men are mortal, Socrates is a man; therefore Socrates is mortal.

This has a valid **Q**-formalization:

  4) $\forall x(Fx \rightarrow Gx), F\alpha \vDash_{\mathbf{Q}} G\alpha$

with "F" corresponding to "is a man", "G" to "is mortal" and "α" to "Socrates".

One early method of formalizing everyday arguments, Aristotle's syllogistic, had particular trouble with arguments like:

  5) All horses are animals; therefore all heads of horses are heads of animals.

This is **Q**-validly formalizable:

  6) $\forall x(Fx \rightarrow Gx) \vDash_{\mathbf{Q}} \forall x \forall y((Fx \ \& \ Hyx) \rightarrow \exists z(Gz \ \& \ Hyz))$

with "F" corresponding to "is a horse", "G" to "is an animal" and

† See Exercise 109, page 358.

"*Hxy*" to "*x* is a head of *y*". We can argue informally for the truth of (6) by showing that there is no counterexample. Suppose some interpretation, *i*, verifies the premise. Then every member of *i*(*F*) belongs to *i*(*G*). For *i* to falsify the conclusion, it needs to verify "*Fα* & *Hβα*", yet there be no *i′* agreeing with *i* on assignments to everything except, perhaps, "*γ*" which verifies "*Gγ* & *Hβγ*". Some *i′* will verify "*Hβγ*" by letting *i′*(*γ*) = *i′*(*α*). This *i′* can fail to verify "*Gγ* & *Hβγ*" only if *i′*(*α*) does not belong to *i′*(*G*). But this is impossible, given that *i′* agrees with *i* on *α*, *F* and *G* and accordingly, *i′*(*α*) belongs to *i′*(*F*) and so to *i′*(*G*). Hence there is no counterexample. Intuitively, the idea is that if we have a horse and a head which satisfy the conclusion's antecedent, satisfying the premise ensures that we *thereby* have an animal and a head which satisfy the conclusion's consequent.

Aristotle's syllogistic has been criticized on the grounds that it counts as valid arguments which are not valid. An alleged example is:

7)   All unicorns are self-identical, all unicorns are non-existent; therefore some self-identical things are non-existent.

Aristotelian logic regarded this as an instance of a valid argument-form because it took the truth conditions of a universal quantification, "All *F*s are *G*s", to require the existence of *F*s. Setting aside the question whether the argument is valid, it is certainly the case that:

8) $\forall x(Fx \rightarrow Gx)$, $\forall x(Fx \rightarrow Hx) \nvDash \exists x(Gx \ \& \ Hx)$.

A counterexample which establishes (8) is, of course, one assigning the empty set to all the predicate-letters.

Consider

9)   Only the brave deserve the fair, Harry is brave and Mary is fair; so Harry deserves Mary.

Formalizing as

10) $\forall x \forall y((Gxy \ \& \ Hy) \rightarrow Fx)$, $F\alpha \ \& \ H\beta$; $G\alpha\beta$

with "*Gxy*" corresponding to "*x* deserves *y*", "*H*" to "is fair", "*F*" to "is brave", "*α*" to "Harry" and "*β*" to "Mary" yields a **Q**-invalid argument. The invalidity can be seen by considering an interpretation which assigns to "F" the set of 1-membered sequences each of whose members is an even number, to "G" the set of ordered pairs such that the first member of each pair is greater by one than the second, to "H" the set of 1-membered sequences each of whose members is an odd number, to "α" 8 and to "β" 3.

A valid argument resembling (**9**) is:

   11)   Only the brave deserve the fair, Mary is fair but Harry isn't brave; so Harry doesn't deserve Mary.

With correspondences as before, this can be **Q**-validly formalized:

   12) $\forall x \forall y ((Gxy \,\&\, Hy) \to Fx), H\beta \,\&\, \neg F\alpha; \neg G\alpha\beta.$

The **Q**-validity is plain if we reflect that all that matters about the premise is its instance "$(G\alpha\beta \,\&\, H\beta) \to F\alpha$", and that the following is true:

   13) $(p \,\&\, q) \to r, q \,\&\, \neg r \vDash_{\mathbf{P}} \neg p.$

Now for some examples of arguments involving identity, definite descriptions, and numeral adjectives.

   14)   Hesperus is a planet, Hesperus is identical to Phosphorus; so Phosphorus is a planet.

With "α" corresponding to "Hesperus", "β" to "Phosphorus" and "F" to "is a planet", we can formalize **Q**-validly:

   15) $F\alpha, \alpha = \beta \vDash_{\mathbf{Q}} F\beta.$

Any interpretation, i, upon which the premises are true assigns the same object to "α" and "β" and that object to "F". Evidently, then, the conclusion is true upon i.

However, compare:

   16)   John believes that Hesperus is a planet, Hesperus is identical to Phosphorus; so John believes that Phosphorus is a planet.

If we could make "Fx" correspond to "John believes that x is a planet", then (**16**) would be formalizable by (**15**), but, intuitively, (**16**) is invalid.

For suppose John does not realize that Hesperus is identical with Phosphorus. He uses "Hesperus" of a heavenly body he sees in the evening. He uses "Phosphorus" of a heavenly body he sees in the morning (never suspecting that these are one and the same). He believes that Hesperus is a planet, but believes that Phosphorus is not a planet but a star. (When you ask, "Is Phosphorus a planet?" he replies, firmly, "No".) So for this case the premises are true and the conclusion false, so (**16**) is not valid.[†]

One moral is that "John believes that x is a planet" should not be allowed to count as a predicate. If a predicate is something adequately formalizable by a predicate-letter, the justice of this ruling can be shown not just by the example considered, but more generally. A predicate-letter is, on any interpretation, assigned a set: intuitively, a set of (1-membered sequences of) things of which the predicate is true. But there is no set of (1-membered sequences of) things of which "John believes that x is a planet" is true. This is shown by the fact that Hesperus ought to be both a member of and not a member of any such set, and this is impossible.

Consider

    17)   Only the fastest walker will reach London. John walks faster than Mary. So Mary will not reach London.

We might offer the formalization

    18) $\exists x(\forall y(x \neq y \rightarrow Fxy) \ \& \ Gx \ \& \ \forall x(Gz \rightarrow z = x)), \ F\alpha\beta; \ \neg G\beta$

with "Fxy" corresponding to "x walks faster than y", "G" to "will reach London", "α" to "John" and "β" to "Mary". The idea is to treat the first premise of (**17**) as saying that someone walks faster than anyone else and will reach London, and no one else will reach London. However, (**18**) as it stands is not **Q**-valid. We need to add to the premises "$\alpha \neq \beta$", and replace "$F\alpha\beta$" by "$\neg F\beta\alpha$"; thus amended, the argument is **Q**-valid. We have interpreted "only" in such a way that "Only α is F" entails "α is F".[‡] We have not needed to use Russell's theory of descriptions to formalize "the fastest walker", and hence we have not included the uniqueness, which "the" imparts, in the formalization. To see this, reflect that if "F" is assigned the set of ordered pairs such that the first loves the second, then

---

† See Exercise 110, page 359.   ‡ See Exercise 111, page 359.

"$\forall y(\alpha \neq y \rightarrow F\alpha y)$" can be true on interpretations differing only in what they assign to "α". That is, more than one person can satisfy the condition of loving everyone else.

The following argument requires the uniqueness to be shown in the formalization, if the formalization is to be valid:

19) Only the fastest walker will reach London. John will reach London; so only John walks faster than anyone else.

Using the style of (**18**), and the same correspondences, we would get:

20) $\exists x(\forall y(x \neq y \rightarrow Fxy) \ \& \ Gx \ \& \ \forall x(Gz \rightarrow z = x)), \ G\alpha;$
$\forall x(\forall y(x \neq y \rightarrow Fxy) \rightarrow x = \alpha) \ \& \ \forall y(\alpha \neq y \rightarrow F\alpha y).$

This is not **Q**-valid, though (**19**) is valid. (**20**) fails to capture the validity through failing to formalize the uniqueness implied by "the" in the premise.[†] This can be captured by:

21) $\exists x(\forall z(x \neq z \rightarrow Fxz) \ \& \ \forall y(\forall z(y \neq z \rightarrow Fyz) \rightarrow y = x) \ \& \ Gx$
$\& \ \forall z(Gz \rightarrow z = x)), \ G\alpha;$
$\forall x(\forall y(x \neq y \rightarrow Fxy) \rightarrow x = \alpha) \ \& \ \forall y(\alpha \neq y \rightarrow F\alpha y).$

The first premise is easier to read if we see it as an instance of the familiar:

$\exists x(Fx \ \& \ \forall y(Fy \rightarrow x = y))$

with "F" replaced by "$\forall z(x \neq z \rightarrow Fxz)$".
Consider

22) Every man has two hands, every hand has a thumb; so every man has two thumbs.

This may strike one as valid, even reading the "two" in the conclusion as "exactly two", but only by making explicit a number of presuppositions can it be formalized as **Q**-valid, with this reading of the conclusion. First, we assume that "two" in the first premise is intended as "exactly two". Secondly, we assume that "a thumb" in the second premise is intended as "exactly one thumb". Thirdly, we assume, and make explicit as a third premise, that the relation of having, as obtaining between a person and his

† See Exercise 112, page 359.

bodily parts, is transitive, so that if the person has a hand, and the hand a thumb, then the person has a thumb. "F" corresponds to "is a man", "Gxy" to "x has y", "H" to "is a hand" and "$\mathcal{J}$" to "is a thumb".

23) $\forall x(Fx \rightarrow \exists y \, \exists z(y \neq z \ \& \ Gxy \ \& \ Gxz \ \& \ Hy \ \& \ Hz \ \&$
$\forall w((Gxw \ \& \ Hw) \rightarrow w = y \lor w = z)))$,
$\forall x(Hx \rightarrow \exists y(\mathcal{J}y \ \& \ Gxy \ \& \ \forall z((Gxz \ \& \ \mathcal{J}z) \rightarrow z = y)))$,
$\forall x \, \forall y \, \forall z((Gxy \ \& \ Gyz) \rightarrow Gxz)$;
$\forall x(Fx \rightarrow \exists y \, \exists z(y \neq z \ \& \ Gxy \ \& \ Gxz \ \& \ \mathcal{J}y \ \& \ \mathcal{J}z \ \&$
$\forall w((Gxw \ \& \ \mathcal{J}w) \rightarrow w = y \lor w = z)))$.

It is worth checking the accuracy of the formalization. I will not undertake the exhausting task of using informal reasoning to establish the **Q**-validity of (**23**).

I conclude this section with a valid argument that is not formalizable as **Q**-valid:

24) Necessarily, if there is a first moment in time, the history of the universe up to now is finite; therefore if there had to be a first moment in time, the history of the universe up to now has to be finite.

In a **Q**-formalization, we cannot reach beyond the non-truth functional sentence connective "necessarily". The deepest **Q**-formalization of the premise which is not inadequate is "p", and it is obvious that however hard we try with the premise we cannot find an adequate **Q**-valid argument.[†]

16 Attitudes

It was suggested in connection with (**15.16**) that an adequate formalization of "John believes that Hesperus is a planet" should not match "John believes that x is a planet" with a predicate-letter. It might seem that the task of providing **Q**-formalizations of sentences of this kind is hopeless. But one should not despair too quickly, since there is a proposal, due to Donald Davidson, for adequately **Q**-formalizing any sentence of the form

† See Exercise 113, page 359.

"John believes that ...", provided that what fills the dots is itself **Q**-formalizable. Indeed, the essence of the proposal applies more widely, to include also sentences of the forms: "John *knows* that ...", "John *said* that ...", "John *wonders* whether ...". The italicized expressions are called *verbs of propositional attitude.*

Davidson called his proposal "paratactic" on the grounds that it sees the relevant sentences as really pairs of sentences. Thus

 1) John believes that Hesperus is a planet

is held to consist of

 2) John believes that. Hesperus is a planet.

"That" is held to be a demonstrative pronoun, referring forward to the subsequent "Hesperus is a planet".[†] The **Q**-formalization is thus:

 3) $F\alpha\beta, G\gamma$

with "Fxy" corresponding to "x believes y", "G" to "is a planet", "α" to "John", "β" to "that" and "γ" to "Hesperus". An intended interpretation of (3) will assign to "β" a certain sentence, namely the second sentence in (2).[25]

This proposal formalizes (**15.16**):

John believes that Hesperus is a planet, Hesperus is identical to Phosphorus; so John believes that Phosphorus is a planet

as follows

 4) $F\alpha\beta, G\gamma, \gamma = \delta; F\alpha\varepsilon, G\delta$

with "F", "α", "β", "γ" and "G" as before, "δ" corresponding to "Phosphorus" and "ε" corresponding to the second "that" (the one referring to "Phosphorus is a planet"). Notice that since the demonstrative pronouns refer to different things, they must be formalized by different name-letters.

Q-validity is strictly speaking undefined for (4), since there are two sentences in the conclusion. However, we will for the moment overlook this, making the obvious *ad hoc* stipulation that (4) is valid only if all

† See Exercise 114, page 359.

interpretations on which all the premises are true are interpretations on which the first conclusion is true. It is obvious that there is no chance whatsoever of (4) being Q-valid. If (15.16) is indeed invalid, this is a point in favour of Davidson's proposal.

The proposal has the merit of giving Q-valid Q-formalizations to intuitively valid English arguments involving propositional attitudes. The valid

> 5)   John believes that Hesperus is a planet, that is true; therefore John believes something true

can be given a valid Q-logical form as follows:

> 6) $F\alpha\beta$, $G\gamma$, $H\beta \vDash_Q \exists x(F\alpha x \,\&\, Hx)$

with "H" corresponding to "is true", and other correspondences as before. Note that we have to regard both occurrences of "that" as referring to the same thing, and thus as formalizable by the same name-letter.[†] The fact that English contains a sentence like the conclusion of (5), which seems naturally formalizable by the conclusion of (6), supports Davidson's proposal by suggesting that "believes" is at least sometimes a predicate of degree two. And if sometimes, why not always?

One would need to analyse the predicate of degree-2 in an appropriate way (and different analyses will be appropriate for the different verbs of propositional attitude). For example, one should not require, as a condition for "x believes y", that x understand English. One might rather, as a start, say something like: x believes y iff x stands in the relation of believing to the proposition expressed by y. It would be a philosophical task, but not one directly related to problems of Q-formalization, to say what this believing relation consists in.[26][‡]

17   Binary quantifiers

There are quantifier expressions that Q does not consider, for example "most" and "few". If we wish to formalize English sentences containing these quantifiers, could we not just add them to Q, to form a new language, call it Q^+? Let us see how we might proceed.

---

† See Exercise 115, page 360.   ‡ See Exercise 116, page 360.

Suppose we introduce the quantifiers "T" and "W" by the syntactic rule that all **Q**-sentences are **Q**$^+$-sentences, and if X is a **Q**$^+$-sentence and $X^\star$ results from X by replacing a name-letter by some variable, v, not in X, then the following are **Q**$^+$-sentences:

$TvX^\star$

$WvX^\star$.

Before raising the question of how one might give rules of interpretation for these quantifiers (along the lines of the rules for "∀" and "∃" given in (**2.2**)), let us see how one might apply them to formalizing English.

For sentences which are (in a respect to be made more precise shortly) like "Everything is physical", there are no problems.

1)   Most things are physical

and

2)   Few things are physical

are **Q**$^+$-formalizable as

3) $TxFx$

and

4) $WxFx$.

Thinking just of such cases, appropriate rules for interpretations could be modelled on the rule (**2.2vi**), simply replacing "upon i, and also upon every other interpretation" by "upon i, and also upon most/a few other interpretations". However, quantifiers thus specified cannot adequately formalize sentences like

5)   Most men lead lives of quiet desperation

(or the corresponding optimistic sentence with "few" for "most"). If we modelled our attempt on the method used to formalize similar sentences starting with "all" we would write:

6) $Tx(Fx{\rightarrow}Gx)$

with "F" corresponding to "is a man" and "G" to "leads a life of quiet desperation". But this properly speaking formalizes not (5) but rather:

7)   Most things are such that: if they are men then they lead lives of quiet desperation.

(7) is true iff most things in the universe have this property: that if they are men then they lead lives of quiet desperation. One way in which this could be true, at least as formalized by (6), is for most things not to be men. Then most things vacuously have the property in question. (Recall the truth table for material implication: a false antecedent is enough for the truth of the whole.) But this is not a condition upon which (5) is true. Hence (6) is not an adequate formalization of (5).

Perhaps only the details of the strategy were wrong. Perhaps we simply used the wrong connective. But what should we use in place of "$\rightarrow$"? Not "&", as we do when formalizing similar existentially quantified sentences. For, with correspondences as before,

8) $Tx(Fx$ & $Gx)$

is true only upon an interpretation which assigns most things to the intersection of what it assigns to "F" and "G". In other words, formalizing (5) by (8) misrepresents the former as requiring for its truth that most things in the universe are both men, and also leaders of quietly desperate lives. This is false since it is not the case even that most things in the universe are men.

It can be shown that no truth functional sentence connective can be inserted in the place marked ¢ in

9) $Tx(Fx$ ¢ $Gx)$

in such a way as to yield an adequate formalization of (5).[†]

This fact motivates a somewhat different conception of quantification, which I shall now introduce. Let us say that an *open sentence* is what

† See Exercise 117, page 361.

results from a $\mathbf{Q}^+$- or $\mathbf{Q}$-sentence by replacing a name by a variable not already contained in the sentence. (Note that, in this terminology, an open sentence is a kind of non-sentence.) We can characterize the quantifiers of $\mathbf{Q}$ as *unary* quantifiers because they take just one open sentence to make a sentence. This is why the formalization of sentences like "Everything is physical" was so straightforward, whereas the formalization of sentences like "All men are happy" was not. This last sentence contains two predicative expressions, "man" and "happy", welded into a sentence by the quantifier and the copula "are". To $\mathbf{Q}$-formalize it, however, we have to find a *single* open sentence for the quantifier to attach to. What seems approximately to do the trick is "$Fx \rightarrow Gx$". The problem with formalizing sentences like (5) was precisely that, with the available resources, we could not find a suitable single open sentence for the "T" quantifier to apply to.

Once the difficulty is put in this way, it becomes natural to respond by introducing a quantifier which takes *two* open sentences to make a sentence. Such a quantifier is called *binary*. So let us now add binary quantifiers "μ" and "ϕ" to $\mathbf{Q}$, to create the language $\mathbf{QB}$, by the stipulation that every $\mathbf{Q}$-sentence is to be a $\mathbf{QB}$-sentence, and if $X\star$ and $Y\star$ result from $\mathbf{QB}$-sentences X and Y, by replacing in both some occurrence(s) of a name-letter by a variable, v, which occurs in neither, then the following are also $\mathbf{QB}$-sentences:

$$\mu v(X\star : Y\star)$$

$$\phi v(X\star : Y\star).$$

The mark "$:$" functions merely as a separator.

Applying this to (5) yields:

10) $\mu x(Fx : Gx)$

with correspondences as for (6). On the appropriate interpretation, this will say that most of the things which are men lead lives of quiet desperation. So now we have an adequate formalization of (5).

A rule of interpretation for "μ" could be phrased on the following lines (after (**2.2vi**)):

11) $\mu v(X : Y)$ is true upon an interpretation i whose domain is D iff most interpretations with domain D upon which $X\frac{n}{v}$ is true, and which agree with i on everything except, perhaps, the assignment to n, are ones upon which $Y\frac{n}{v}$ is also true.

Notice that the English "most" used in the statement of the rule is naturally seen as a binary quantifier.[†]

It is hard to resist the thought that "every", "all" and "most" belong to the same linguistic category, and so should be treated in the same way. Any sentence containing "most" is still grammatical if that quantifier is replaced by "every" or "all"; and conversely. So it is natural to conclude that if "most" is a binary quantifier, so are "every" and "all". Since we have the antecedent of this conditional, its conclusion comes naturally.

It is straightforward to add a binary universal quantifier to **QB**. Let us write it "λ". Then, with the obvious correspondences, we could formalize "all men are happy" as

12) $\lambda x(Fx:Gx)$,

which looks somewhat more like the English. On the formalization by a unary universal quantifier, an "$\rightarrow$" appeared which was invisible in the English, and there is no such distortion in (**12**).

How should we treat those sentences, like "everything is physical", which appeared well adapted to the unary treatment? The English itself suggests a way, which serves to confirm the view that it involves a binary quantifier. We cannot say "every is physical". "Thing" appears to function precisely as a first term to the quantifier, so that the appropriate formalization is still (**12**), but now with "F" corresponding to "is a thing" and "G" to "is physical". An intended interpretation will, of course, assign every thing in its domain to "F".[‡]

We saw earlier that there is room for doubt concerning whether the standard unary account correctly specifies the truth conditions of English quantifications. Suppose we were convinced that the binary treatment is correct. Would that resolve the doubts? The doubts concerned whether sentences with empty first terms like (**3.4**), (**3.5**), and (**3.6**),

All John's children are asleep
All bodies acted on by no forces continue in a uniform state of rest or motion
All bodies acted on by no forces undergo random changes of velocity

are true just in virtue of the fact that their first terms are empty. The affirmative answer which is unequivocally given by the standard unary treatment is rendered dubious by such sentences as the above.

---

† See Exercise 118, page 361.   ‡ See Exercise 119, page 361.

However, there is nothing in the nature of the binary treatment which pronounces either way on this issue. The obvious rule of interpretation for "λ" is

13) $\lambda v(X:Y)$ is true on an interpretation i whose domain is D iff every interpretation with domain D upon which X^n_v is true, and which agrees with i on everything except, perhaps, the assignment to n, is one upon which Y^n_v is also true.

The question of whether "$\lambda x(Fx:Gx)$" is true on an interpretation, i, which assigns to "F" the empty set turns on the question of whether it is true that "every interpretation upon which '$F\alpha$' is true is one upon which '$G\alpha$' is also true", given that *no* interpretation is one upon which "$F\alpha$" is true. Presumably, those who are most struck by the suggestion that "All John's children are asleep" is not true if John has no children will answer the last question negatively; whereas those who are impressed by the truth of (**3.5**) ("All bodies acted on by no forces ...") will incline to answer it positively, laying themselves open to the charge that perhaps in that case they ought also to assign truth to (**3.6**). Thus binary quantification as such does not address the debate.[†]

*18 Substitutional quantifiers

The **Q**-quantifiers are called "objectual". The reason is that whether or not a quantification is true upon an interpretation depends on how things are with the *objects* in the domain of interpretation. For example, the rule for "$\exists$" entails that "$\exists xFx$" is true upon an interpretation, i, iff some *object* in the domain of i belongs to $i(F)$.

An alternative style of quantifier is called "substitutional". The idea behind the appellation is that the rule for such a quantifier makes whether a quantification is true upon an interpretation depend on whether sentences resulting from the quantification by deleting the quantifier and *substituting* a name for the variable of quantification are true. Thus for an existential substitutional quantifier, written "E", the rule might be:

1) $EvX\star$ is true upon an interpretation, i, iff for some name, N, X^N_v is true upon i.

† See Exercise 120, page 361.

Analogously, for a universal substitutional quantifier, written "A", the rule might be:

> **2)** $AvX^\star$ is true upon an interpretation, i, iff for all names, N, X^N_v is true upon i.

Here, $X^\star$ results from a sentence X by replacing a name-letter by some variable, v, not already in X, and X^N_v results from $X^\star$ by replacing each occurrence of v by some name, N. Note that we do not require that N be absent from $X^\star$.[†]

Q, as defined so far, contains no names, but only name-letters. Names should be related to name-letters as predicates to predicate-letters. Whereas a predicate is assigned the same set on every interpretation, a predicate-letter may be assigned different sets upon different interpretations. Likewise, a name, as opposed to a name-letter, will be assigned the same object in every interpretation. Let us call a language "**QS**" if it adds to **Q** the two substitutional quantifiers recently mentioned, and also adds names, specifying, for each name, the object which any interpretaion must assign to it.

QS is still underspecified, since we have not said what names it contains. Let us suppose, first, that it contains just the name "Reagan", and we stipulate that in every interpretation, i, i(Reagan) is Ronald Reagan, the famous American political film star. In conjunction with (**1**) and (**2**), this would ensure that "$AxFx$" and "$ExFx$" would alike be true upon an interpretation, i, iff ⟨Ronald Reagan⟩ belongs to $i(F)$. There is nothing incoherent here, but equally there is no discernible utility in such quantifiers.

Could we specify **QS** in such a way that a substitutional quantification in **QS** is true upon an interpretation iff the corresponding objectual quantification in **Q** is true upon the corresponding interpretation?[27] A necessary condition, obviously, is that **QS** contain a name for every object in the domain of interpretation. If it did not, it would be "easier" for "$AxFx$" than for "$\forall xFx$" to be true upon an interpretation. Equally, a further necessary condition is that **QS** should not contain a name for any object not in the domain of interpretation.

The two necessary conditions are jointly sufficient for coincidence of objectual quantifiers of **Q** and the substitutional quantifiers of **QS**. If we held to these conditions, there would be no interest in substitutional quantification. There are two ways of modifying **QS** which would make

† See Exercise 121, page 361.

substitutional quantification of interest. One is to allow **QS** to contain empty names; the other is to allow it to contain opaque contexts.

Considerations relevant to the first way are, in effect, discussed in §20 below, so here I shall concentrate only on considerations relevant to the second.

An *opaque context with respect to names* is one in which there is no guarantee that two co-referring names can be substituted *salva veritate*; that is, it is a context in which the substitutivity of identicals – (**8.5**) –fails; that is, the context " ... ..." is opaque with respect to names iff there is no guarantee that "... N_1 ..." and "... N_2 ..." have the same truth value, despite the fact that "N_1" and "N_2" have the same bearer. Thus the context

3)   John believes that ... is a student of Christ Church

is opaque with respect to names, since it may be that the first but not the second of the following is true

4)   John believes that Charles Dodgson is a student of Christ Church

5)   John believes that Lewis Carroll is a student of Christ Church

even though Dodgson is Carroll.

Arguably, we cannot regard (3) as a predicate, for reasons given in §15 (see (**15.16**)). Let us instead call it a "quasi-predicate". To formalize sentences like (4) and (5) we need what I shall call "quasi-predicate-letters". These letters cannot be interpreted by being assigned a set of objects. We will not consider how they are to be interpreted, but let us assume that somehow or other they can be, so that there are **QS**-formalizations of (4) and (5) which, upon an intended interpretation, are true and false respectively. Then by the interpretation rules for the substitutional quantifiers, both of the following are true upon an intended interpretation (where "ψ" corresponds to the quasi-predicate (3)):

6) $\mathbf{E}x\psi x$

7) $\mathbf{E}x\neg\psi x$.

We can now see why these quantifiers are justly regarded as non-objectual. The truth of (6) and (7) on an intended interpretation does not turn on how things are with some object, for the same object is involved in the verification of both, yet no one object can be both ψ and not-ψ.

Substitutional quantification is intelligible in contexts in which objectual quantification would not be. Consider the quasi-predicate "was so-called on account of of his size" and the putative English sentence

    8)   Someone was so-called on account of his size.

The objectual quantification

    9) $\exists x(x$ was so-called on account of of his size$)$

is nonsense. What does the "so" refer back to? On the other hand, if

    10)   Giorgione was so-called because of his size

is true, and "Giorgione" is included in the substitution class of names with respect to which the quantifier is defined, then

    11) $\mathrm{E}x\psi x$

is an adequate formalization of (**8**) (with "ψ" corresponding to the quasi-predicate "was so-called on account of his size"), and is true upon an intended interpretation.

Substitutional quantifiers are perfectly intelligible. It is quite unclear whether any English quantifiers are substitutional. If there is such a quantifier in English, then what (**11**) formalizes would be expressible in English, and true. But it is doubtful whether (**8**) is any more intelligible than (**9**): we ask again "What does the 'so' refer back to?" At best, (**8**) could be counted as a less than accurate expression of

    12)   Someone was called what he was called on account of his size.

Notice, however, that any doubts about the intelligibility of (**8**) do not extend to (**11**), which has been given entirely intelligible truth conditions.[†]

† See Exercise 122, page 361.

*19 Predicate quantifiers

The **Q**-quantifiers are "name" quantifiers. That is, the variables they bind occupy a position in a sentence appropriate to a name. Can one add to **Q** quantifiers binding variables which occupy a different kind of position, in particular, predicate position?

It may seem that in English we can generalize from a sentence like

1) Reagan and Thatcher are both powerful

in two kinds of way. One way is the way of familiar name quantification:

2) Someone, x, and someone, y, are such that x is powerful and y is powerful.

A **Q**-formalization, with obvious correspondences, would be

3) $\exists x\, \exists y (Fx \,\&\, Fy \,\&\, Gx \,\&\, Gy)$.

The other way of generalizing from (**1**) is:

4) There is something which both Reagan and Thatcher are.

To formalize sentences involving the kind of quantification (**4**) exemplifies, we might augment **Q** to a language – call it **QP** – in which there are quantifiers, say "$\triangle$" for the universal, "$\triangledown$" for the existential, which bind variables occupying predicate position – we will use italic letters, $f, g, \ldots$. Then (**4**) could be **QP**-formalized by

5) $\triangledown f (f\alpha \,\&\, f\beta)$.

Interpretation rules for a name quantifier make truth upon an interpretation depend upon all the objects which can be assigned to name-letters, that is, on all the objects in the domain. Extending this to predicate quantifiers, we would say that appropriate interpretation rules would make truth upon a **QP**-interpretation depend upon all objects which can be assigned to predicate-letters, that is, on all sets of n-tuples whose members belong to the domain. We could therefore write rules for $\triangle$ and $\triangledown$ which exactly parallel the rules for $\forall$ and $\exists$. With this change in the interpretation rules, the definition of **QP**-validity can exactly match that of **Q**-validity.

(5) would be true upon a **QP**-interpretation, i, iff there is a **QP**-interpretation, i', agreeing with i except, perhaps on what it assigns to "F",[28] such that "$F\alpha$ & $F\beta$" is true upon i', that is, such that $\langle i'(\alpha)\rangle$ and $\langle i'(\beta)\rangle$ both belong to $i'(F)$; that is, iff there is a set of 1-tuples containing both $\langle i(\alpha)\rangle$ and $\langle i(\beta)\rangle$. The predicate variables (f, g, ...) are seen as standing to the predicate-letters just as the individual variables (x, y, ...) stand to the name-letters. Whether there is or is not such a set in part depends upon facts of set theory. The domain settles which objects are available for an interpretation to assign to name-letters; but what sets are available for assignment to name-letters, relative to a given domain, is a matter of set theory.

On this account, (5) is equivalent to the claim that there is some set of n-tuples to which both Thatcher and Reagan belong. Perhaps we originally wanted something rather meatier, some talk of Reagan and Thatcher having a common *property*. For the moment, we will turn aside from that desire. However, seeing predicate quantifiers as ranging over sets of n-tuples satisfies part of the desire, for properties divide the world into such sets.

QP is, approximately, the language of "second order logic", in contrast to **Q**, the language of "first order logic". Second order logic has a number of different properties from first order logic.[29] From our present point of view, the most important difference is that some intuitively valid sentences are formalizable as **QP**-valid, but are not formalizable as **Q**-valid. An example is:

6)   Everything has at least one property.

Setting aside for the moment the difference between a property and a set of n-tuples, this is formalizable as the **QP**-valid

7) $\forall x \triangledown f(fx)$.

Its validity depends upon the fact that for any unary predicate letter and any object, o, in the domain of interpretation, there is an interpretation which assigns to the predicate letter a set to which $\langle o\rangle$ belongs.

By contrast, a first attempt at **Q**-formalizing (6) would give merely "$\exists x(Fx)$", which would be false upon an interpretation which assigned the empty set to "F".

We can adequately (in the technical sense of this book) formalize any sentence by a sentence-letter. However, there are valid arguments whose validity corresponds to **Q**-validity but not to **P**-validity. **Q**-enables us to provide deeper formalizations than any available in **P**. There is a parallel contrast between **Q** and **QP**. In particular, **QP** can provide formalizations of mathematical statements so as to capture validity which is not matched by **Q**-validity.

However, one can for certain non-mathematical purposes simulate some of the features of **QP** in **Q**. Although a **Q**-domain is standardly called a set of *individuals*, the logician does not mean this in any metaphysically loaded way: an individual is simply a member of a **Q**-domain. Hence if we include some sets of n-tuples among the individuals, we might think to mirror some features of **QP** in **Q**. Thus we might **Q**-formalize (4) as follows:

8) $\exists x(G\alpha x \,\&\, G\beta x)$

with "Gxy" corresponding to "x belongs to y" and "α" and "β" as before. The idea is to use name quantifiers in the place of predicate quantifiers, but achieve the effect of predicate quantification by requiring that, upon intended interpretations, the name quantifier will range over, *inter alia*, the sort of thing a predicate quantifier ranges over. Upon an intended interpretation, (8) will be true iff there is some set to which both ⟨Thatcher⟩ and ⟨Reagan⟩ belong, which is, near enough, the gloss given of the **QP**-sentence (5).

If we try to develop this **Q**-simulation of **QP**, in such a way as to try to capture intuitive validities as **Q**-validities, we need to formalize what are intuitively one-place English atoms, like "Bill jogs", as two-place. For suppose we do not. Then the intuitively valid

9)   John does everything that Bill does, Bill jogs; therefore John jogs

would be formalized along the lines of:

10) $\forall x(G\beta x \rightarrow G\alpha x)$, $F\beta$; $F\alpha$,

which is **Q**-invalid.[30] (The correspondence scheme is: "Gxy" for "x belongs to y", "F" for "jogs", "β" for "Bill" and "α" for "John".)†

However, if "Bill jogs" is formalized as

11) $G\alpha\gamma$

† See Exercise 123, page 362.

with "Gxy" corresponding to "x belongs to y" and "γ" to "the joggers", thus treating (11) as equivalent to "John is among the joggers",[31] then (9) could be formalized as the **Q**-valid:

    12) $\forall x(G\beta x \rightarrow G\alpha x),\ G\beta\gamma;\ G\alpha\gamma.$

    Even with such devices, we are no nearer formalizing (6) as **Q**-valid, so **QP** can reveal depth which **Q** cannot.[†]

Second order logic is sometimes characterized as the logic of properties, whereas we have taken the predicate quantifiers to range over sets. The connection between sets and properties is this: every (coherent) property determines a set (albeit in some cases the empty set), but two properties may determine the same set. Historically, there have been doubts about the existence of properties which have not extended to sets.[32] The semantics of **Q** already mention the entities, sets of n-tuples whose members belong to D, needed for interpreting the **QP** predicate quantifiers. An alternative to **QP** which saw properties as the range of the predicate quantifiers would be less continuous with **Q** and would raise a new range of issues.

*20 Freedom from existence

We recalled in connection with (14.7), viz.

    $X \vDash_{\mathbf{Q}} \exists v X^\star,$

that we have required any **Q**-interpretation to assign to each name-letter an object. The central formal feature of "free logic" is that this requirement should be dropped. Before looking at languages which could implement this idea, let us try to understand the philosophical motivation.

English contains names lacking bearers, like "Vulcan" and "Pegasus", and such names, as we have seen, cannot be appropriately formalized by name-letters in **Q**. There is no syntactic test for whether a name has a bearer or not. It could be argued that whether a name has a bearer is a fact not available *a priori* to a user of the language. In this case, the validity of arguments containing names should be independent of whether they have bearers, for validity is supposed to be discernible *a priori*.

---

† See Exercise 124, page 362.

Thinking on these lines could lead to a number of specific claims. First, consider the argument

   1)   Everything is physical; therefore Vulcan is physical.

If "Vulcan" were simply treated as a name, the argument would be **Q**-formalized as valid, yet arguably (**1**) is invalid. Even though there may be disagreement about whether the conclusion is false, no one could suppose that it is true; yet the premise is arguably true.

Secondly, consider the arguments:

   2)   Pegasus does not fly; so Pegasus exists.

   3)   Pegasus does not exist; so something does not exist.

Arguably, these have true premises but false conclusions. Everyone agrees that "Pegasus flies" is not (genuinely and non-fictionally) true; so if there is a way of denying whatever "Pegasus flies" asserts, the result should be true. Arguably, the premise of (**2**) expresses this denial, and so is true, yet the conclusion is false. For the same reason, the premise of (**3**) is true, yet, unless Meinong's theory is true, its conclusion is false. However, as we have already seen (in §11), the obvious formalizations of (**2**) and (**3**) using name-letters are **Q**-valid.

Thirdly, one might have doubts about whether

   4)   Vulcan is identical with Vulcan

is true. If it is true, it had better not entail that something is identical with Vulcan. These intuitions are not straightforwardly accommodated in **Q**.†

Fourthly, it might be doubted whether

   5)   Something exists,

though indisputably true, is a truth of *logic*. Yet a possible formalization of (**5**),

   6) $\exists x \exists y \; x = y$,

is **Q**-valid. While this fourth point does not concern English names lacking bearers, it does relate to the basic formal feature of free logic: if we

† See Exercise 125, page 362.

changed the semantics of **Q** just by dropping the requirement that every interpretation assign an object from the domain to every name-letter, (**6**) would not be valid.

These intuitions, or alleged intuitions, fall naturally into two pairs. The first pair shares the feature that the presence of bearerless names in a language invalidates arguments concerning quantifiers that are the English analogues of **Q**-valid arguments. For this point, it is essential that there be atomic sentences containing bearerless names which are false (for the invalidity of (**1**)), and negated sentences containing bearerless names which are true (for the invalidity of (**2**) and (**3**)).

The second pair of alleged intuitions raises the question of what should count as logically true. An argument for regarding (**4**) as valid (that is, as logically true) might be based on these assumptions: (a) that a logical truth should be recognizable as such *a priori*; (b) that whether or not a name has a bearer is not a fact accessible *a priori* to a competent user of the language. This blocks any discrimination among sentences of the form of (**4**): all or none must be valid. Repugnance for the latter option might persuade one to adopt the former.[33]

Some logicians believe that (**5**) is not logically true, and it turns out to be fairly easy to make room for this in one version of free logic.

The first pair of intuitions, or alleged intuitions, are central to free logic. The second pair are optional, jointly and severally. Either one, or both, are adopted by some but not all free logicians.

We will keep to the syntax of **Q**, that is, we will discuss just the same sentences, but as we are going to introduce different semantics, that is, different rules for interpretation, let us regard ourselves as introducing a new language, **QF**, to mark the fact that it is Free of existence-assumptions with respect to name-letters.

In place of the stipulation on interpretations in (**2.1**), we say that for every name-letter in **QF** an interpretation with respect to a domain, D, may, but need not, assign it an object from D. This means that when we consider the totality of interpretations with respect to a given name-letter, we must not forget those which assign no object to that letter.

Our first revision of the interpretation rules will speak just to the first pair of intuitions. A subsequent revision will show how the other intuitions can be accommodated. First, the new rule for atoms, replacing (**2.2v**), should read:

7)   an atom $\phi v_1 \dots v_n$ is true upon i iff (a) i assigns an object from D to each of $v_1 \dots v_n$ and (b) the sequence $\langle i(v_1) \dots i(v_n) \rangle$ belongs to the set $i(\phi)$; $\phi v_1 \dots v_n$ is false upon i iff either (a) or (b) fails.

This ensures that the conclusion of (1), **QF**-formalized as "$F\alpha$", is false upon an intended interpretation (one which, among other things, assigns nothing to "Vulcan"). It also ensures that the premises of (2) and (3), formalized using a name-letter for "Pegasus", are true upon an intended interpretation. An appropriate modification of the quantifier rules, replacing (**2.2**vi and vii) is as follows:

8) $\forall v X$ is true upon an interpretation, i, with respect to domain, D, iff X_{v}^{n} is true upon every interpretation whose domain is D which assigns something to n and agrees with i except, perhaps, in point of what is assigned to n; otherwise $\forall v X$ is false upon i.

$\exists v X$ is true upon an interpretation, i, with respect to domain, D, iff X_{v}^{n} is true upon some interpretation over D which assigns something to n and agrees with i except, perhaps, in point of what is assigned to n; otherwise $\exists v X$ is false upon i.

The rulings ensure that quantification is genuinely over objects, by considering only the interpretations which assign something to the name-letters replacing the variables of quantification. They have the effect that the obvious formalizations of (1) and (4) are **QF**-invalid.[†] The closest valid argument to the formalization of (1) is:

9) $\forall x F x, \exists x x = \alpha; F\alpha.$

The free logician will urge that if an argument of the form of (1) strikes us as intuitively valid, it is because we are presupposing the truth of the suppressed existential premise, made explicit in (**9**).

The premise of (3) can be adequately **QF**-formalized without quantification and with name-letters as "$\neg F\alpha$", with "F" corresponding to "exists" and "α" to "Pegasus". Upon an intended interpretation, i, "$F\alpha$" is false, and so the **QF**-formalization is true, which is the desired result. The same sentence can also be adequately formalized with quantification and with name-letters as

10) $\neg \exists x\, x = \alpha,$

for "$\exists x\, x = \alpha$" is not **QF**-valid. (**10**) is true upon an intended interpretation.[‡]

† See Exercise 126, page 362.   ‡ See Exercise 127, page 362.

The original intuition was that an atom with an empty name is false, and so its negation true. However, it is plain that there could be a tension within this view, since we could regard, for example, "Pegasus is not happy" either as a denial of "Pegasus is happy", and so as true (given that "Pegasus" has no bearer), or as an affirmation of "is not happy" with respect to Pegasus, and so as false. From the perspective of **QF**, it would seem that we have to regard a sentence like "Pegasus is not happy" as ambiguous.

The problem with not recognizing such ambiguity can be brought out in the following way. It should not be incorrect to formalize, for example, "John is not happy" as "$F\alpha$", letting "F" correspond to the negative predicate "is not happy". If we allowed this general policy in **QF**, we would have to say that "Vulcan does not exist" could be adequately formalized as "$F\alpha$", with "F" corresponding to "does not exist". However, this formalization would treat the English as entailing that Vulcan does exist, since

11) $F\alpha \vDash_{\mathbf{QF}} \exists x Fx.$

The basis for (11) is the fact that an *atom* is false upon an interpretation that fails to assign an object to one of its name-letters.

Thus the rationale of **QF** demands recognition of ambiguity in sentences like "Pegasus is not happy". This ambiguity can be represented as scope differences with respect to name-letters. The distinction we want to capture might be represented in English by parentheses: it is between "Pegasus (is not happy)" and "not (Pegasus is happy)", the former being a negative predication upon Pegasus, and therefore false, the latter the denial of a predication upon Pegasus, and therefore true. We could make a similar distinction in **QF**, writing

12) $[\alpha](\neg F\alpha)$

to match the first alternative, and keeping to the familiar

13) $\neg F\alpha$

for the second. We can read an occurrence of "$[\alpha]$" as "it is true with respect to α that:", so that (12) is read "it is true with respect to α that: α is not F". A formalization of "Vulcan does not exist" which is true upon an intended interpretation will be (13).

The proposed convention is that unless otherwise indicated by an occurrence of "$[\alpha]$", "α" will take narrowest scope in a sentence. A sentence entirely within the scope of an occurrence of a name-letter in square brackets will be said to be *dominated by* that name-letter. Thus "α" *dominates* (12) and does not dominate (13). Implementing this idea in revised interpretation rules in effect means requiring that a sentence of the form

14) $[v]X$

is true upon an interpretation only if it assigns something to v.[†]

QF is a perfectly coherent and sensible language, which gives effect to the first two intuitions mentioned, whose essence is that some sentences containing bearerless names are true and some are false. This makes names more like definite descriptions, and entails that the condition for understanding a name cannot be that one knows who or what its bearer is, since some names which can be understood do not, on this view, have bearers. The view is therefore closer to our ordinary classification of names in English, marked by their being capitalized non-complex subject expressions. The question of whether all names are in this respect like descriptions is thus not to be resolved purely by considerations about formalizability, unless one has some independent reason for thinking that **Q**-formalizability is a necessary condition for intelligibility.

The second pair of intuitions are not accommodated within the present version of **QF**, and so raise further questions. They are that the so-called Law of Identity – the validity of every sentence of the form "$\alpha = \alpha$" – should be respected, and that the truth of (5) – "something exists" – should not be a matter of logic. Yet the Law of Identity fails in **QF**, and (6) is **QF**-invalid. Respecting these intuitions is made more difficult by the fact that the Law of Identity standardly entails (6): we have to restore the validity of the Law of Identity without restoring its standard consequences.

The obvious way to attend to the first intuition is to revise the rule for atoms as follows:

15)   an atom $\phi v_1 \ldots v_n$ is true upon i iff either (a) $\phi v_1 \ldots v_n$ is true upon every interpretation over D which assigns an object to every name-letter in the atom, or (b) i assigns an object from D

to each of $v_1 \ldots v_n$ and the sequence $\langle i(v_1) \ldots i(v_n) \rangle$ belongs to the set $i(\phi)$; otherwise $\phi v_1 \ldots v_n$ is false upon i.

This, which ensures the **QF**-validity of "$\alpha = \alpha$", is inconsistent with (7), and springs from an inconsistent motivation. (7) was motivated by the thought that any *atom* containing a bearerless name should be false; (15) by the thought that classical validities should be preserved as far as possible. However, (6) remains valid, and it cannot be rendered invalid by any obvious changes in the quantifier rules.† To reflect the envisaged combination of views, the validity of "$\alpha = \alpha$" and the invalidity of (6), we need to make some explicit stipulations about domains.

QF differs from **Q** in point of the stipulation that a **Q**-interpretation must, but a **QF**-interpretation need not, assign an object from the domain to each name-letter; it differs from **Q** also in point of the rules for atoms and quantifiers; and these are the only differences. Hence the **Q**-requirement that every domain be non-empty remains in force. Once we make the relaxation upon interpretations, however, it becomes natural to investigate the further step of relaxing the requirement that all domains of interpretation be non-empty. Once this step is taken, achieving the invalidity of (6) requires no modification of the rules for atoms or quantifiers. A universal quantification will be true upon any interpretation with respect to the empty domain, since there are no interpretations over that domain which assign anything to any name-letter, and so every such interpretation meets any condition at all. An existential quantification will be false upon any interpretation with respect to the empty domain, since there are no interpretations over that domain which assign anything to any name-letter, and so no interpretation meets the relevant condition. Hence no existential quantification will be valid.

Though it may not be immediately obvious, merely relaxing the stipulation about the non-emptiness of domains, in addition to achieving the **QF**-invalidity of (6), does nothing to undermine the **QF**-validity of "$\alpha = \alpha$".‡

It is useful to be reminded that there are coherent alternatives to **Q**. We must not become fixated by **Q**, thinking of it as the sole repository of logic. For example, whether "Something exists" is valid (in the sense of ⊨) is not decided by the **Q**-validity of (6). We must choose the best language for logic on philosophical grounds: these should tell us what a correct logic should validate, and what it should not.

† See Exercise 129, page 362.   ‡ See Exercise 130, page 362.

21 Depth

A **Q**-formalization may be adequate, yet not reveal all the logical structure of the English. Thus

1) $F\alpha$

is an adequate **Q**-formalization of

2)   John is not happy

(with "*F*" corresponding to "is not happy"). The truth conditions of (2) coincide with the truth-upon-an-intended-interpretation conditions of (1). Yet (1) does not reflect the fact that (2) contains an occurrence of a logical constant. We shall say that, relative to (2), (1) is *a less deep* **Q**-*formalization* than

3) $\neg F\alpha$

with "*F*" corresponding to "is happy".
Sometimes lack of depth in a formalization may prevent the validity of an argument becoming manifest. For example, the following is valid:

4)   John will enjoy any book about cosmology. *The First Three Seconds* is a book about cosmology. Therefore John will enjoy *The First Three Seconds*.

The following meets the standard for being an adequate **Q**-formalization:

5) $F\alpha\beta$, $G\gamma\beta$; $H\alpha\gamma$

with "*Fxy*" corresponding to "*x* will enjoy any book about *y*", "*Gxy*" to "*x* is a book about *y*", "*Hxy*" to "*x* will enjoy *y*", "*α*" to "John", "*β*" to "cosmology" and "*γ*" to "*The First Three Seconds*". Yet (5) is plainly **Q**-invalid.
Lack of depth may not prevent the manifestation of validity. The following is **Q**-valid and an adequate formalization of (4):

6) $\forall x(Kx \rightarrow H\alpha x)$, $K\gamma$; $H\alpha\gamma$

with "K" corresponding to "is a book about cosmology" and the other correspondences as before. Yet the following is adequate and deeper:

7) $\forall x(Gx\beta \rightarrow H\alpha x)$, $G\gamma\beta$; $H\alpha\gamma$

(correspondences as before). The crucial point is that if X and Y are adequate formalizations of A, and Y is deeper than X, and X is **Q**-valid, then Y is also **Q**-valid.

How far should depth go? One extreme idea, to be explored in chapter **6.2**, is that wherever there is validity in the English, then some adequate **Q**-formalization is valid. The execution of this idea would require one, for example, to formalize the argument

8)   Tom is a bachelor, therefore Tom is not married

along the lines of

9) $F\alpha \ \& \ \neg G\alpha$; $\neg G\alpha$

(with "F" corresponding to "is a man" and "G" to "is married"). And it would require one to formalize

10)   Socrates is mortal, therefore he is or will be dead

along the lines

11) $\exists x(Fx \ \& \ G\alpha x)$; $\exists x(Fx \ \& \ G\alpha x)$

(with "F" corresponding to "is a time", "Gxy" to "x dies at y" and "α" to "Socrates").

We will confine ourselves here to noting two points. First, as in the case of **P**, that an adequate formalization of an argument is not **Q**-valid does not in itself enable us to infer anything about the validity of the argument. Perhaps a deeper **Q**-formalization would reveal it as valid – and valid in virtue of its **Q**-logical form. Or perhaps it is valid, but not in virtue of its **Q**-logical form (*pace* the proponents of the extreme idea just mooted).

Secondly, standard practice shrinks not at all from introducing in a **Q**-formalization logical constants that do not visibly appear in the English. Examples are the formalizations of two-term universal and existential quantifications (e.g. "All (some) men are happy"), the former seen as introducing an occurrence of "$\rightarrow$", the latter an occurrence of

"&". Indeed, it is not unknown for practitioners to go further, and introduce predicate-letters not corresponding to predicates visible in the English. Some versions of Davidson's theory of adverbs, for example, see a sentence like "John runs" as formalizable by a **Q**-sentence containing a predicate-letter which, on an intended interpretation, will be assigned the set of all 1-membered sequences each having a run as member (see §6). No such predicate is visible in the English, for "runs" appears to be true of running people rather than of runs.

22 From *Q*-validity to validity

If an adequate formalization of an English argument is not **Q**-valid, what can be inferred about the validity of the English? Answer (given in the last section): Nothing.

Suppose, however, that an English argument is adequately formalized by a **Q**-valid one? The exercise would have been pointless unless we can infer that the English argument is valid – valid, as we shall say, in virtue of its **Q**-logical form. How can this inference be justified?

We have to show that if a **Q**-argument ϕ is **Q**-valid and is an *adequate* formalization of an English argument ψ, then ψ is valid.

The adequacy of the formalization ensures that, where ϕ' is the argument recovered from ϕ by applying the relevant correspondences, ϕ' says the same as ψ. The notion of "saying the same" that is required here is that the sentences of ϕ' should have the same truth conditions as the corresponding sentences in ψ. Validity can be defined in terms of truth conditions, and if two arguments are related as ψ and ϕ' then, necessarily, both or neither are valid. So it will be enough if we can show that if ϕ is **Q**-valid then ϕ' is valid.

A necessary condition for there being any adequate **Q**-formalizations is that the **Q**-operators make the same contribution to truth conditions as the corresponding English expressions. Let us suspend any doubts on this score for the moment. We will use a phrase like "corresponding name", "corresponding predicate", etc, to refer to a name (predicate) in ϕ' to which, by the correspondence scheme, a name-letter (predicate-letter) in ϕ corresponds. The following argument establishes what is needed:

1) (i)   Suppose ϕ is **Q**-valid.
    (ii) Then every interpretation upon which the premises of ϕ are true is one upon which the conclusion is true.

(iii) Hence whatever the corresponding names in ϕ' refer to, whatever the corresponding predicates in ϕ' are true of, and whatever the truth values of the corresponding sentences, if the premises of ϕ' are true, so is the conclusion.

(iv) Hence, necessarily, if the premises are true so is the conclusion.

One crucial step is from (ii) to (iii). This essentially requires that every English expression in ϕ' corresponding to a constant in ϕ expresses the same as the **Q**-constant.

Another crucial step is that from (iii) to (iv). The former makes no explicit mention of logical necessity, the latter does. How can the intrusion of this notion be justified? The basic idea is this:

2)   the truth of a sentence which is adequately **Q**-formalizable will of necessity depend on nothing except the reference of any corresponding names it contains, what any corresponding predicates are true of, and whether any corresponding sentences are true or false.

If all logical possibilities for what the corresponding names refer to, what the corresponding predicates are true of, and what truth value the corresponding sentences have, are ones upon which the conclusion is true if the premises are, as the **Q**-validity of ϕ assures us is the case, then we can legitimately infer that, necessarily, if the premises are true so is the conclusion.

The property of a sentence specified by (2) is known as *extensionality*. I discuss this notion in more detail in chapter **5.6**.

Finally, it remains to note that $\vDash_{\mathbf{Q}}$ shares the formal properties of $\vDash$.[†]

† See Exercise 131, page 362.

Notes

§1–4

Strawson [1952], chs 5 and 6, provides useful comparisons between **Q**-quantifiers and English; see also McCawley [1981]. For a claim that **Q**-quantifiers cannot represent everything we want to say in English by the corresponding quantifiers, see Quine [1970] pp. 89–91 (branched quantifiers). For the claim that other quantifiers (e.g. "most") cannot be represented in the style of the **Q**-quantifiers, see below, §17. On alleged contrasts between universal generalizations in English and **Q**: Barwise and Cooper [1981].

§5

For more detailed discussions of adjectives, see Platts [1979], ch. 7; Kamp [1975].

§6

For Davidson's proposal see his [1967b] and [1970a]. For discussion see Platts [1979] pp. 190–201 and Taylor [1985]. For a different approach, see Clark [1970] and Parsons [1972]. For a proposal which accepts much of the spirit of Davidson's, while rejecting the full appropriateness of **Q**-formalization, see Wiggins [1985].

§7 and §12

Mill [1879], book 1, held that names and descriptions differ: names denote but do not connote. Frege [1892b] p. 59, footnote, suggests that the sense of a name might be given by a definite description, but it is certain that this is not a consequence of his general doctrines on the subject (see Dummett [1973] pp. 97–8: "there is nothing in what [Frege] says to warrant the conclusion that the sense of a proper name is always the sense of some complex description"), and doubtful whether it is even consistent with them (Evans [1982] pp. 22–38). None the less, the theory that names are descriptions is frequently attributed to Frege, for example by Kripke [1972], esp. p. 27, and Searle [1958]. Russell thought there are two kinds of names: "logically" proper names formalizable by name-letters, but in natural language consisting in only "this" and "that"; and "ordinary" proper names, like "Aristotle", which are "really" abbreviations for descriptions. See Russell [1912], ch. 5, and [1918] pp. 200ff. Russell does not intend this as a doctrine about the meaning of names, if meaning is what is in common between speaker and hearer in communication, but rather as contributing to an account of "the thought in the mind of a person using a proper name correctly" ([1912] p. 29).

A gentle introduction to opposing views is Searle [1958]. Kripke [1972] attacks description theories of names, and his arguments have been widely discussed, for example by Dummett [1973], appendix to ch. 5. Some subtle distinctions are brought to bear by McDowell [1977]. See also Platts [1979], ch. 6, Linsky [1977], Davies [1981], esp. pp. 103ff., Pollock [1982], chs 2–3, and Evans [1982], chs 1–3 and 11. A recent contribution which takes the topic of names as a basis for a very wide-ranging discussion is McCulloch [1989].

§8

See Quine [1960], section 24.

§9

This standard treatment of numeral adjectives derives from Frege [1884], esp. §55ff.

§10

Russell's theory of descriptions was first presented in his [1905], along with a general theory of quantification. A much clearer presentation, detached from the general theory of quantification, is in Russell [1919], ch. 16. A classic criticism is by Strawson (see his [1950], and his [1952] pp. 184–90), and a source of much debate is Donnellan [1966]. For further discussion, see Peacocke [1975], Sainsbury [1979], ch. 4, Davies [1981], ch. 7, Pollock [1982], ch. 4, McCulloch [1989], Salmon [1989].

§11

For Russell's account, together with his attack on Meinong, see his [1905] and [1918], lecture 5. For Meinong's theory see Meinong [1904]. For a recent philosophical discussion see Lambert [1983], ch. 5. For a formal defence of a Meinongian position, see Parsons [1974]. For a recent philosophical plea for non-existents, see Yourgrau [1987].

For an ingenious account of negative singular existential truths which treats "exists" as a predicate, see Evans [1982], ch. 10.

§13

For discussions of pronouns see, for example, Evans [1977a], [1977b], and K. Bach [1987], chs 11 and 12.

§16

For Davidson's proposal, see Davidson [1969]. For discussion, see Platts [1979], ch. 5, Burge [1986], Schiffer [1987], ch. 5, Segal [1989].

§17

See Platts [1979] pp. 100–6, Davies [1981], esp. pp. 123ff., and Wiggins [1980a].

§18

A classic text on substitutional quantification is Kripke [1976]. A gentler introduction can be found in Quine [1969a]. See also Davies [1981] pp. 142ff.

§19

For philosophical discussion of predicate quantification, see Strawson [1974], Davies [1981] pp. 136ff., and, for a sceptical view, Quine [1953b] and [1970] pp. 66–8. For second order logic, see Boolos [1975] and Boolos and Jeffrey [1974], ch. 18, and van Benthem and Doets [1983]. Both Frege [1879] and Russell [1908] used predicate quantification in their formalizations of mathematics.

§20

For a version of free logic motivated by the first pair of intuitions, see Schock [1968]. See Lambert [1965] for the alternative motivation. See also Leonard [1956], Routley [1970], Lambert [1983], esp. ch. 5, and references therein; and Bencivenga [1985] (which includes a 60-item bibliography). For an interesting application, see Evans [1979], from whom I have drawn the idea of scope distinctions for names. For the empty domain, see Quine [1950] p. 96. For a philosophical discussion of whether "Something exists" is a logical truth, see Cohen [1962], section 33.

Among extensions of **Q** not treated here, see Quine [1970] pp. 89–90 for a discussion of branched quantifiers. For a survey of many variants of **Q**, see Haack [1974].

---

1 There is no need for the interpretation to assign to predicates of degree 1 sets of sequences rather than sets of objects, except for attaining uniformity with predicates whose degree exceeds 1. The example which follows in the text shows why sequences are needed for the latter.
2 The standard approach is Tarski's, for which the classic reference is his [1937]. For a discussion of the alleged philosophical difference between Tarski's approach and that adopted here, see Evans [1977a] pp. 81ff., and for a contrasting view Hodges [1983], §10. There are several alternatives. One very clear example is Lewis [1970].
3 By Strawson [1952] pp. 173–6.
4 Cf. Geach [1956].
5 To say that a quantifier is implicitly present is just to say that it is present in the logical form. A justification for this sort of claim is discussed in ch. **6**.
6 This phrase and other points in this section come from Taylor [1985].
7 It was Davidson himself who first insisted on this point: see his [1967b] p. 121.
8 Cf. Wiggins [1985].

9 There would be a decisive objection, if there were no properties, that is, if nominalism were true (see Armstrong [1978]).

10 But see Hintikka [1973] p. 195 for a formalization of (20) based on the "exactly *n*" quantifier.

11 These are the only alternatives in **Q**. Some have found both inadequate, and seen in this a reason for thinking that English is not adequately **Q**-formalizable: cf. Strawson [1952] pp. 184ff.

12 There are many alternative treatments of descriptions, but they are not available within **Q**.

13 Cf. Frege [1892b] p. 70.

14 See Russell [1918] p. 201.

15 I do not think that this approach is adequate as it stands. For a recent sophisticated view, see Evans [1982], ch. 10.

16 Notice that, contrary to what is sometimes affirmed (e.g. by Thomson [1967] pp. 104–5), this shows that, if a predicate is an expression which takes one or more names to form a sentence, **Q** contains a predicate which expresses existence. Those who hold that **Q** contains no predicate for existence may be either remarking on the fact that formalizing "exists" by a predicate letter is not always adequate, or else observing, correctly, that **Q** contains no *simple* predicate which has to be interpreted as existence. (But it also contains no simple predicate which has to be interpreted as *mountain*.)

17 Quine [1948] proposes this approach. For a critical discussion, see Hochberg [1957].

18 The phrase "truncated description" comes from Russell [1918] p. 243 in connection with "Romulus". Cf. also Russell [1912] p. 29.

19 Cf. that attributed by Kripke [1972] p. 27. But see Russell [1912] pp. 29–31.

20 Cf. Frege [1892b] p. 59. It is not clear that his own account avoided these subjective sources. See the footnote concerning "Aristotle" on that page.

21 By Evans [1982], ch. 10.

22 The point is stressed by Kripke [1972], esp. pp. 53ff., the distinction between specifying the meaning and fixing the reference.

23 It is the "∃¹" of (**9.15**). See Exercise 100a.

24 Strictly speaking, I should say: "assign to '*F*' a set whose only member is the empty sequence". *Mutatis mutandis* when I speak of the assignment of a non-empty set. In this case, pedantic accuracy seemed hopelessly indigestible.

25 Or should one say, rather: a **Q**-formalization of the second sentence of (2)? Neither answer would be quite right, for reasons not discussed until ch. **6.6**.

26 On the contrast between logical form proposals and philosophical analyses, see Davidson [1970a], and below ch. **6.2**.

27 That is, one just like the interpretation for **QS**, except not assigning anything to names (for **Q**-interpretations have no truck with names, as opposed to name-letters).

28 Here I assume that "*F*" is the first predicate letter (relative to some standard ordering) which does not appear in (5). Cf. (**2.2**).

29 It is not complete, the compactness and Löwenheim–Skolem theorems fail,

and it is rich enough to express arithmetic. For a discussion, see Boolos and Jeffrey [1974], ch. 18.

30 Cf. Higginbotham [1986].

31 Properly speaking, "γ" should correspond to "the set of singletons whose members are joggers", for it is ⟨John⟩, not John, which would be said to belong to it.

32 See e.g. Quine [1970], ch. 5, esp. pp. 66–9.

33 I can discover no reasonable ground for this repugnance. Whatever one might want from the "Law of Identity" (the validity of every sentence of the form "$\alpha = \alpha$") can be obtained from a conditionalized version of the law (the validity of every sentence of the form "$(\exists x\ x = \alpha) \rightarrow \alpha = \alpha$").

5

Necessity

If you are a physicalist, you may wish to assert not merely that everything is physical – which is consistent with this important fact being accidental – but also, more strongly:

1) Necessarily, everything is physical.

This chapter explores how one might augment **Q** so as to capture the distinctive contribution of English expressions like "necessarily". Superficially at least, (**1**) appears to be composed of "everything is physical", dominated by the non-truth functional sentence operator "necessarily". One way to proceed is to take appearances at face value, and add suitable non-truth functional sentence operators to **Q** (stipulating that these be counted among the logical constants). This augmented language will be called **QN**.

(**1**) appears to be equivalent to

2) It is necessarily the case that everything is physical.

It is natural to regard such a sentence as ascribing something – the property of being necessarily true – to a proposition. In these terms, it is easy to see how possibility and necessity are interdefinable. A proposition is possible iff it is not impossible, iff its negation is not necessary. Possibility and necessity are called "modal" notions, and studying them is part of the study of *modality*.

Modal notions surface in idioms other than "necessarily" and "possibly", notably in "has to" and "must" as used in sentences like the following:

3) You have to make adequate financial provision for your children.

4) You must leave now or you'll miss your plane.

5) If you press down on one end of a rigid rod, freely balanced at its centre, the other end must go up.

6) This just has to be a herring.

Notice that these examples exploit different standards or criteria for the necessity they invoke: (3) is naturally heard as invoking moral necessity, (4) prudential necessity, (5) natural necessity, and (6) epistemic necessity. We will discuss necessity of the broadest kind, so-called logical or metaphysical necessity. We make no attempt to give any account of what this is. It is exemplified in (1), and there will be many subsequent examples.

1   Adding "□" to *P*

It will be convenient to approach **QN** by an intermediate stage, a language, to be called **PN**, obtained from **P** by adding a new sentence operator, "□", pronounced "box", with the intended meaning of "necessarily". **PN** is called a "propositional modal language".

Syntactically, "□" is just like "¬": that is, if X is a **PN**-sentence, so is $□X$. A sentence of the form $□X$ is called a "necessitation".

Intuitively, we want $□A$ to be true iff A has to be true, iff A is true however things might have been or may be, iff A is true in every possible world.† We will for the moment take the notion of a possible world for granted, glossing it, in the style of David Lewis, as a way things might have been.

What is it for a sentence to be true "in", "at" or "with respect to" a possible world? Roughly, it is for the sentence to be one which would have been true, had the world in question been actual. Thus had snow been black, "snow is black" would have been true. In other words, "snow is black" is true with respect to a world in which snow is black.

This suggests a wholesale revision of the semantics applicable to **P**. An *interpretation of PN* will be a distribution of truth values to all sentence-letters with respect to each world. That is, a **PN**-interpretation fixes for each sentence-letter, and with respect to each world, the truth value of the

† See Exercise 132, page 363.

sentence-letter at that world. This corresponds to the intuitive idea that the meaning of a sentence determines, among other things, in what circumstances it would be true, and in what circumstances it would be false. The interpretation rules likewise require that truth be relativized to a world as well as to an interpretation. Our original rules from chapter **2.1** have to be rewritten; one way to do so is as follows:

1) For any set of worlds, W, any world, w, in W, and any interpretation i of **PN**,
    $\neg X$ is true at w upon i iff X is false at w upon i;
    $(X \& Y)$ is true at w upon i iff X is true at w upon i and Y is true at w upon i;
    $(X \vee Y)$ is true at w upon i iff X is true at w upon i or Y is true at w upon i;
    $(X \rightarrow Y)$ is true at w upon i iff X is false at w upon i or Y is true at w upon i;
    $(X \leftrightarrow Y)$ is true at w upon i if either both X and Y are true at w upon i or both X and Y are false at w upon i.
    $\Box X$ is true at w upon i iff for every world w' in W, X is true at w' upon i.

We can add a symbol for "possibly" by the following definition:

2) $\Diamond X =_{df} \neg \Box \neg X$.

"$\Diamond$" is pronounced "diamond". Validity is defined as follows:

3) $X_1, \ldots X_n \vDash_{PN} Y$ iff: for all interpretations, i, and all sets of worlds, W, if, for any world w in W, all of $X_1, \ldots X_n$ are true at w upon i, Y is true at w upon i.

These semantics determine a system of propositional modal logic known as S5.[1] In the rest of this section we will explore its properties. At the end of the section I will briefly indicate how the semantics can be varied to give alternative propositional modal logics.

Consider the following argument:

4) Possibly, no one will come. Possibly, many people will come. Therefore it's possible that both no one and many people will come.

The argument is plainly invalid, rather in the way that

> 5) Some numbers are odd and some are even, therefore some numbers are both odd and even

is invalid.

With "p" corresponding to "no one will come" and "q" to "many people will come", the obvious **PN**-formalization of (4) is

> 6) $\Diamond p, \Diamond q; \Diamond(p \,\&\, q)$.

To show that (6) is **PN**-invalid, we need to find an interpretation and a set of worlds such that the premises are true at a world in the set upon the interpretation, whereas the conclusion is false at that world upon the interpretation. It will be useful to spell out, from the definition of "$\Diamond$", together with the interpretation rule for "□", the derived interpretation rule for "$\Diamond$":

> 7) $\Diamond X$ is true at w upon i iff for some world w' in W, X is true at w' upon i.[†]

Let i assign truth to "p" and falsehood to "q" with respect to w_1, falsehood to "p" and truth to "q" with respect to w_2; and falsehood to both "p" and "q" with respect to all other worlds in W. At w_1, "$\Diamond p$" is true upon i, since there is a world, viz. w_1, at which "p" is true upon i; and at w_1, "$\Diamond q$" is true upon i, since there is a world, viz. w_2, at which "q" is true upon i; but at w_1, "$\Diamond(p \,\&\, q)$" is false upon i, since there is no world at which "$p \,\&\, q$" is true upon i. Hence i is a counterexample to the validity of (6). The essential point is structurally like what needs to be said to identify the fallacy of (5): that there is a world at which "p" is true and one at which "q" is true does not entail that there is one at which both are true.

The point can be made diagrammatically. Let W consist of just w_1 and w_2, and let us represent them by ellipses. The truth values assigned to the relevant sentence-letters by some interpretation are represented in the ellipses by writing the sentence letter itself, if the letter is assigned true by the interpretation at that world, or its negation if it is assigned false by the interpretation at that world. Below we add relevant complex sentences whose truth upon the interpretation follows from the assignments to the sentence letters together with the features of the worlds. The first two

† See Exercise 133, page 363.

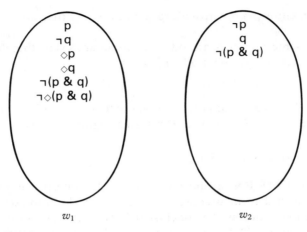

Figure 5.1

lines in each ellipse in figure 5.1 represent the assignments made by the interpretation in question to "p" and "q" at the worlds w_1 and w_2. The justification for adding "$\Diamond p$" to w_1 is provided by the presence of "p" somewhere in the diagram; likewise for the justification for adding "$\Diamond q$". The justification for adding "$\neg \Diamond (p \& q)$" is that no world in W contains "$p \& q$".[†]

The following is also invalid:

    8) $p, \Box(p \rightarrow q); \Box q$.[‡]

This can be established by the diagram (figure 5.2) of the set W of worlds containing just w_1 and w_2. The justification for adding "$\neg \Box q$" to w_1 is that "q" does not appear in every world in the diagram, corresponding to the fact that, at some world, "q" is false upon the interpretation represented.

In contrast to (**8**), the following is a cardinal principle of modal logic:

    9) $\Box X, \Box(X \rightarrow Y) \vDash_{\mathbf{PN}} \Box Y$.

The truth of (**9**) can be established informally by reflecting that an interpretation upon which both premises are true at some world in some arbitrary set of worlds, W, assigns truth at *every* world in W to both X and

† See Exercise 134, page 363.   ‡ See Exercise 135, page 363.

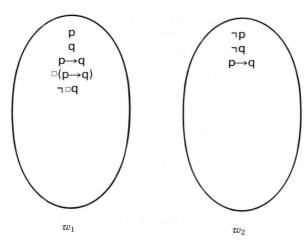

Figure 5.2

$X \to Y$, and so, by the interpretation rule for "→", must also assign truth to Y at every world in **W**. Hence for any set of worlds, **W**, and any world in **W**, any interpretation upon which the premises are true at that world is one upon which the conclusion is true at that world.

Since the interpretation rules of **P** are mirrored in those for **PN**, we have:

10) If [⊨**P** X] then [⊨**PN** X].

In other words, any **P**-valid sentence is also a **PN**-valid sentence. We also have the stronger

11) If [⊨**P** X] then [⊨**PN** □X].

This says that the result of prefixing a **P**-valid sentence by a box is **PN**-valid. It reflects the thought that a valid **P**-sentence corresponds to a necessary truth. If ⊨**P** is enough for ⊨, in the way argued in chapter **2.10**, and box corresponds to "it is logically necessary that", then (**11**) must hold. For "⊨A" is by abbreviation equivalent to "it is logically necessary that A" (cf. chapter **1.6**).

The intuitively natural thought that what is necessarily true (at any world) is true (at that world) is verified by the interpretation rule for "□". It can be expressed by the truth of the generalization:

12) ⊨**PN** □$X \to X$.

This says that every conditional whose antecedent consists of its consequent preceded by "$\Box$" is **PN**-valid.

By the definition of "$\Diamond$", (12) ensures

13) $\vDash_{PN} \neg \Diamond \neg X \rightarrow X$

and thus, by contraposition and the equivalence of X with $\neg \neg X$:

14) $\vDash_{PN} \neg X \rightarrow \Diamond \neg X.$

Since X stands for any formula, this entails

15) $\vDash_{PN} X \rightarrow \Diamond X.$[2]

This accords with the intuition that anything actual is possible: what is in fact true *can* be true.

More controversially, **PN** has it that

16) $\vDash_{PN} \Box X \rightarrow \Box \Box X.$

This says that any conditional is valid in **PN** if its antecedent starts with an occurrence of "$\Box$" and its consequent consists of the antecedent prefixed by an occurrence of "$\Box$". This corresponds to the claim that anything that is necessarily true is of necessity necessarily true. It is no accident which truths are necessary.

It is not entirely uncontroversial that (16) accurately reflects our intuitive views about necessity. Suppose, for example, that necessity is not an objective feature of the world, but is a product of human thinking. Suppose, further, that our patterns of thinking are determined by evolutionary pressures, but in a way that is not necessary. (Random mutations might be involved.) Then we do not *necessarily* think as we do, so something which is in fact a necessary truth might not have been. I do not endorse this line of thought, but I do show later how those who regard (16) as inappropriate to an account of "necessarily" can rewrite the interpretation rule for "$\Box$" to reflect their views.

There is also in **PN** a stronger version of (15):

17) $\vDash_{PN} X \rightarrow \Box \Diamond X.$

An interpretation upon which X is true at a world, w, is one upon which $\Diamond X$ is true at that world and every other world, and so is one upon which $\Box \Diamond X$ is true at that world.

Again, it is not uncontroversial that this correctly reflects our modal notions. You might agree that the fact that something is true guarantees that it is in fact possible, but not agree that it had to be possible. This is certainly correct for some restricted notions of necessity. For example, it is true (let us suppose) that you will catch the train, and this shows that it is possible for you to catch it. But it did not have to be possible: something could have happened to delay you, and then it would not have been possible for you to catch it (even though you will in fact catch it). So we have both "You will catch the train" and "It is not necessarily possible for you to catch the train", contrary to (17). The modality here, however, is related to time, and it would seem that the claim that it did not have to be possible for you to catch the train relates to an earlier time than the claims that you will catch it and that it is possible for you to catch it.

A further fact about **PN** which might be regarded as only dubiously appropriate to our intuitions about modality is that

18) $\vDash_{\textbf{PN}} \Diamond \Diamond X \rightarrow \Diamond X.$

This holds because for any world w, any interpretation upon which the antecedent is true at w is one upon which $\Diamond X$ is true at some world, and thus one upon which X is true at some world; this last condition is enough to ensure that $\Diamond X$ is true at w.

Do we intuitively believe that if something is possibly possible it is actually possible? A case for a negative answer is as follows. This table on which I am writing, call it α_0, could not have been made out of entirely different parts, for a table made out of entirely different parts would not have been α_0. However, it could have been made out of slightly different parts. Let us express this as follows: a table, α_1, made out of slightly different parts from α_0 would have been α_0. We will no doubt also hold that α_1 could have been made out of parts differing from the parts it is made of. Call such a table α_2. If we have chosen the degree of difference appropriately, we may reasonably deny that α_0 could have been constructed out of the components of which α_2 is constructed. We might be prepared to affirm that *if*, as we allow is possible, α_0 had been constructed out of α_1's parts, then it *could* have been constructed out of α_2's parts; yet deny that α_0 could have been constructed out of α_2's parts. We might express this by the combination of claims, inconsistent with (18):

19) $\Diamond \Diamond (\alpha_0$ is constructed out of α_2's parts).
         $\neg \Diamond (\alpha_0$ is constructed out of α_2's parts).

PN has the following interesting property, called "modal collapse": given a non-modal sentence, X, that is, one containing no boxes or diamonds, there are only two non-equivalent fresh sentences you can form from it just by adding modal operators. They are, simply, $\Box X$ and $\Diamond X$. However many other boxes and diamonds you may stick in front of X, in whatever order, the result will be equivalent to one of these two sentences. To take two examples, it is easy to see, from (**12**) and (**16**), that

20) $\vDash_{\mathbf{PN}} \Box\Box X \leftrightarrow \Box X,$

which means that any pair of boxes can be collapsed to a single box. In addition,

21) $\vDash_{\mathbf{PN}} \Box\Diamond X \leftrightarrow \Diamond X,$

which shows that a box followed by a diamond can be collapsed to just a diamond.[†]

Those who think that some **PN** validities are inappropriate to our ordinary conception of modality can devise weaker modal languages by restricting in various way the worlds which are relevant to truth-upon-an-interpretation. The standard way to do this is to introduce a relation between worlds, called the "accessibility" relation, here abbreviated to "R". The rule for box will be revised to

22) $\Box X$ is true at w upon i iff for every world w' in W such that $Rw'w$, X is true at w' upon i.

Only worlds R-related to the world of evaluation will be relevant to the truth-upon-an-interpretation of a necessitation.

The original rule for box is equivalent to letting R be an entirely unrestricting relation, so that any pair of worlds are R-related.

Instances of (**12**) hold only if every world is R-related to itself. If you want a language in which (**12**) does not hold, stipulate that some worlds are not R-related to themselves (that R is non-reflexive).

Instances of (**16**) hold only if R is transitive, that is, only if for all worlds w_1, w_2, and w_3 such that w_1 is R-related to w_2 and w_2 to w_3, w_1 is R-related to w_3. For a set of worlds containing three worlds lacking this property, we can construct, as in figure 5.3, a counterexample to (**16**). Here arrows represent what the worlds are R-related to. Thus we have Rw_1w_2 and

† See Exercise 136, page 363.

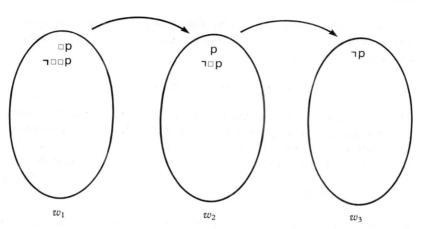

w_1 w_2 w_3

Figure 5.3

Rw_2w_3, and no other instances of the R-relation. Notice that the description of w_1 as containing "$\Box p$", on the interpretation represented in the diagram, would be inconsistent if w_1 were R-related to w_3.[†]

One could restrict the R-relation in different kinds of ways if one wanted to represent more limited kinds of necessity. For example, it is plausible to say that it is morally necessary that A iff A is true in all morally perfect worlds, those in which what morality requires obtains. One could use the framework provided here to construct an appropriate language, stipulating that $Rw'w$ holds iff the requirements of morality which obtain at w are satisfied in w'.

We will not pursue these more restrictive notions, keeping to the broadest conception of necessity. Notice, however, that we now have some account of what this broadness consists in.[‡]

*2   Subjunctive conditionals

We saw in chapter 2 that a distinction could be drawn between indicative and subjunctive conditionals. A rough guide is that a subjunctive conditional is one expressed by means of the subjunctive mood, e.g.:

   1)   If Oswald hadn't shot Kennedy, someone else would have done.

† See Exercise 137, page 363.   ‡ See Exercise 138, page 363.

We also noted that there is a case for saying that some conditionals expressed in the indicative might well be best classified with subjunctive ones. Our example was (**2.4.39**):

> If John dies before Joan, she will inherit the lot.

It seems clear that subjunctive conditionals belong with modal idioms like "necessarily". It might even seem that **PN** offers the resources for adequately formalizing such conditionals. Can they not be rendered as necessitated material implications? Then (**1**) would be formalized

$$\textbf{2)} \quad \Box(\neg p \to q)$$

with "p" corresponding to "Oswald shot Kennedy" and "q" to "Someone else shot Kennedy". (We will ignore the fact that **PN** obviously cannot express the quantificational structure of (**1**).)

In favour of the formalization, you might argue that all possible worlds divide into two classes: those in which Oswald did shoot Kennedy, and those in which he did not. "$\neg p \to q$" is true (upon an intended interpretation) at all members of the first class of worlds, which will include the actual world (on the assumption of Oswald's guilt), in virtue of the falsity of the antecedent; and if (**1**) is true, it is natural to suppose that "$\neg p \to q$" is true (upon an intended interpretation) at every world in the second class as well. Part of the point of (**1**) seems to be to say that any world in which Oswald didn't shoot Kennedy is one in which someone else did.

However, it soon becomes apparent that (**2**) is much too demanding a formalization of (**1**). For example, (**1**) is not falsified by a world in which Kennedy never existed, nor by a world in which he never came before the public eye, and so was not a target for assassination, nor by a world in which the most stringent security precautions were invariably taken.

The defect may seem easy to remedy. Let us restrict the worlds that are relevant to the truth of a formalization of (**1**) to worlds that are similar, in certain contextually determined respects, to the actual one. For (**1**), the only relevant worlds are those in which Kennedy exists, became President, and was not always subject to the most stringent security precautions.

No **PN**-operator carries this restriction to similar worlds. However, let us use "$\Box\!\!\rightarrow$" to formalize subjunctive conditionals, introduce a special interpretation rule for them, and use "**PNS**" to stand for the result of adding this symbol and this rule to **PN**.

The syntax of "$\Box\!\!\rightarrow$" (pronounced "box arrow") is that it takes two (indicative) sentences to form a sentence. Thus with the correspondences of (2), (1) will be formalized:

3) $p\,\Box\!\!\rightarrow q$.

The appropriate interpretation rule for "$\Box\!\!\rightarrow$", on the basis of the present suggestion, is:

4) $X\,\Box\!\!\rightarrow Y$ is true at w upon i iff for every world w' in W which is similar to w, $X\rightarrow Y$ is true at w' upon i.

This version of **PNS** will validate intuitively invalid inferences using subjunctive conditionals. For example, we have

5) $X\,\Box\!\!\rightarrow Y \vDash_{\textbf{PNS}} (X\ \&\ Z)\,\Box\!\!\rightarrow Y$.

The premise is true at a world, w, upon an interpretation, i, iff $X\rightarrow Y$ is true at all worlds in W which are similar to w. But any world at which $X\rightarrow Y$ is true upon i is one at which $(X\ \&\ Z)\rightarrow Y$ is true upon i. So the conclusion is true at w upon i. However, we do not accept that arguments like the following are valid (compare (**2.5.28**)):

6)   If I had put sugar in this cup of coffee, it would have tasted good. So if I had put sugar and diesel oil in this cup of coffee, it would have tasted good.

Reflection on this example suggests that there cannot be a fixed standard of similarity of worlds, applicable to all subjunctive conditionals. For example, considering evaluations with respect to the actual world, if we set the standard high, then "(I put sugar and diesel oil in this cup of coffee)→(it tastes good)" is true at every similar world, since every world which is *very* similar to ours is one in which the antecedent is false. This high standard would formalize the conclusion of (6) as true upon an intended interpretation, which is not what we want. If we set the standard low, then "(I put sugar in this cup of coffee)→(it tastes good)" is false at some similar world, for example a world in which diesel is added as well as sugar. This low standard would formalize the premise of (6) as false upon an intended interpretation, which is not what we want. What is needed, therefore, is variability in how great the similarity must be, as a function of the content of the antecedent.[†]

One theory which has this property is Stalnaker's, discussed in chapter **3**. In our current terminology, he suggests that the rule for "□→" should be:

> 7) X□→Y is true at w upon i iff Y is true upon i at the most similar world to w at which X is true upon i, if there is one.

In chapter **3.3**, we noted one of David Lewis's objections to this proposal: that (7) incorrectly assumes that there will never be more than one most similar X-world (i.e. world at which X is true). Recall, however, the competing subjunctives:

> 8)   If Bizet and Verdi had been compatriots, Bizet would have been Italian
>
> 9)   If Bizet and Verdi had been compatriots, Verdi would have been French.

A world in which both composers are French is as similar to ours as a world in which both composers are Italian. Hence there is no such thing as *the* world at which "Bizet and Verdi are compatriots" is true which is most similar to the actual world. This verifies the standard reading of the right hand side of the relevant instance of the biconditional in (7): it is itself a conditional, and has a false antecedent. So (8) and (9) and, for that matter,

> 10)   If Bizet and Verdi had been compatriots they would have been Ukrainian,

will all be **PNS**-formalized as true upon an intended interpretation. Since these sentences are intuitively not true, the formalizations are inadequate.[3]

Lewis offers an analysis of counterfactuals which does not suffer from this alleged defect. Adapted to our terminology, it would give the following interpretation rule for "□→":

> 11) X□→Y is true at w upon i iff some world at which X & Y is true upon i is more similar to w than any world at which X & ¬ Y is true upon i, if there are any worlds at which X is true.

† See Exercise 139, page 364.

This makes formalizations of (8)–(10) false upon an intended interpretation, which seems acceptable.[†]

The account has various merits of structure and detail. One pleasing feature is that it makes room for a unified account of "if". The material conditional features in the analysis of the subjunctive conditional.[‡] So if we could accept the material conditional analysis of indicative conditionals, we would see a unity between all English conditionals: either they simply are material conditionals, or else they are material conditionals embedded in a modal construction, expressed in English by the subjunctive mood. Further, like Stalnaker's account, it leaves room for indeterminacy: context may settle the required standards of similarity in different ways on different occasions, and this corresponds well to the context sensitivity of many subjunctive conditionals.[¶]

The account is also able to explain why certain intuitively invalid arguments using subjunctive conditionals are invalid, for example:

12)   If Hoover had been born a Russian, he would have been a communist.
        If he had been a communist, he would have been a traitor.
        So if he had been born a Russian, he would have been a traitor.

Formalizing this as

13) $p \, \square\!\!\rightarrow q, \; q \, \square\!\!\rightarrow r; \; p \, \square\!\!\rightarrow r$

with obvious correspondences, we can see that the fallacy has something in common with that noted in (1.4) and (1.5). Upon an intended interpretation with respect to the actual world, there are some $p \, \& \, q$ worlds more similar to ours than any $p \, \& \, \neg q$ world, and some $q \, \& \, r$ worlds more similar to ours than any $q \, \& \, \neg r$ world; but this does not guarantee that there is a $p \, \& \, r$ world more similar to ours than any $p \, \& \, \neg r$ world.[‖]

One feature of Lewis's theory which has attracted criticism is that it entails that inserting "$\square\!\!\rightarrow$" between truths yields a truth. Suppose that Jones had a narrow escape while driving along a mountainous road in icy weather. He approached a corner too fast, realized his mistake, and slammed on the brakes. Luckily for him, the brakes failed: had they not failed, he would have skidded out of control and crashed. As things were,

† See Exercise 140, page 364.   ‡ See Exercise 141, page 364.
¶ See Exercise 142, page 364.   ‖ See Exercise 143, page 364.

he was able to take an escape road, which he had not noticed when he applied the brakes, and so stop safely. Lewis's theory verifies:

> 14) If Jones had slammed on the brakes, he would not have crashed.

But one may feel that (14) is not true: under the circumstances, slamming on the brakes was very likely to produce a crash, and failed to do so only thanks to a freakish intervention.[4]

The objection is not decisive, for even if it is agreed on all hands that there is something unsatisfactory about (14), it would seem quite unclear whether this amounts to its being false, or, rather, true but misleading. The misleadingness could be analysed as being based on two general features of subjunctive conditionals. First, they implicate that the indicative corresponding to the antecedent is false; secondly, they implicate that there is a connection, typically causal, between what would have made the antecedent true and what would make the consequent true. No intuition appears to guide the choice between treating (14) as false and treating it as true but misleading, so we could well agree with Lewis that the choice should be guided by theory instead, in particular, by Lewis's theory.

Another criticism of the theory is that if we use intuitive accounts of similarity it yields intuitively incorrect assignments of truth value. Suppose we believe that Oswald was acting alone, and so we deny (1). Compare it with:

> 15) If Oswald hadn't shot Kennedy, Kennedy would not have been shot.

This we should accept (given our assumptions). However, a world in which Oswald didn't shoot Kennedy, and no one else did either, is extremely unlike the actual world: all the repercussions of the assassination would be absent from it. A world in which Oswald didn't shoot Kennedy but someone else, acting independently, did shoot Kennedy, would be much more similar to our world. So it looks as if Lewis is committed to holding that (15) is false.

This suggests a genuine problem of detail, a problem indeed, to which Lewis gives some attention. The problem is that of specifying the ways in which different similarities and dissimilarities count when assessing a subjunctive conditional. For example, one might propose disregarding dissimilarities occurring after the time of an event verifying the antecedent, and this would allow one to hold that (15) is true: on our assumptions, a world in which, say, Oswald fired and missed, so that Kennedy was not shot, is more like ours up to the time of the firing than one in which some other assassin shot Kennedy.[5][†]

3 Adding "□" to Q

QN results from adding box to **Q**, treating it as having the syntactical properties of tilde. This gives rise not only to sentences like

    1) $\Box \forall x(Fx \rightarrow Fx)$

which, with "F" corresponding to "is a mathematician", seems a reasonable formalization of

    2) Necessarily, all mathematicians are mathematicians;

but also to sentences like

    3) $\forall x(Fx \rightarrow \Box Fx)$

which, with correspondences as before, would formalize

    4) All mathematicians are necessarily mathematicians.

Perhaps (4) is ambiguous, but the reading formalized by (3) intuitively strikes one as false. People who become mathematicians don't have to turn to mathematics. In western systems of education, there is an element of choice. Consider, for example, Cantor. He didn't *have* to become a mathematician (though in fact he was one). He might have died in infancy, and so become nothing at all. He might have met at an early age a charismatic botanist, and followed in his footsteps. So Cantor constitutes a counterexample to the truth of (4).

If we try to apply our earlier interpretation rules to (3), we can see roughly what should happen. Some interpretation of "$F\alpha \rightarrow \Box F\alpha$", which assigns to "$F$" the set of all 1-membered sequences whose members are mathematicians, will assign Cantor to "α"; on this interpretation the

† See Exercise 144, page 365.

antecedent of the conditional is true with respect to the actual world and the consequent false, so the conditional is false with respect to the actual world; hence upon no interpretation agreeing with this one on the assignment to "*F*" is (**3**) true with respect to the actual world. This story requires a necessitation of the form "$\Box F\alpha$" to be true or false upon an interpretation, with respect to a world.

To arrive at **QN**, the interpretation rules for **Q** need to be revised rather as those for **P** were revised to obtain **PN**. In that case, we were guided by the thought that one approaches an interpretation of a sentence by specifying the worlds in which it is true. Extending that thought to **Q**-constructions, one would approach the interpretation of a predicate by specifying, for each world, the things of which it is true at that world, and of a name by specifying, for each world, its bearer at that world.

There are various ways in which this can be done: later we shall indicate some roads not taken. First, we need to relativize interpretations to worlds. We stipulate that for each world w there is a (possibly empty)[6] domain D^w (the entities in w). In chapter **4**, we used expressions like "$i(F)$" to designate the set assigned by a **Q**-interpretation, i, to the predicate letter F. Here we extend that notation, using expressions of the form "$i_w(X)$" to designate what a **QN**-interpretation, i, assigns to an expression, X, with respect to a world, w.

5) For any set of worlds, W, any world, w, in W, and any interpretation i of **QN**:

for each sentence-letter P, $i_w(P)$ is a truth value, either T or F;

for each name-letter n, for some world w', $i_{w'}(n)$ is an object, α, in $D^{w'}$ and for every world, w'', if α belongs to $D^{w''}$, $i_{w''}(n)$ is α and if α does not belong to $D^{w''}$ there is no $i_{w''}(n)$ (e.g. "α" might be assigned Ronald Reagan with respect to any world at which Reagan exists, and assigned nothing with respect to the other worlds);

for each predicate-letter ϕ of degree n, $i_w(\phi)$ is a set of n-tuples (n-membered sequences) all of whose members belong to D^w (e.g. $i_w(F)$ might be a set of ordered pairs such that the first member of the pair loves the second in w);

$i_w(=)$ is the set of ordered pairs of members of D^w such that in each pair the first object is the same thing as the second in w.

An intended interpretation of a **QN**-sentence will assign to each name-letter in the sentence, with respect to each world w, the actual bearer of the corresponding name, if that thing exists at w, and otherwise will assign nothing to the letter with respect to w; it will assign to each n-ary predicate-letter in the sentence, with respect to each world w, those n-tuples of members of D^w which possess the property associated with the corresponding predicate.

6)   For any set of worlds, W, any world, w, in W, and any interpretation i of **QN**,

   (i) $i_w(\neg X)$ is T iff $i_w(X)$ is F.
   (ii) $i_w(X \ \& \ Y)$ is T iff $i_w(X)$ is T and $i_w(Y)$ is T.
   (iii) $i_w(X \vee Y)$ is T iff $i_w(X)$ is T or $i_w(Y)$ is T.
   (iv) $i_w(X \rightarrow Y)$ is T iff $i_w(X)$ is F or $i_w(Y)$ is T.
   (v) $i_w(\phi n_1 \ldots n_k)$ is T iff all of $i_w(n_1) \ldots i_w(n_k)$ belong to D^w and $\langle i_w(n_1) \ldots i_w(n_k) \rangle$ belongs to $i_w(\phi)$; otherwise $i_w(\phi n_1 \ldots n_k)$ is F.[7]
   (vi) $i_w(\forall v \ X)$ is T iff for all i', for the name letter n as specified in $X^n_{\bar{v}}$, if there is an assignment $i'_w(n)$ and i' agrees with i on everything except, perhaps, this assignment, then $i'_w(X^n_{\bar{v}})$ is T.
   (vii) $i_w(\exists v \ X)$ is T iff for some i', for the name letter n as specified in $X^n_{\bar{v}}$, there is an assignment $i'_w(n)$ and i' agrees with i on everything except, perhaps, this assignment, and $i'_w(X^n_{\bar{v}})$ is T.
   (viii) $i_w(\Box X)$ is T iff for all worlds w' in W, $i_w(X)$ is T.

QN-validity can be identified exactly like **PN**-validity. What makes the difference is the richer notion of interpretation for **QN** as compared with **PN**.

Is (viii) appropriate to the ordinary notion of necessity? We have given an example in which a sentence of the form "$\Box F \alpha$", "Necessarily, Cantor was a mathematician", is false. Are there any examples of truths of this form? One candidate is

7)   Necessarily, Socrates is human.

You might well think that this is true: anything non-human – a stone or a crocodile – simply could not be Socrates. Even if there is room for doubt on this point, it would seem that an appropriate language for necessity should not preclude the truth of a sentence like (7). However, the **QN**-formalization of (7) as "$\Box F \alpha$" will not be true upon an intended interpretation, i. With respect to each world, w, i assigns to "F" the set of all 1-membered sequences each having a human member at w; i assigns Socrates to "α" with respect to every world at which Socrates exists. Accordingly, for some reasonably rich set of worlds W, i does not assign truth with respect to every world to "$F\alpha$", for we all agree that Socrates might not have existed (and would not have done if the world had ended 10 years before the year in which he was born), so that there are worlds with respect to which i assigns nothing to "α", hence worlds at which "$F\alpha$" is false upon i, so "$\square F\alpha$" is false upon i.

There are at least two ways of responding to this difficulty. One involves weakening the interpretation rule for $\square$ so that $\square X$ is true on an interpretation i iff X is true on i at every world at which all the objects i assigns to any name-letters in X exist. Then we could formalize (7) straightforwardly, as "$\square F\alpha$", and it would not be false on an intended interpretation merely in virtue of the fact that Socrates might not have existed.

Alternatively, we could keep to the original interpretation rule for $\square$, and formalize sentences like (7) as

8) $\square((\exists x\; x = \alpha) \to F\alpha)$

with "F" corresponding to "is human". The semantics for **QN** do not preclude (8) being true upon an intended interpretation. Worlds at which Socrates does not exist will be worlds at which "$(\exists x\; x = \alpha) \to F\alpha$" is true upon an intended interpretation in virtue of the falsity of the antecedent at such worlds.

The first response introduces the so-called "weak" interpretation of necessity; the interpretation rule (6viii) expresses the "strong" interpretation.

Necessitated existential sentences themselves provide a reason for preferring the strong interpretation, and thus for adopting the more complex formalizations, in the style of (8), of sentences like (7). We might debate whether the following is true:

9)   Necessarily, the number 7 exists.

It is clear how we could formalize this using strong necessity, in such a way that the question of its truth upon an intended interpretation is not foreclosed by the formalization:

10) $\square \exists x\; x = \alpha$

with "α" corresponding to "the number 7".[8] But if we use this formalization and interpret box as expressing weak necessity, (10) will express a trivial truth upon an intended interpretation: it will in effect say that α exists at every world at which it exists.

On the strong interpretation of box, (10) is false upon some interpretations at some worlds and so is not trivial. Consider an interpretation assigning Socrates to "α". "$\exists x \, x = \alpha$" will be false upon this interpretation at a world at which Socrates does not exist, so upon this interpretation (10) will be false (at every world).

This virtue of the strong interpretation carries with it a source of worry. For the fact that (10) is not **QN**-valid means that we cannot accept for **QN** the correlate of (1.11) for **PN**. That is, the following is false:

11)   If $\vDash_Q X$ then $\vDash_{QN} \Box X$.

For, as we noted in connection with (4.11.20), $\vDash_Q \exists x \, x = \alpha$. Given the philosophical motivation for (1.11), this should be genuinely disturbing. It would not be satisfactory to try to reinstate (11) by insisting that every name latter is always assigned something by every interpretation with respect to every world, for then, intuitively, a name letter cannot adequately formalize a name for a contingent being, one which actually exists but which might not have done.[†]

This raises deep issues about the role of names.[9] At the more superficial level, the strong version of necessity has the advantage noted of giving natural formalizations both of claims like (7) and claims like (9), so I shall persist with it.

A sentence of the form "$\Box F\alpha$" can be read as ascribing a property to an object: ascribing to α the property of being necessarily F. Sentences like (3) speak generally of things being necessarily thus-and-so. For a number of different reasons, philosophers have held that such ways of talking are illegitimate. If they are right, then it is bootless to investigate much further the properties of **QN**, since it is committed to this supposedly illegitimate way of talking.[‡]

4   Necessity *de re* and *de dicto*

Let us say that a sentence expresses "necessity *de re*" iff it is adequately **QN**-formalizable by a sentence in which either there is a name within the

† See Exercise 145, page 365.   ‡ See Exercise 146, page 365.

scope of some modal operator or a modal operator within the scope of a quantifier. Let us say that a sentence expresses "necessity *de dicto*" just on condition that it expresses necessity but does not express necessity de re. The doubts alluded to at the end of the last section are doubts concerning the coherence of the notion of de re necessity.

As I have defined the difference between de re and de dicto necessity, it is simply a matter of scope. The definitions do not introduce two concepts of necessity. We could with our available resources make an exactly parallel distinction between "negation de re" and "negation de dicto", but the distinction would give no support to the thought that there are two distinct concepts of negation.[†]

However, many philosophers associate the distinction between de re and dicto necessity with some or all of the following further theses:

1) In ascribing de re necessity we attribute a property to a non-linguistic object; in ascribing de dicto necessity we attribute a property to a sentence.

2) A de re necessary truth records how things are in the world; a de dicto necessary truth records only linguistic facts.

3) Whether a de re necessary sentence is true depends on which objects exist at which worlds.

One who believes that there are de re necessary truths is called an "essentialist". Optionally, he may also subscribe to some or all of the above theses. Attacks on the intelligibility of de re necessity are thus sometimes expressed as attacks on essentialism. In the following three sections, I shall consider three such attacks.

5 The number of the planets

Some years ago, Quine advanced two related arguments against essentialism.

First, he asked us to consider the following argument:

1) 9 is necessarily greater than 7
    9 = the number of the planets
    Therefore, the number of the planets is necessarily greater than 7.

† See Exercise 147, page 366.

Quine claimed that the argument has true premises and a false conclusion, yet that it ought to be valid if essentialism is true. For if essentialism is true, the first premise will constitute the ascription to an object (the number 9) of a property, that of being necessarily greater than 7, and the conclusion will constitute the ascription of the same property to the same object.

Notice that Quine does not use the distinction between de re and de dicto given in §4. Rather he identifies a de re statement by a claim of (4.1): it is one which ascribes a property to an object. However, a sentence judged by this last standard to be one of de re necessity is also de re by our official standard. If the premise of (1) ascribes the property of being necessarily greater than 7 to the number 9, then it is appropriately **QN** formalizable as:

2) $\Box F\alpha$.

Hence we must follow Quine in counting it de re.

We can also agree with Quine that an essentialist may view (1) as sound. The following formalization is **QN**-valid:

3) $\Box F\alpha, \alpha = \beta; \Box F\beta$.

The correspondence scheme is: "F" to "is greater than 7", "α" to "9" and "β" to "the number of the planets". The essentialist will allow that the premises are true upon an intended interpretation. So he must allow that the conclusion is also true upon an intended interpretation. So he should hold that the conclusion of (1) is true. Why does Quine say that it is false?

His ground is that there might not have been nine planets. If there had been only 5, then the number which numbers them would not have been greater than 7, let alone necessarily greater than 7. Although this is correct, it is not inconsistent with the truth upon an intended interpretation of the conclusion of (3). Let us see what an intended interpretation, i, assigns to "β", "F" and "$F\beta$" with respect to a world, w, at which there are just 5 planets. First, $i_w(\beta)$ must be the very object that i assigns to "β" with respect to the actual world, viz. the number 9 (assuming it belongs to D^w).[10] $i_w(F)$ must be the set of all 1-membered sequences whose member belongs to D^w and, at w, is greater than 7. $\langle 9 \rangle$ will thus be a member of $i_w(F)$, and so $i_w(F\beta)$ is T. That there are worlds in which there are only 5 planets is thus irrelevant to the truth of the conclusion of (1), as it will be understood by one who takes the argument to be valid.

The conclusion of (1) is, perhaps, ambiguous. If (1) is a valid argument, its conclusion must be read in the de re fashion brought out by the formalization (3). The de dicto reading of the conclusion would be formalized:

$$\Box\exists x\ (Gx\ \&\ \forall y\ (Gy\rightarrow x=y)\ \&\ Fx)$$

with "G" corresponding to "numbers the planets". For the argument to have a chance of being valid, we must make an analogous adjustment to the formalization of the phrase "the number of the planets" as it occurs in the premise, so that the overall formalization would be:

4) $\Box F\alpha,\ \exists x\ (Gx\ \&\ \forall y\ (Gy\rightarrow x=y)\ \&\ x=\alpha);$
     $\Box\exists x\ (Gx\ \&\ \forall y\ (Gy\rightarrow x=y)\ \&\ Fx).$

(4) is invalid: the premises do not ensure even that every world has a unique G, let alone one which is F. However, the essentialist has no more reason than anyone else to suppose that (4) is valid or has a true conclusion.

The failure of Quine's argument against essentialism can be summarized as follows: if the conclusion of (1) is read in such a way that it is false, the essentialist has no more reason than the rest of us for supposing the argument to be valid. If the conclusion of (1) is read in such a way that the argument is valid, then no reason has been given to suppose that the conclusion is false.

Quine used another argument to make a similar point, and a similar criticism applies.[11] He says that essentialists will subscribe to the following:

5)   All bachelors are necessarily unmarried but not necessarily army officers.

6)   All majors are necessarily army officers but not necessarily unmarried.

Consider Major Smith, who is both a bachelor and a major: qua bachelor, he is not necessarily an army officer, but qua major he is necessarily one; qua bachelor he is necessarily unmarried, but qua major he is not necessarily unmarried. So there is no question of an *object* possessing a property necessarily or contingently: what matters is how you refer to the object. Hence essentialism – the claim that necessity attaches to objects – is false.

Appropriate **QN** formalizations of (5) and (6) are all de dicto:

7) $\Box\forall x(Fx\to Gx)$ & $\neg\,\Box\forall x(Fx\to Hx)$

8) $\Box\forall x(\mathcal{J}x\to Hx)$ & $\neg\,\Box\forall x(\mathcal{J}x\to Gx)$

with "F" corresponding to "is a bachelor", "G" to "is unmarried", "H" to "is an army officer" and "$\mathcal{J}$" to "is a major". However, absolutely no conclusions of the de re form "$\Box G\alpha$" or "$\neg\,\Box H\alpha$" follow from (7) or (8) or their conjunction. The essentialist should say that Major Smith is neither necessarily unmarried nor necessarily an army officer. The crucial point is that the following is consistent – that is, true on at least one interpretation:

9) $\Box\forall x\,(Fx\to Gx)$ & $F\alpha$ & $\neg\,\Box G\alpha$.

Bachelors don't have to be bachelors (denial of de re necessity), even though, necessarily, everyone who is one is unmarried (affirmation of related de dicto necessity).†

Quine's arguments do indeed show that not all necessity is de re. However, no one should deny this. Essentialism is not the claim that all necessity is de re, but only that some is.

6 "Frege's argument"

An argument has been ascribed to Frege which, if sound, would establish the absurdity of de re necessity. Without committing myself to the accuracy of the ascription, I shall refer to it as "Frege's argument".[12]

De re necessities formalizable as "$\Box F\alpha$" satisfy the substitutivity of identicals: in conjunction with a premise formalizable as "$\alpha=\beta$" you validly get a conclusion formalizable as "$\Box F\beta$". Moreover, "$\Box$" is plainly a non-truth functional sentence connective. The truth value of $\Box X$ is not determined uniquely by the truth value of X. The conclusion of Frege's argument is that these features are incompatible: where the

† See Exercise 148, page 366.

substitutivity of identicals holds, there you must have truth functionality. I will introduce some terminology which will place this point in a wider setting, and facilitate its discussion.

Let us say that the *extension* of a sentence is its truth value, the extension of a name, its bearer, and the extension of a predicate of degree n the set of n-tuples of which it is true. In other words, extensions of expressions are the sorts of objects a **Q**-interpretation assigns to corresponding letters.

A sentence is *extensional with respect to a position for an expression (sentence, name or predicate)* iff replacing the expression in that position with any other expression having the same extension leaves the truth value of the whole sentence unchanged.

A sentence is *extensional, tout court,* iff it is extensional with respect to all the positions for sentences, names and predicates it contains.

A language is extensional iff all its sentences are extensional.

Truth functionality is a special case of extensionality: to say that a language is truth functional is just to say that all its sentences are extensional with respect to all positions for sentences.

The sentence, which I stipulate to be true,

1)   John believes that Charles Dodgson is a mathematician

is not extensional with respect to the position of the sentence "Charles Dodgson is a mathematician". If it were extensional with respect to that position, then the result of inserting any truth into that position would be a truth. We know that this is not so, since John, being of only finite intellect, does not believe every truth.

(1) is also not extensional with respect to the position occupied by the name "Charles Dodgson".[13] I can without inconsistency stipulate that John does not believe that Lewis Carroll is a mathematician, despite the fact that

2)   John believes that Lewis Carroll is a mathematician

results from (1) by placing in a position occupied by a name another name having the same extension.

Finally, (1) is not extensional with respect to the position occupied by the predicate "is a mathematician". For, let us suppose, this predicate is true of just the same things as those of which the predicate "can rattle off a proof that there is no greatest prime" is true. Yet I can without inconsistency stipulate that

3)   John believes that Charles Dodgson can rattle off a proof that there is no greatest prime

is false, despite the fact that it results from (1) by placing in a position occupied by a predicate another predicate having the same extension.[14]

Notice that the non-extensionality of (1) with respect to the position occupied by "Charles Dodgson" does not entail that it is non-extensional with respect to the position occupied by "John". Moreover, if the same name (or predicate or sentence) occurs twice, the sentence may be extensional with respect to the position of one occurrence but not extensional with respect to the position of another. For example,

4)   Charles Dodgson believes that Charles Dodgson is a mathematician

is extensional with respect to the first but not the second position occupied by "Charles Dodgson". To illustrate the non-extensionality, imagine that some other name, say N, for Dodgson is in general currency, though Dodgson wrongly thinks that N refers to someone else, a botanist, and so does not believe that N is a mathematician.

Modal contexts uncontroversially give rise to non-extensionality with respect to sentence-positions (that is just to say that modal sentence connectives are not truth functional). They also give rise to non-extensionality with respect to predicate-positions.

5)   Necessarily, all humans are humans

is true, but presumably

6)   Necessarily, all featherless bipeds are humans

is false, since it is only contingently true that there are no non-human bipeds without feathers. Yet (6) results from (5) by replacing a predicate occupying the relevant position by another with the same extension.

We can extend the definition of extensionality to the language **Q** by relativizing to an interpretation throughout, and replacing talk of name-position by talk of name-letter-position. Thus defined, **Q** is extensional, a fact ensured by the interpretation rules which attend only to the extensions (upon an interpretation) of the various expressions. Any English sentences adequately **Q**-formalizable are extensional with respect to the positions of expressions corresponding to **Q**-letters.

By specifying *intensions* for sentences, names and predicates we can arrive, by exactly analogous definitions, at various notions of intensionality. We stipulate that the intension of an expression is the set of all ordered pairs whose first member is a possible world and whose second member is the extension of that expression with respect to that world. A **QN**-interpretation thus fixes an intension for every category of **QN**-letter.

QN is non-extensional, but it is *intensional*. This is because the interpretation rules in effect attend just to the intensions of the various expressions. Some English sentences dominated by "necessarily" are certainly intensional with respect to some contained positions for sentences and predicates. Obviously, an English sentence "necessarily, *A*" is intensional with respect to the *A*-position: if *A* and *B* have the same intension, that is, are true in just the same worlds, "necessarily *A*" will have the same truth value as "necessarily *B* ".

Again, an English sentence "necessarily, all *F*s are *G*s" is intensional with respect to the positions of "*F*" and "*G*".

Some English sentences contain positions which are neither intensional nor extensional.[15] (**1**) is an example:

John believes that Charles Dodgson is a mathematician.

"Charles Dodgson is a mathematician" has the same intension as "Either Charles Dodgson is a mathematician or there is a largest prime", but one may consistently suppose that John believes the first and not the second.

Again, "Charles Dodgson" and "Lewis Carroll" have the same intension: with respect to each world, they name the same person. Yet we have already seen that their substitution in (**1**) may fail to preserve truth value.

Frege's argument claims that if a sentence is extensional with respect to all its name-positions, then it is extensional with respect to all its sentence-positions; that is, it is truth functional. Sentences formalizable in **QN** are extensional with respect to name-positions, as will shortly be made plain. Hence it should follow that **QN** is truth functional, which it evidently is not. If one is convinced of the soundness of Frege's argument, one can but conclude that there is something incoherent about the semantic notions which animate **QN**.

Built in to the interpretation rules for **QN** is the requirement that a name-letter be assigned, with respect to every world w, the same object, providing that the object exists at w. An intended interpretation of an English sentence was said to be one which assigned to a name-letter in the formalization the object which bears the corresponding name. So if English is adequately **QN**-formalizable, an English name which in fact names o must also name o with respect to each world at which o exists. This is turn ensures that names having the same extension also have the same intension. Thus, the intensionality with respect to name-positions of a language formalizable in **QN** ensures the extensionality with respect to name-positions of that language.

To take an example. The fact that "Lewis Carroll" and "Charles Dodgson" have the same extension is enough to ensure that the following sentences have the same truth value:

7) Necessarily, Charles Dodgson is human

8) Necessarily, Lewis Carroll is human.

We thus have the most important premise that Frege's argument requires: **QN** is extensional with respect to all its name-positions.

The argument requires a further uncontroversial assumption:

9) if any sentence adequately formalizable as $\Box(A \leftrightarrow B)$ is true, then the truth value of any larger **QN**-formalizable sentence containing A is the same as the truth value of the result of replacing A by B.

We know that (9) would be false without the restriction to adequately **QN**-formalizable sentences containing A and B;[†] but its truth with respect to **QN**-formalizable sentences is secured by the intensionality of **QN**, and thus by the intensionality of the sentence-position occupied by A.

Now for the argument itself. We suppose that "p" and "q" are two sentences with the same truth value, and we use "δ" as a device to form from a sentence an expression able to occupy name-position, with the stipulation that for any sentence, s, "$\delta s = 1$" is true iff s is true, and "$\delta s = 0$" is true iff s is false. Since this is constitutive of the meaning of "δ", we can infer that

10) $\Box(\delta p = 1 \leftrightarrow p)$, and $\Box(\delta q = 1 \leftrightarrow q)$.

† See Exercise 149, page 366.

Then the Fregean claim is that the following series of transformations preserves truth value:

11)   (i)   ... p ...
        (ii) ... $\delta p = 1$ ...
        (iii) ... $\delta q = 1$ ...
        (iv) ... q ...

(9) and (10) together ensure that (i) and (ii) have the same truth value and so do (iii) and (iv). The assumption of extensionality with respect to name position, together with the obvious fact that the co-extensiveness of "p" and "q" ensures the co-extensiveness of "δp" and "δq", establishes that (ii) and (iii) have the same truth value. Hence a context which is extensional with respect to positions for names must also be extensional with respect to positions for sentences, that is, truth functional.

The argument appears to be valid and to have true premises. Must we, therefore, conclude that **QN** is incoherent?

We need to distinguish a broad and a narrow conception of a "name", corresponding to which there are broad and narrow conceptions of extensionality with respect to name-positions. We shall say that on the broad conception, Frege's argument is valid but not sound, since it is not the case that a language formalizable in **QN** is broadly name-extensional. We shall say that on the narrow conception, Frege's argument is not valid, since the crucial expressions, those which stand to the left of the identity sign in (11ii) and (11iii), are not, on this conception, names.

A name, broadly conceived, will be any expression which in some language or other can combine with an n-ary predicate and $n-1$ other names to form a sentence. (The circularity will not matter for our purposes.) The broad conception of a name will include definite descriptions, like "the value taken by s", and its symbolic equivalent "δs". However, no expressions of this kind belong to **Q** or to **QN**. In these languages, there is no expression which is both complex (that is, contains other expressions as proper parts) and also capable of occupying the position occupied by name-letters. We have already seen (chapter **4.10**, especially (**4.10.13**) and (**4.10.14**)) that sentences like those to the left of the biconditionals in (10) will not be **Q**-classified as identity sentences, but rather as existential generalizations; likewise in **QN**. If we try to formalize "$\delta s = 1$" in **QN**, the result would be something like:

12) $\exists x (Fx\beta \ \& \ \forall y (Fy\beta \rightarrow y = x) \ \& \ x = \alpha)$

where "*Fxy*" corresponds to "*x* is a value taken by *y*", "*β*" corresponds to "*s*" (which here has to be thought of as the *name* of a sentence), and "*α*" corresponds to "1". (12) says that there is a unique thing, *x*, which is a value taken by *s*, and *x* = 1. This will not serve as a premise to the application of the substitutivity of identicals. If "name" is defined widely enough to include definite descriptions, then a language formalizable in **QN** is not extensional with respect to name-positions. Consider the pair:

13) Necessarily, the author of *Waverley* (if there is a unique author of *Waverley*) wrote *Waverley*

14) Necessarily, Scott (if there is a unique author of *Waverley*) wrote *Waverley*.

The first is true and the second false, despite the fact that the second results from the first by replacing an occurrence of a "name" ("the author of *Waverley*") by a co-extensive name ("Scott").[16]

The **QN**-formalizations of (13) and (14) are, respectively:

15) $\Box(\exists x(Fx \,\&\, \forall y(Fy \to x = y)) \to \exists x(Fx \,\&\, \forall y(Fy \to x = y) \,\&\, Fx))$

16) $\Box(\exists x(Fx \,\&\, \forall y(Fy \to x = y)) \to F\alpha)$

with "*F*" corresponding to "wrote *Waverley*" and "*α*" to "Scott". (15) is true upon any interpretation, having almost the form "$\Box(p \to p)$". (16) is arguably not true upon an intended interpretation, since there are worlds in which someone other than Scott uniquely wrote *Waverley*, worlds *w* such that the antecedent of the conditional is true upon the interpretation at *w* and the consequent false upon the interpretation at *w*.

If we keep to the narrower conception of a name, according to which only an expression appropriately formalizable by a **QN**-name-letter is a name, then Frege's argument is invalid, since there is no justification for the claim that, regardless of the context, (ii) and (iii) have the same truth value.

The very different behaviour of names and descriptions in these contexts might well ground the view that they should not be treated as belonging to a common semantic category. We will not pursue this thought, but we will always think of a name in accordance with the narrower conception, which excludes complex expressions like definite descriptions.

The upshot of this discussion is that Frege's argument does not show that the name-extensionality of **QN**, and the associated expressions of de re modality, are inconsistent with **QN**'s non-truth functionality.

7 Trans-world identity

Evaluating de re sentences by the interpretation rules of **QN** requires attention to the *identity* of objects at various possible worlds. The rule for interpreting name-letters requires an interpretation to assign the same object to any name-letter with respect to each world at which the object exists. Hence this rule already incorporates the notion of trans-world identity: an interpretation must settle which object in, say, w is the object, say o, which it has assigned to "α" with respect to some other world.

The assumption of trans-world identity also emerges in connection with de re quantifications. Compare

1) $\Box \forall x F x$

2) $\forall x \Box F x$.

In interpreting (**1**), a de dicto sentence, we do not have to consider how the entities at various worlds are related to the ones at the world of evaluation. For any interpretation, i, any world, w, any n-ary predicate, F, all we have to consider is whether all n-tuples formed from the members of D^w (whatever they may be) belong to $i_w(F)$. In interpreting (**2**), a de re sentence, with respect to w, we have to consider which objects at other worlds belong to D^w. Suppose the question is whether (**2**) is true upon an interpretation, i, with respect to a world w. In effect, the quantifier of (**2**) ranges just over D^w, since the interpretation rule for "$\forall$" makes the truth of (**2**) turn on the truth of "$\Box F \alpha$" with respect to interpretations which assign a member of D^w to "α" and agree as far as possible with i. The upshot is that (**2**) is true on i at w iff, for all w', all the members of D^w belong to $D^{w'}$, and are members of members of $i_{w'}(F)$. So i has to determine, for each world, what objects from w that world contains. It thus has to settle questions of "trans-world identity".

Imagine two types of atheistic views about God's existence. They are both "soft" atheisms, in that they agree that although there is no God, there could have been one. One view is formalizable:

3) $\Diamond \exists x F x$

with "*F*" corresponding to "is omnipotent, benevolent etc." (here insert a complete list of the attributes appropriate to God). The other view is formalizable:

4) $\exists x \Diamond F x$

with the correspondence as before. The first view is de dicto: it is true upon an intended interpretation, i, with respect to w, iff for some world, w', $i_{w'}(F)$ has at least one non-empty member. The second view is de re: it is true upon an intended interpretation, i, with respect to w, iff for some interpretation i' for which there is an assignment $i'_w(\alpha)$ and which agrees as far as possible with i except, perhaps, in respect of this assignment, there is a world w' such that $i'_{w'}(\alpha)$ belongs to $D^{w'}$ and $\langle i'_{w'}(\alpha) \rangle$ belongs to $i'_{w'}(F)$. Since, by the stipulation of (**3.5**), $i'_w(\alpha) = i'_{w'}(\alpha)$, (**4**) corresponds to the claim that someone exists who *could* have been God. The interpretation will settle whether someone who exists at the world of evaluation, w, is God at some possibly distinct world, w', so it involves trans-world identity.

We must therefore agree that de re sentences, as **QN**-formalized, involve trans-world identity. If Quine is right, trans-world identity is unintelligible, and so **QN**, and all languages adequately formalizable therein, would likewise be unintelligible.

Let us set aside a question that is not in dispute. We are concerned with a metaphysical issue, not an epistemic one. The question of how, if at all, we *know*, concerning a possible world, that it does or does not contain a given individual is not at issue. If there is fact of the matter which we do not know, then de re modality is intelligible, but permits the expression of propositions whose truth values we do not know.

It seems as if the intelligibility of trans-world identity is ensured by the fact that we can introduce a counterfactual situation by referring to a particular object. We might say: envisage a situation in which Nixon tells the truth. If this introduces a possible situation at all, there seems no room for doubt about whether it is a situation in which Nixon exists. We have specified the situation in terms of an object in it. On the face of it, then, Nixon belongs to at least two worlds, the actual world and a world where he is honest. How can there be some hidden difficulty here?

A difficulty can be concocted by holding the following two theses: (a) the only legitimate specification of a possible world is purely qualitative; and (b) a purely qualitative specification is insufficient to determine which actual individuals are present at a non-actual world. If the only facts which could determine trans-world identity, namely purely qualitative facts, fail to do so, then, of course, the trans-world identities are not determined. But theses (a) and (b) are an unattractive combination. If you believe that all facts are determined by qualitative facts, then, fair enough, you will hold thesis (a); but, by the same token, you will reject thesis (b). If you believe thesis (b), then you have a reason to reject the view that all facts are determined by qualitative facts, and thus a reason not to agree that only qualitative specifications of worlds are legitimate.

That is the end of the discussion, unless we can find some arguments against the intelligibility of trans-world identity. I shall mention one, and briefly discuss another.

There is a powerful argument in Lewis [1968], but it has two features which make it inappropriate for discussion here. First, it depends upon Lewis's preferred method of constructing non-actual worlds (as mereological sums of genuinely existing, though non-actual, spatio-temporally related things), and is therefore only available to one who is familiar with, and accepts, that construction. Secondly, it in no way impugns the intelligibility of de re modality, but simply justifies Lewis's preferred method of representing it.

The argument I shall briefly discuss is due to Quine. He compares trans-world identity with trans-moment identity:

> our cross-moment identification of bodies turned on continuity of displacement, distortion and chemical change. These considerations cannot be extended across worlds, because you can change anything to anything by easy stages through some connecting series of possible worlds. ([1976] p. 861)

Quine is claiming that we can make sense of identity across moments, because this is determined by various kinds of continuity, but that we cannot make sense of identity across worlds, because it is not determined by anything. For example, this table on which I am writing, α_0, could have been made of slightly different parts. So there is a world, w_1, and a w_1-object, α_1, such that α_1 is identical with α_0 yet is made of slightly different parts. α_1 could also have been made of slightly different parts. So there is a world, w_2, and a w_2-object, α_2, such that α_2 is identical with α_1 but is made of slightly different parts. Continue this process through a hundred or a hundred million stages, making small variations to design as well as to parts, and you will end up with a submarine, or anything else you choose. So there are no limits on how something could be at another world. This is a *reductio ad absurdum* of the view that there are facts of the form: α at w is the same object as β at w' and is distinct from γ at w'.

Quine underplays the paradoxical nature of the argument upon which he relies. We do *not* happily accept that α_0 could have been made of the parts that some remote successor in the series is made of. So we are not happy to accept that "anything can be changed into anything by easy stages". The reasoning is on a par with the reasoning that seems to force us to accept that a heap of sand can never be destroyed by one-by-one removal of grains (for taking away one grain can never turn a heap into a non-heap). The reasoning is powerful, yet we all know that there must be something wrong either with it or with the premises, for we all know that the conclusion is unacceptable.

He also underplays the problem-free nature of identity through time. There are well-known puzzles. For example, we are drawn to a continuity account of identity through time, as is shown by the fact that we allow that a ship that has been endlessly repaired over many years, in gentle stages, is the very same ship, even if in its later years it is composed of none of its original parts. In addition, we are drawn to a compositional account of identity through time, as shown by the fact that if we imagine a ship's parts being successively replaced, but the old parts kept and finally reassembled into a ship, we have *some* inclination (of varying strength depending upon context)[17] to hold that this is really the original ship, the one bearing a continuity relation to the original being merely a replica.

Finally, Quine assumes that identity through worlds must be qualitatively determined, in the way that he takes identity through time to be. This assumption requires justification.

Quine can indeed be justly read as presenting a believer in trans-world identity with a challenge to give a systematic account of it, but he cannot be said to have shown that it is any more incoherent than our talk of ships and heaps.

One way to take up the challenge has been proposed by Lewis. He does not, as I said, allow trans-world identity in the sense that Quine intends, though he does provide a substitute that is said to yield everything for which trans-world identity was needed, and is governed by fairly well-articulated principles. This theory is discussed in §**9**.

Kripke has suggested that some of the distrust of trans-world identity may be fostered by a faulty picture: thinking of a possible world as like something viewed through the wrong end of a telescope. The question of whether Socrates could have been an alligator is not to be addressed by envisaging an alligator-infested world, and reviewing the individuals therein to "see" if one of them is Socrates. Rather, it is to be answered by connecting it with other questions, like: must an individual of a species be propagated by individuals of that species? Could anyone have had different parents (propagators) from the ones he actually had? Kripke seems to have in mind an epistemic version of the problem of trans-world identity. He is certainly right to say that possible worlds are not going to *supply* the answers to questions like the one about Socrates. Rather, worlds provide a way of expressing the answers, once found.

We have seen that a version of the third thesis about de re modality mentioned in §4 is correct as applied to **QN**: **QN**-formalizations expressing de re necessity do involve, in their **QN**-interpretation, questions of trans-world identity. (We will see in §9 that there is an alternative approach.) We now very briefly consider the other two theses, viz.:

4.1)   In ascribing de re necessity we attribute a property to a non-linguistic object; in ascribing de dicto necessity we attribute a property to a sentence.

4.2)   A de re necessary truth records how things are in the world; a de dicto necessary truth records only linguistic facts.

We have seen that de re necessity does indeed involve the ascription of properties to non-linguistic objects, but we have seen no reason at all to suppose that de dicto expressions of necessity attribute it to a sentence. It is unquestionably a property of all people that they are, necessarily, unmarried if bachelors. Likewise, though our discussion sits happily with the thought that a de re necessary truth records how things are in the world, we have seen no reason to suppose that a de dicto necessary truth is made true by linguistic facts. There is a *prima facie* case against any such view. As far as we have seen, the de re/de dicto distinction is merely one of scope, so that the very same concept of necessity is involved in both cases. Hence one would expect it to be true in both cases or neither that they are "made true by linguistic facts".

8   "∀" for "□"

The semantics for **QN** have been given in terms of possible worlds. Roughly, "□" is associated with universal quantification over worlds, "◇" with existential quantification. The connection suggests that we could have proceeded in a different way: instead of enriching **Q** with the non-truth functional sentence connectives we could instead have enriched it with a further predicate constant, "W", stipulating that every interpretation assign to "W" the set of all 1-membered sequences whose member is a possible world. Let us call the result of enriching **Q** in this way **QW**. We might then hope to express a **QN**-sentence of the form "□ ... " by a corresponding **QW**-sentence of the form "$\forall x(Wx \rightarrow ...)$". Let us refer to any treatment on these lines as a quantifier treatment of necessity.

Intuitively, the idea is to exploit the equivalence between it being necessarily true that A and it being true at every possible world that A. As soon as we look at the details of **QW**, however, some difficulties arise. What, in **QW**, should correspond to the **QN**-sentence "$\Box p$"? We cannot simply write

1) $\forall x(Wx \rightarrow p).$

For this severs the connection between "p" and the possible worlds: the idea was to say that "p" is *true at* each possible world. There are a considerable number of different tacks one might take, of which I shall consider only two. Both have been influential, and both are due to Lewis. In both cases, the base language (corresponding to our **Q**) is envisaged to have no sentence-letters, so let us imagine that modification of **Q** to have been made. Our original problem emerges in the same way if we ask: how should we fill the dots in "$\forall x(Wx \rightarrow ...)$" when formalizing what in **QN** is formalized as "$\Box F\alpha$"?

The first suggestion I shall call the "extra argument place" treatment. Suppose we are to formalize

2)   Socrates is necessarily human.

The proposal is that we use

3) $\forall x(Wx \rightarrow F'\alpha x)$

where "$F'xy$" corresponds, not to "human", but to "x is human at y". On an intended **QW**-interpretation, "F'" will be assigned the set of ordered pairs σ such that, for each world, w, and each object o, $\langle o,w \rangle$ belongs to σ just on condition that o is human at w.

It is easy to see that the idea could be generalized. Every n-ary predicate in an English necessitation will be formalized by an $n + 1$-ary predicate-letter of **QW**, the extra argument place being filled by a variable bound by a quantifier in the phrase "$\forall x(Wx \rightarrow ...)$".

Let us note three connected features of this treatment. First, it provides an extensional treatment of modality. The extensionality of **QW** follows from the fact that its interpretation rules attend only to the extensions of expressions. To see how apparent evidence of non-extensionality disappears on the account, consider the invalidity of the argument:

4)   Necessarily Socrates is human.

Socrates is human iff Socrates is snub-nosed.

Therefore necessarily, Socrates is snub-nosed.

Notice that (4) is evidence for non-extensionality only if we can construe the first premise as consisting in the application of a sentence connective to a sentence. On the extra argument place treatment, the premise is not construed in this way, but rather as having the logical form of (3) in which there is no sentential component. The invalidity of (4), as formalized in **QW** on the extra argument place approach, is quite consistent with **QW**'s extensionality.[†] In his [1968], Lewis held this to be a virtue.

The second, related, feature is that the approach provides a vivid example of how proposed logical forms may differ from the way a sentence would intuitively be supposed to be constructed.

The third feature is that this divergence will hinder **QW** in formalizing as valid intuitively valid arguments. For example, intuitively the following argument is valid:

5)   Necessarily Socrates is human. Therefore Socrates is human.

The present **QW** approach cannot, on the face of it, even discern a common constituent in premise and conclusion corresponding to "Socrates is human". "Human" in the premise has to be matched with a 2-place predicate letter, whereas in the conclusion it will, for all that has been said, be matched with a 1-place one.[‡]

There are ways of patching up the damage. For example, it might be suggested that we can use a 2-place predicate-letter in formalizing the conclusion, filling the additional argument place by a name for the actual world. The justification might be that that is the world with respect to which it is intended that the conclusion be evaluated. However, this would seem to justify equally well matching "human" in the premise with a 3-place predicate-letter. The sentence as a whole is to be evaluated

† See Exercise 150, page 366.   ‡ See Exercise 151, page 366.

for truth with respect to the actual world, w∗, so should we not think of it as abbreviating "For all worlds, w, Socrates is human-at-w-with-respect-to-w∗"? Again, "′" might be used to introduce a special semantic relation between predicate letters, the interpretations being constrained to behave according to the following rule: if an interpretation assigns to "F" the set of all 1-membered sequences meeting a condition, C, then it must assign to "F'" the set of all ordered pairs $\langle o,w \rangle$ such that o meets condition C with respect to w. The patches are at best *ad hoc*.

This treatment, as I said, derives from Lewis [1968]. Recently, Lewis has, without explicitly alluding to this earlier treatment, given a different kind of reason for rejecting it.

The extra argument place treatment takes all of a thing's properties as really relations between it and a world. You may think that this page has the intrinsic property of being rectangular, but on the proposed treatment, we have to say that there is no such intrinsic property. To permit **QW**-formalization of, say

6)   This page is rectangular

we regard it as abbreviating something like

7)   This page is rectangular-at-w

where the relevant world w is contextually determined (presumably as the actual world in a self-standing utterance of (**6**)). Lewis's recent criticism is that this implicit repudiation of intrinsic properties is unjustified. Being white or rectangular are properties which objects have in themselves, and are not relations the objects bear to other things.†

Let us therefore consider an alternative way of implementing the quantifier approach. I follow Lewis [1986], though as he is not very explicit on certain details I may not be entirely faithful. I call the approach the "open-sentence former" approach.

Lewis suggests, in effect, that a phrase like "In Australia ..." can be seen as a one-place sentence operator. He suggests that it works by restricting the quantifiers in what follows. Thus evaluating

8)   In Australia, all beer is good

requires one to attend only to a proper subset of all the beer there is, viz.

† See Exercise 152, page 366.

that which is in Australia. One could think of a phrase like "At the actual world ..." as working in a similar way, so that an affirmation of

9)   At the actual world, all beer is good

would be intelligible, and would be read as a specially emphatic way of refraining from committing oneself to the necessity of all beer being good.

Lewis's suggestion is that a phrase like "at v", where "v" is a variable ranging over possible worlds, can function in a similar way to "in Australia" and "in the actual world". The main difference is that such a phrase contains a variable of quantification that can fall in the scope of a quantifier, thus:

10) $\forall x(\text{at } x(p))$.

Syntactically, "at x" forms from "p" an *open* sentence "at $x(p)$", and hence something from which a quantifier can form a closed sentence. I call any phrase "at v" (v any variable) an "open-sentence former": it takes a sentence, open or closed, to make an open sentence. Semantically, the basic idea is that "at $x(p)$" is true iff p is true with respect to x (a condition which will fail if x is not a world).

Let us take the syntax of **QW** to be enriched by the addition of the open-sentence former "at". How should the interpretation rules be modified? It is plain that we cannot escape a thoroughgoing relativization of all **QW**-interpretations to worlds. The reason is that a sentence like "at $\alpha(p)$", which will be involved in interpreting a sentence which quantifies over worlds (assuming we leave the quantifier rules unchanged), should be true upon an interpretation iff "p" is true upon an interpretation *with respect to* whatever the interpretation assigns to "α".

Moreover, to be true to Lewis's ideas, we cannot simply take over the **QW**-interpretation rules as they stand. Lewis wants quantifiers to range over absolutely everything there is, and he takes this to include non-actual objects as well as actual ones. Thus he takes there to be a reading of

11)   There are talking donkeys

upon which it is true (with respect to the actual world): a reading which treats the quantifier as ranging over *everything*, actual and possible. Context can implicitly restrict or derestrict quantifiers, and the same can be done explicitly by the "at v" operators. Lewis sees the effect of replacing "are" by "could be" in (11) as that of unambiguously ensuring that the quantifier will be completely unrestricted, and thus range over non-actual as well as actual objects.

Implementing Lewis's idea is slightly complicated. We will need two styles of name-letter, neither of which behaves as in **Q**. A variable of quantification, say "x", occurring within the immediate scope[18] of an "at v" operator, needs to be thought of as quantifying over objects in the world assigned to "v". Hence one style of name-letter replacing variables like "x" for the purposes of interpretation requires a corresponding relativity: the interpretation must assign it something in the world assigned to "v". Unrestricted quantifiers are thought of as quantifying over the totality of actual and non-actual objects (including the actual and all non-actual worlds). A corresponding distinct style of name-letter is required, and to such a name-letter an interpretation is free to assign anything from any world. The quantifier rules will need to use name-letters of the unrestricted sort to mark positions occupied by variables whose quantifier does not fall in the scope of any "at v" operator, and name-letters of the restricted sort to mark positions occupied by variables whose quantifier does fall within the scope of an "at v" operator. The full details are not necessary for present purposes. The general idea is that "at n, X" is true upon an interpretation, i, with respect to a world, w, iff X is true upon i with respect to whatever i assigns to "n".

I pass over without comment Lewis's claim that there are non-actual objects; see, however, §11 below.

To confirm our understanding of **QW**, let us see how it deals with what, superficially, appears to be a difficulty. In **Q**, and therefore in **QW**, the order of quantifiers of the same sort is irrelevant. For example, the prefixes "$\exists x \exists y$" and "$\exists y \exists x$" are equivalent. But "$\exists x \Diamond$" and "$\Diamond \exists x$" are far from equivalent prefixes, the first expressing de re, the second de dicto, possibility. (Compare (**7.3**) and (**7.4**).)

The difference is not brought out in **QW** by the following pair of formalizations:

12) $\exists x\ \exists y(Wy\ \&\ \text{at } y(\dots x \dots))$

13) $\exists y\ \exists x(Wy\ \&\ \text{at } y(\dots x \dots))$

for these are equivalent. Rather, Lewis will see the de re English phrase "There is something which could be … " as involving an implicit restriction of the quantifier to the actual world. To bring this out in **QW**, we will need to add a name for the actual world, say "$w*$". (The rule will be that every interpretation, with respect to every world, w, assigns the actual world to "w*". "At w*" thus forms closed sentences from closed sentences.) The de re sentence could be formalized:

14) $\exists x$ (at w*($\exists y \, y = x$) & $\exists y(Wy$ & (at $y(... \, x \, ...)$)))).

(12) or (13) serve to formalize the de dicto sentence. There is no question of the de re/de dicto distinction being submerged.

QW (henceforth understood in its open-sentence former version) and **QN** use pretty similar semantic resources, despite their syntactic differences. For example, in the semantics for **QN** there is quantification over worlds, including non-actual worlds, and their domains, including domains containing non-actual objects. Is there any reason to prefer one of these languages to the other? Lewis's earlier preference for extensionality does not bear on the choice, since "at v" is a non-truth functional open-sentence former.[†]

One might initially be tempted to suppose that there could be little to choose between the languages. However, Lewis has argued that **QW** has greater expressive resources than **QN**, revealed in the greater depth of the formalizations of English the former can provide.

As **QN** and **QW** stand, it is true that the expressive resources of the latter outstrip the former. One reason for this is connected with the addition made to **QW** of the name "w*". There is no corresponding device in **QN**. The difference can be brought out by considering the following sentence.

15)   It is possible that everything that is actually red should also have been shiny.

This is straightforwardly formalizable in **QW** as

16) $\exists x(Wx$ & $\forall y$ (at w*(Fy)$\rightarrow$at $x(Gy)$))),

with "F" corresponding to "is red" and "G" to "is red and shiny". However, (15) cannot be adequately **QN**-formalized as either of

17) $\forall x(Fx \rightarrow \Diamond Gx)$

18) $\Diamond \forall x(Fx \rightarrow Gx)$

with correspondences as before.[‡]

† See Exercise 153, page 366.   ‡ See Exercise 154, page 366.

A suitable supplementation of **QN** that keeps to the sentence connective approach to modality, and does for **QN** something like what "w∗" does for **QW**, is a one-place sentence connective, say "Ⓐ", and the interpretation rule:

19) For any interpretation, i, any world, w, $i_w(ⒶX)$ is T iff X is true upon i with respect to the actual world.

One should then **QN**-formalize (15) as

20) $\Diamond \forall x(ⒶFx \rightarrow Gx).$[19]

Similarly the claim that there are non-actual objects can be **QN**-formalized as:

21) $\exists x(\Diamond \exists y\, y = x\ \&\ \neg\, Ⓐ\exists y\, y = x),$

or **QW**-formalized as

22) $\exists x \exists y(Wy\ \&\ \text{at } y(\exists z\, z = x)\ \&\ \text{at } w∗ (\neg \exists z\, z = x)).$

If we build in to our definition of **QW**-interpretation that there are non-actual objects, then (22) will come out **QW**-valid; otherwise not. Similarly, one has a choice whether or not so to engineer the interpretation rules of **QN**, in particular one's account of the worlds and their domains, as to make (21) valid, or even to make its negation valid. A common prejudice among logicians would be to prefer neutrality on this issue, which might well be seen to belong to "metaphysics".

Lewis gives various examples of natural claims formalizable in **QW** but allegedly not formalizable (to any reasonable depth) in **QN**. If even one allegation is correct, this would constitute a powerful reason for preferring **QW**. I shall consider three of the examples.

23) It might happen in three different ways that a donkey talks.

In this example, we appear to talk of, indeed count, various different possibilities. These, says Lewis, can be represented by possible worlds, but not by boxes and diamonds. Thus (23) is **QW**-formalizable as

24) $\exists x\, \exists y\, \exists z(Wx\ \&\ Wy\ \&\ Wz\ \&\ x \neq y\ \&\ x \neq z\ \&\ y \neq z\ \&\ \text{at } x(\exists vFv)\ \&\ \text{at } y(\exists vFv)\ \&\ \text{at } z(\exists vFv))$

with "*F*" corresponding to "is a talking donkey". One might doubt if this is entirely adequate, since the differences between the worlds might be ones irrelevant to the way the donkey talked, whereas intuitively what is required for the truth of (23) is, for example, that a donkey might be made to talk by special training, or by the injection of a chemical, or by genetic engineering. However, it may seem that in **QN** one cannot get even as close as this, since there would seem to be no way in which numeral adjectives can be conjured out of boxes and diamonds.

In fact, one can use resources going at most a very little beyond **QN** to produce a better formalization of (23) than (24). The idea will be to understand the English as saying that there are three different properties a donkey could have, and a possessor of any of these properties is a talker. I shall treat the quantification over properties in the way that simulates predicate quantification within a first order language, and so minimizes alterations to **QN** (compare the discussion surrounding (**4.19.8**)): I shall assume that properties are among the objects in some (or all) domains of interpretation.[20] A **QN**-formalization is:

25) $\exists x \, \exists y \, \exists z (Fx \,\&\, Fy \,\&\, Fz \,\&\, x \neq y \,\&\, x \neq z \,\&\, y \neq z \,\&$
$\forall v (Gvx \rightarrow Hv) \,\&\, \forall v (Gvy \rightarrow Hv) \,\&\, \forall v (Gvz \rightarrow Hv) \,\&$
$\Diamond \exists v (\mathcal{J}v \,\&\, Gvx) \,\&\, \Diamond \exists v (\mathcal{J}v \,\&\, Gvy) \,\&\, \Diamond \exists v (\mathcal{J}v \,\&\, Gvz))$

with "*F*" corresponding to "is a property", "*Gxy*" to "*x* possesses (the property) *y*", "*H*" to "talks" and "*$\mathcal{J}$*" to "is a donkey". (25) is better than (24), since it connects the different ways with different ways of *being a talker*. It may not be perfect, since the connections between having any one of the properties and being a talker are rather weak, but the fact remains that we have yet to find any expressive superiority of **QW** over **QN**.

This suggestion, like some later ones, could be criticized for its treating properties as among the individuals. From a perspective upon which there are no properties, this is clearly unsatisfactory. However, one who would use this reply as part of a case for preferring **QW** to **QN** (as opposed to a case simply against **QN**) would need to show that, though properties do not exist, worlds do. I do not say that no such case could be devised, but it would form a very controversial basis for a preference for **QW**. A separate point is that the main feature of the suggestion which led to (25) would be retained if properties were replaced by sets.

Another example which Lewis uses to ground a preference for **QW** is:

26) A red thing could resemble an orange thing more closely than a red thing could resemble a blue thing.

Lewis suggests that the obvious **QN**-formalization

27) $\Diamond \exists x \exists y \exists z \exists v (Fx \,\&\, Gy \,\&\, Fv \,\&\, Hz \,\&\, \mathcal{J}xyvz)$

with "F", "G", and "H" corresponding to the three colour predicates "red", "orange" and "blue", and "$\mathcal{J}xyvz$" to "x resembles y more than v resembles z" is inadequate. There does not have to be a single world containing all the objects. It is enough that there be two worlds, one containing one pair, another the other. Lewis sums up his case by saying that English essentially involves cross-world comparisons of similarity. His own formalization is simply:

28) $\exists x \exists y \exists z \exists v (Fx \,\&\, Gy \,\&\, Fv \,\&\, Hz \,\&\, \mathcal{J}xyvz)$

the force of the English "could" being reflected simply in the fact that the quantifiers are quite unrestricted, each ranging over the totality of actual and non-actual objects.

There is certainly a formal difference between the claim that there could be things x, y, z, v, such that … and the claim that there could be things x, y and there could be things z, v, such that … . But I can discover no difference of substance that could make the truth-upon-an-intended-interpretation-conditions of (**27**) differ from the truth conditions of (**26**). If there is a world with a closely similar red-orange pair, and a world with a less closely similar red-blue pair, then surely there is a world where both pairs co-exist. Indeed, that there is such a world follows from Lewis's own principles determining what worlds exist. I can find no sentence of idiomatic English whose truth is differentially sensitive to the distinction Lewis makes.[†]

A kind of example upon which Lewis places a good deal of weight are supervenience claims. These have the general form:

29)   There could be no differences of one sort without differences of another sort.

Let us take a specific example, the claim that the mental supervenes upon the physical. In Lewis's words:

> The idea is that the mental supervenes on the physical … [i.e.] there could be no mental difference between two people without there being some physical difference, whether intrinsic or extrinsic.

† See Exercise 155, page 367.

Reading the "could" as a diamond, the thesis becomes this: there is no world ... wherein two people differ mentally without there being some physical difference, whether intrinsic or extrinsic, between them. That is not quite right. We have gratuitously limited our attention to physical differences between two people in the same world, and that means ignoring those extrinsic differences that only ever arise between people in different worlds. ([1986] p. 16)

"Reading the 'could' as a diamond" goes over into my terminology as "supposing that the English is **QN**-formalizable".

Let us be clear what the rules of the game are. The neutral territory consists in ordinary idiomatic English sentences and our intuitive judgements of whether these are true or false. The two players are to formalize these in the languages **QW** or **QN**. If the truth-upon-an-intended-interpretation value of one of the players' formalizations does not match that of the English sentence (according to our intuitive judgements) that player loses a point. (He might regain it if he can convince his opponent that our intuitive judgements are faulty.)

The paragraph quoted from Lewis should thus identify some feature of the neutral territory not matched by a **QN**-formalization: a way in which there could be a mismatch between the truth value of the English and the truth-upon-an-intended-interpretation value of the **QN**-sentence. What we in fact discover is some remarks couched in the quite non-neutral idioms of Lewis's preferred account, including reference to special features that he attributes to worlds. For example, on his construction of worlds, a person in a Riemannian spacetime cannot inhabit the same world as a person in a Lobachevskian spacetime. Such details, however, do not belong to the neutral territory. Only within the framework for which Lewis is arguing is there a gratuitous limitation of attention in the **QN** approach.

One might base a **QN**-formalization of the supervenience of the mental upon the physical upon the idea that the claim amounts to: necessarily, any things differing in mental properties necessarily differ also in physical ones. This suggests:

30) $\Box\forall x\Box\forall y\Box(\exists z(Mz \,\&\, Fxz \,\&\, \neg Fyz) \rightarrow \exists z(Pz \,\&\, Fxz \,\&\, \neg Fyz))$

where "M" corresponds to "is a mental property", "Fxy" to "x possesses (the property) y" and "P" to "is a physical property". In what respect is this, upon an intended interpretation, weaker than we want? The answer must be couched in ordinary English idioms, and must not assume any thesis about the structure of possible worlds, unless this has an independent justification. I cannot find such an answer. Note that in Lewis's own formulation of the supervenience claim, there is quantification over differences, so if this is legitimate, so should quantification over properties be.

Lewis allows that the alleged failure of **QN** with respect to this case makes little odds, but he suggests that it is more serious with respect to another, structurally similar, thesis, the supervenience of laws. The thesis is that two worlds could not differ in their laws without also differing in local qualitative character. Here is the supposed problem with **QN**-formalization:

> ... if we read the "could" as a diamond, the thesis in question turns into this: it is not the case that, possibly, two worlds differ in their laws without differing in their distribution of local qualitative character. That's trivial – there is no world wherein two worlds do anything. At any one world W, there is only the single world W. ([1986] p. 16)

"World" here is playing two roles which should be kept distinct. One role is the special one which Lewis is developing and defending in his book: a world, in this *technical* sense, is the sort of thing quantification over which will be said to translate English modal idioms. As we might put it in our terminology: it is open to the defender of the expressive advantages of **QW** to add whatever further constraints he feels are necessary to his interpretation rule for "W". In the technical sense, it is certainly correct for Lewis to say: "At any one world w, there is only the single world w", for on his construction, for all worlds w, the only world which exists at w is w.

The other role for "world" is non-technical, and it cannot be taken for granted that any Lewisian thesis holds with respect to it. The non-technical sense is that used in stating the thesis that two *worlds* could not differ in their laws without also differing in local qualitative character. A suitable **QN**-formalization is:

31) $\Box \forall x \, \Box \forall y \, \Box((Gx \,\&\, Gy) \to (\exists z(Mz \,\&\, Fxz \,\&\, \neg Fyz) \to \exists z(Pz \,\&\, Hxz \,\&\, \neg Hyz)))$,

with "G" corresponding to "is a (non-technical) world", "M" to "is a law", "Fxy" to "x is governed by y", "P" to "is a local qualitative property" and "Hxy" to "x possesses (the property) y". No doubt a full appreciation of this thesis will require a more detailed understanding of what sort of entity will be assigned to "G" upon an intended interpretation. But (a) one must not assume that these entities are the possible worlds used in giving the semantics for **QN**; and (b) a formalization can be entirely adequate even if it gives no analysis of the concepts employed. (One does not criticize the **Q**-formalization of "Socrates is human" as "$F\alpha$" on the grounds that it offers no analysis of the concept of humanity.) (**31**) is far from trivial: there are endless interpretations upon which it is false.

While we have not yet encountered an ordinary English sentence better formalizable in **QW** than in **QN**, it is certainly true that the expressive resources of **QW** outstrip those of **QN**. The existence of the predicate constant, "W", in **QW** but not **QN**, is enough to establish this. (For example, "There are worlds" is adequately formalizable as **QW**-valid, but not as **QN**-valid) Further, a **QW**-sentence like

32) $\exists w(\mathrm{W}w\ \&\ \forall w'(\mathrm{W}w' \rightarrow (\forall z\ (\text{at } w'(\exists y\ y = z) \rightarrow \text{at } w(\exists y\ y = z)))))$[21]

corresponds to an important claim about the structure of worlds (that some world contains all possible and actual individuals), yet has no obvious **QN**-correlate.

33) $\Diamond \forall x(\Diamond \exists y\ y = x \rightarrow \exists y\ y = x)$

is not an adequate formalization, for it corresponds to the trivial claim that it is possible for all possible existents to exist.

One way to progress is to introduce a device which will enable "▣" so to speak to refer back to what would have been actual if the possibility introduced by "◇" were realized. We can achieve this effect in **QN** by indexing the operators. ▣X will, by default, be true on an interpretation, i, with respect to a world, w, iff X is true upon i with respect to the actual world; but if it is indexed and occurs in the scope of an operator with the same index, ▣$_n X$ will be true upon an interpretation, i, with respect to any world, w, iff X is true upon i with respect to w', where w' is the world introduced, according to the semantics, by the previous n-indexed operator.[22] Using this device, what corresponds to (**32**) is:

34) $\Diamond_1 \forall x(\Diamond \exists y\ y = x \rightarrow ▣_1 \exists y\ y = x)$.

The interpretation of "▣$_1$" will pick up the variable attached to the existential quantifier introduced, by the interpretation rules, in the interpretation of "$\Diamond_1$".

These considerations suggest that the framework of **QN** can be extended to achieve expressive adequacy, without departing from the original idea that modal idioms are essentially sentence operators. Two questions remain: would there be any point in making these additions to **QN**, given the availability of **QW**? And do these additions keep to the spirit, as well as the letter, of the operator account?

The most promising basis for a positive answer to the first question is that there is a case for saying that **QN** but not **QW** can avoid committing itself to non-actual objects, for example, non-actual worlds. This view will be considered in §11.

In answer to the second question, it is worth noting that a distinctive feature of quantification is the possibility of back reference that can be achieved by associating quantifiers with variables. If this feature is simply being mirrored by indexing, then it looks as if indexed operators are really quantifiers in all but name. What is certain is that the indexed operators are explicitly linked, by the envisaged rules of interpretation, to variables of quantification. It is natural to conclude that unless these rules mislead, the indices function as variables. If this conclusion is justified, then it would seem that we could also conclude that even the unindexed operators are "really" quantifiers over worlds. The proponent of the suggestion just mentioned, that **QN** can avoid commitment to non-actual entities, will need to speak to this question (see §11).

9 Counterpart theory

Lewis's quantifier treatment of modal operators is combined with another distinctive, but theoretically separable, view. He holds that nothing exists at more than one world. In considering whether, for example,

1) Socrates is necessarily human

is true, we must ask not whether "Socrates is human" is true with respect to every world at which he exists, for he exists only at the actual world. What we have to ask is whether every counterpart of Socrates is human. A counterpart of Socrates at a world is an object such that nothing else at that world is more like Socrates than it is.

To express counterpart theory, we need to add to **QW** a further predicate. Let us call such an augmentation **QC**. One possibility is to add a predicate of degree 3, "$Cxyz$", to be understood "x is a counterpart in z of y". We would then need to specify an appropriate rule of interpretation, for example, the following:

  2) For any interpretation, i, any world, w, $i_w(Cn_1n_2n_3)$ is T iff $i_w(n_3)$ is a world, w', $i_w(n_1)$ exists at w, and $i_w(n_2)$ exists at w' and nothing in w' is more similar than $i_w(n_2)$ to $i_w(n_1)$.

We could then formalize (1) as:

  3) $\forall x \forall y (Cx\alpha y \rightarrow \text{at } y(Fx))$

with "α" corresponding to "Socrates" and "F" to "is human".[23] This takes us directly to a weak reading of "necessarily" in (1), since worlds at which Socrates has no counterpart will verify rather than falsify (3). We could formalize the claim that the number 7 exists necessarily (on the reading which makes it non-trivial) as:

  4) $\forall y \exists x Cx\alpha y$.

Given the predicate "C", we no longer need "W".

Adopting the above proposal leads to a problem about contingent existence. If we use the method of (3) to formalize

  5) Socrates necessarily exists

which, intuitively, should be false upon an intended interpretation, we get

  6) $\forall x \forall y (Cx\alpha y \rightarrow \text{at } y(\exists z \; z = x))$

which is true upon every interpretation. One response is simply to refuse to follow the pattern of (3) and instead, as was done in **Q**, treat existence-sentences as a special case. Then (4) can be used to formalize (5), and is, as desired, false upon an intended interpretation.

Further features of counterpart theory will emerge in connection with some objections. A well-known one, which has been put both by Plantinga and Kripke,[24] is that it fails to do justice to our intuition that when we say, for example,

  7) Socrates might have been foolish

we mean to ascribe a property, that of being possibly foolish, to Socrates himself and not to someone else, however similar.

Lewis [1986] replies, entirely justly, that an appropriate formalization of (7), e.g.

8) $\exists y \exists x (Cx\alpha y$ & at $y(Fx))$

(with "α" corresponding to "Socrates") is, on an intended interpretation, "about" Socrates. To the extent that it is "about" someone else, it does not ascribe to them the property of being possibly foolish, but rather the property, at y, of being foolish. Showing the formalization to be inadequate would require it to be shown that the truth-upon-an-intended-interpretation value of (8) might differ from the truth value of (7), and this has not been done.

Lewis explicitly allows an object in one world to have two counterparts at another. (If this were not allowed, one might wonder whether the counterpart relation differed from identity.) He is committed to this view by founding the counterpart relation upon overall similarity. Consider a red circle, a red square and a blue circle. The last two can be equally similar to the first, even if very different from one another. Similarity comes in respects (our cases involve similarity with respect to colour and shape), and even within a single respect there are different ways of judging it. A uniformly grey object is similar in respect of colour to an object giving the impression of being grey at a distance through being chequered with small black and white squares; it is also similar to a white object covered in black dots. There is no absolute fact of the matter about which it is more similar to. These facts about similarity are not exclusive to counterpart theory. They are recorded in the consistency of such mundane beliefs as "x and y are very alike, and also very different".

The consequence within at least one version of counterpart theory is that there are cases in which $o_1 = o_2$, but this is only contingent.[25] In terms of **QC**, the following is true upon some interpretations with respect to some worlds:

9) $\alpha = \beta$ & $\exists x \exists y \exists z (Cx\alpha z$ & $Cy\beta z$ & at $z(x \neq y))$,

and this appears to correspond to the following **QN**-sentence:

10) $\alpha = \beta$ & $\Diamond \alpha \neq \beta$.

This last, however, is false upon every **QN**-interpretation, with respect to every world, so there is a head-on collision between **QN** and **QC**.

There are arguments designed to show that any English sentence formalizable as (**10**) is false. Here is one:

11)   (i) Leibniz's Law: identicals have all their properties in common. More formally, every instance of the following (obtained by replacing "Π" by any predicate) is true: $\forall x\ \forall y(x = y \rightarrow (\Pi x \rightarrow \Pi y))$.

    (ii) $\forall x \square x = x$ (assumption).

    (iii) $\forall x \forall y(x = y \rightarrow (\square x = x \rightarrow \square x = y))$ from (i), replacing "Π" by "$\square x = $" (the predicate ascribing the property of being necessarily identical to x).

    (iv) $\forall x \forall y(x = y \rightarrow \square x = y)$ from (ii) and (iii).

This appears to establish quite generally that identicals are necessarily identical.[26]

In material added to his [1968], Lewis objects to step (iii): he denies that Leibniz's Law is correctly applied. He holds, in effect, that there is no unequivocal property

being necessarily identical to x

applicable now to x, now to y. To see the justification for this, we need to see what the counterpart-theoretic representation would be. (iii) would be **QC**-formalized:

12) $\forall x \forall y(x = y \rightarrow (\forall z \forall w(Czxw \rightarrow z = z) \rightarrow$
$\forall z \forall z' \forall w((Czxw\ \&\ Cz'yw) \rightarrow z = z')))$.

But this does not result from (i) by substitution of a predicate (the same one on both occurrences) for "Π". On its first occurrence, "$\Pi\xi$" (I use "ξ" to mark the gap in the predicate, the position to be filled by a name or variable) is replaced by

13) $\forall z \forall w(Cz\xi w \rightarrow z = z)$.

On its second occurrence it is replaced by

14) $\forall z \forall z' \forall w((Cz\xi w\ \&\ Cz'yw) \rightarrow z = z')$.

Since (13) and (14) are distinct, (iii) as interpreted in **QC** is not an instance of Leibniz's Law. This shows that the argument of (11) cannot be used as a basis for rejecting counterpart theory.

Lewis holds that there is a positive advantage in relaxing the usual condition of the necessity of identity (see his [1986a], 4.5).[27] Intuitively, we in fact believe, or at least should believe, something that could be formalized as (9) or (10). Here is a case to support this view. Think of a plastic utensils factory. The various utensils are manufactured by filling moulds with the precursors of plastics. The plastic itself is synthesized in the mould, so there is no gap between a certain lump of plastic coming into being, and some plastic utensil coming into being. Take in particular a bowl made in this way, and suppose that, after a few days, it is incinerated, so that both it and the lump of plastic are destroyed. At every moment of time, both or neither of the bowl and the lump of plastic exist, and when they exist, they do so in the same place, weigh the same, and so forth. This gives us reason to hold that the bowl *is* the lump of plastic, so the first conjunct of (10) is true upon an intended interpretation. However, it is possible that the factory should have received a different order that morning. Suppose the precursors were already divided up into utensil-sized heaps, and that the heap which in fact became the bowl was instead made into a waste-basket. Suppose that the mould from which the bowl was made went unused that day, but was used on the next day to make a bowl (a bowl, say, fulfilling the special order which our actual bowl fulfilled, so we can properly speak of *the* bowl). The possibility we are describing appears to be one in which the original bowl and the original lump of plastic both exist, but are distinct, the lump of plastic being a different utensil, a waste-basket, and the bowl being a different lump of plastic. If so, this will make the second conjunct of (10) true upon an intended interpretation.

In Lewis's scheme, what makes the waste-basket a counterpart of the original bowl is that it is made out of the same stuff; what makes the other bowl a counterpart of the original bowl is that both fulfilled the same order, were made in the same mould, and were the nth bowl to be made by the factory. There are different dimensions of similarity, and this is one way in which there can be more than one counterpart.

The case shows at least that **QN** is not uncontroversially an appropriate language in which to formalize our modal idioms. It raises a methodological issue: is the dispute one which *logic* should attempt to resolve? It is tempting to say that logic ought to allow room for the debate, without resolving it.[28]

Another objection to counterpart theory is that it fails to validate the inference from "$\Box F\alpha\beta$" to "$\Box\exists x F\alpha x$". A candidate **QC**-formalization of the premise is:

    15) $\forall x\forall y\forall z((C\alpha xz \ \& \ Cy\beta z)\to Fxy)$.

(**15**) will be true upon an interpretation with respect to a world at which there is no counterpart of what the interpretation assigns to "β": then the antecedent of the conditional will be false, and so the conditional true. However, such an interpretation could falsify the conclusion of the argument, if it is **QC**-formalized:

    16) $\forall y\forall z\exists x(Cy\alpha z\to Fyx)$.

A world in which there is a counterpart of α but not of β, and not of anything else to which α is related by what the interpretation assigns to "F", verifies the antecedent of the conditional but falsifies the consequent, thus falsifying the whole.[†]

    This could well seem a serious defect, for the following reason. We have

    17) $\vDash_{\mathbf{Q}} F\alpha\beta\to\exists x F\alpha x$,

and this seems plainly to correspond to a valid (in the sense of "$\vDash$") English sentence. But a valid sentence (in the most general sense) must be true of logical necessity. Hence we ought to have, and by our interpretation rules do have:

    18) $\vDash_{\mathbf{QN}} \Box(F\alpha\beta\to\exists x F\alpha x)$.

PN, and also **QN**, ensures that if (**18**) is true so is

    19) $\vDash_{\mathbf{QN}} \Box F\alpha\beta\to\Box\exists x F\alpha x$.

This, finally, ensures that

    20) $\Box F\alpha\beta \vDash_{\mathbf{QN}} \Box\exists x F\alpha x$.

So an argument that is invalid as formalized in **QC** is something which we have powerful reason to count as valid, reasons springing from features of **QN** that appear irresistible.

† See Exercise 156, page 367.

Arguably, a person could not have had different (biological) parents. On this view, necessarily it was Judy Garland (rather than anyone else) who was Liza Minnelli's mother. If we formalize this as the premise of (20), with "Fxy" corresponding to "x begat y", "α" to "Judy Garland" and "β" to "Liza Minnelli" the argument has a premise arguably true upon an intended interpretation, and a conclusion unquestionably false on such an interpretation: Judy Garland did not have to be a mother (and would not have been if certain misfortunes had occurred). There is therefore some case for thinking that **QC** is closer than **QN** to our intuitive thinking about necessity.

There is room for doubt about whether the premise of (20) does adequately formalize what we want to say about Judy Garland and Liza Minnelli. The corresponding English does best with some sort of special stress (as in the second sentence of the previous paragraph.) This can be represented in **QN**, not as the premise of (20), but rather as

21) $\Box(\exists x\ x = \beta \rightarrow F\alpha\beta)$.

The intuitive falsehood would then be **QN**-formalized

22) $\Box(\exists x\ x = \alpha \rightarrow F\alpha\beta)$.

But plainly (22) is not a **QN**-consequence of (21). Despite initial appearances, **QN** can handle the case adequately.

Relations pose other problems for **QC**. Intuitively Caesar is essentially related to his death in that none but he could have died it. Formalizing in **QN** as

23) $\Box F\alpha\beta$

with "α" corresponding to "Caesar", "β" to "Caesar's death" (or, if preferred, to "Fred", where this is a name introduced especially to refer to Caesar's death) and "Fxy" to "x died y", there is no problem about the sentence being true upon an intended interpretation. The same does not hold for the **QC**-formalization:

24) $\forall x \forall y \forall z((Cx\alpha z\ \&\ Cy\beta z) \rightarrow Fxy)$.

For if at a world, z, there are two counterparts of Caesar, c_1 and c_2, and two counterparts of his death, d_1 and d_2, (24), upon an intended interpretation, requires that c_1 be related by the dying relation both to d_1

and d_2, which, intuitively, is impossible: in no world does Caesar literally die two deaths.

However, this very example also cuts the other way. Intuitively, Caesar didn't have to die the death he did (Brutus's courage might have failed), and this fact is not well represented by (23).[†]

Plantinga objects that, intuitively, Socrates could have been very much like Xenophon actually is, and Xenophon could have been very much like Socrates actually is. We can envisage a situation in which extreme versions of these possibilities both obtain: a world in which Socrates is just like Xenophon actually is and Xenophon is just like Socrates actually is. Suppose we make "F" correspond to some predicate which gives a reasonably comprehensive account of Socrates's features (those that Xenophon would have to possess to be "just like" Socrates), and "G" to some predicate which does the same for Xenophon's features. Then

25) $\Diamond (G\alpha \ \& \ F\beta)$,

with "α" corresponding to "Socrates" and "β" to "Xenophon", is an appropriate **QN**-formalization, and, if Plantinga's claim is correct, will be true upon an intended interpretation. However, no **QC**-formalization, the objection continues, can be true upon an intended interpretation since, as counterparts are determined by similarity, the possessor of the features associated with "G" must be a counterpart of Xenophon and not of Socrates, and the possessor of the features associated with "F" must be a counterpart of Socrates and not of Xenophon.

Lewis's reply is already contained in the thought that there are different respects of similarity, which is what enables us consistently to hold that two things can be both very similar and very different. If, in the case Plantinga describes, the intended similarities are, say, in point of life history and personality, then the appropriate counterpart relation will be determined by features like origins (who the parents were). This will make for a counterpart of Socrates who is very like Socrates (say, in point of origin) but also very unlike Socrates, because "just like" Xenophon, in point of life history and personality. Lewis holds that what respect of similarity determines the appropriate counterpart relation is sensitive to context. In interpreting another's speech, the right maxim is to pick, if possible, a counterpart relation which makes what he says true. That is, implicitly, just what we do when we interpret Plantinga's example (it could have been that Socrates is just like Xenophon and Xenophon just like Socrates) as true.

† See Exercise 157, page 367.

In this discussion, we have found no decisive objection to **QC**, and a reason to doubt whether **QN** does justice to our modal judgements: its adherence to the necessity of identity as in (11iv). However, the issues are considerably more complicated than I have indicated, and the present discussion would not be an adequate basis for preferring **QC** to **QN**.

*10 Necessity and vagueness

Vagueness gives rise to borderline cases. Think, for example, of a colour spectrum. There are definite cases of red and definite cases of orange, and in between there are borderline cases: shades neither definitely red nor definitely orange. This feature of vagueness has its analogue in the case of the possible original constitution of artefacts. There are definite cases of possible differences in a thing's original parts; definite cases of impossible differences; and, in between, cases which are neither definitely possible nor definitely impossible.

It is natural to suppose that, if an area on the spectrum is red, then if you move a tiny distance in either direction, the area you get to must be red too; but if you move a large distance the area may not be red. In other words, a small difference does not matter to the correctness of applying "red" but a large difference does. The first amounts to what has been called a "tolerance principle": if two objects differ minutely in shade, then the predicate "red" applies to both or neither. This principle is in tension with the fact that large differences do make a difference to whether "red" is applicable, because you can create large differences out of a number of small differences.

Reasoning that makes this tension manifest, by delivering an explicit contradiction, is called "sorites reasoning", and the contradiction a "sorites paradox". In the case of red, we would construct a sorites argument by naming adjacent areas on a spectrum (imagine the spectrum spread out over a kilometre, and the area strips across it just 1 mm wide), $\alpha_0, \alpha_1, \ldots \alpha_n$, where α_0 is a definitely red area, and α_n definitely orange. Using "F" to correspond to "is red" we have:

1) $F\alpha_0$
$F\alpha_0 \rightarrow F\alpha_1$
$F\alpha_1 \rightarrow F\alpha_2$
...

$F\alpha_1$ (by modus ponens from the first two premises, followed by successive applications to yield:)
$F\alpha_n$.

The conditional premises are licensed by the tolerance principle. Using the fact that what is orange is not red, we get our contradiction from the hypothesis that $\neg F\alpha_n$, which accords with the principle that a big difference does make a difference to the applicability of "red".

It seems as if strictly analogous reasoning yields a similar paradox in the case of artefact concepts. The two principles that appear to be in tension are:

> **2)** An artefact could have been constructed out of somewhat different parts from those actually used.

> **3)** An artefact could not have been constructed out of totally different parts from those actually used.

(3) sounds more controversial than I intend it to be. For example, as it stands it is flatly inconsistent with Lewis's example of the plastic bowl which, he says, might have been made today out of one lot of plastic or tomorrow out of a completely different lot.[29] So let us use "parts" to stand in for all the relevant features of an artefact's construction: not merely the components but also what order the construction fulfilled, when it was made, with what implements, by whom, according to what design, and so on. It is obvious that it is not possible for there to be an arbitrarily large variation in these features, consistently with its being the very same artefact that got made.

Consider an artefact α_0, and let "F_0" represent the relevant facts of its actual construction. Let "F_i" represent a property possessed by anything having one part different from anything possessing the property represented by "F_{i-1}". Since (2) entails that α_0 could have had one part different, we allow that, possibly, $F_1\alpha_0$. We also allow that had α_0 actually possessed F_1, it could have possessed F_2. This appears to allow us to infer that α_0 could have possessed F_2. This begins a slippery slope, which will have us in the end saying that α_0 could have had F_n, where a possessor of F_n has no part in common with a possessor of F_0. But this is inconsistent with (3).

The most natural way to formalize the reasoning just informally presented is as follows:

> **4)** $\Diamond F_1\alpha_0$
> $F_1\alpha_0 \ \Box\!\!\rightarrow \ \Diamond F_2\alpha_0$[30]
>
> ...
>
> $\Diamond F_2\alpha_0$ (from the first two premises, and then further applications to yield):
> $\Diamond F_n\alpha_0$.

(4) is invalid. Unlike (1), the principle used to detach the conclusion is not modus ponens. To show that it is not a truth-preserving principle, suppose we stipulate that a ship★ is something which could have been made only out of a set of parts differing in at most one member from the set it was actually made of, and further stipulate that α is a ship★. We will still hold to the first two premises. We will hold that α could have been F_1, and that if α had been F_1, then it could have been F_2. But the stipulation ensures that α could not have been F_2. (4) fails to represent the reasoning as sorites reasoning, for sorites reasoning is valid, according to classical principles of reasoning.[31]†

Forbes has suggested that the relevant reasoning can be represented in **QN** as follows:

5) $\Diamond F_1\alpha_0$
    $\Diamond F_1\alpha_0 \rightarrow \Diamond F_2\alpha_0$
    $\Diamond F_2\alpha_0 \rightarrow \Diamond F_3\alpha_0$
    ...
    $\Diamond F_{n-1}\alpha_0 \rightarrow \Diamond F_n\alpha_0$
    $\Diamond F_2\alpha_0$ (by modus ponens, followed by $n-2$ similar applications to yield)
    ...
    $\Diamond F_n\alpha_0$.

This gives the reasoning precisely the form of (1) (replacing "F" by "$\Diamond F$" throughout). The conclusion formalizes "α_0 could have possessed F_n", that is, α_0 could have been made of entirely different parts. It is unclear whether this suggestion as it stands does justice to the original reasoning. (5) is, indeed, classically valid, but doubt must attend the truth-upon-an-intended-interpretation of "$\Diamond F_1\alpha_0 \rightarrow \Diamond F_2\alpha_0$". At first sight, it seems that nothing in our intuitions commits us to saying that a ship could have differed in original construction by two parts, merely *because* we are committed to saying that it could have differed by one. If we believe a conditional like the second premise of (5), it will be because we already accept its consequent. Whereas sorites arguments appear to have true premises, there is room for doubt whether this feature is preserved by the formalization of (5).

For (5) to appear to be sound, we have to think of the conditionals as sustained by something like the following tolerance principle:

† See Exercise 158, page 367.

6)   Small differences in membership of two sets makes no difference to the applicability of "α could have been constituted out
of ..." to them.

If this is intuitively acceptable, then (5) genuinely constitutes a sorites
paradox: it is valid, according to classical principles, has intuitively true
premises (upon an intended interpretation), and an intuitively false
conclusion (upon an intended interpretation). Sorites paradoxes are due
to vagueness. The question I want to raise is whether in the present case
the vagueness attaches to our modal concepts or rather to our artefact
concepts. There is a *prima facie* case for the latter option.

Replace our actual artefact concepts by precise ones, and the paradox
disappears. For example, "ship★" is a precise artefact concept: any given
ship★ could have been made out of just one different part, but not out of
more than one different part. The tolerance principle (6) will obviously
fail for ships★, and so there is no reason to think of an argument with the
form of (5) as sound. There is now no whiff of a paradox. A complete
explanation of the paradoxes has not alluded to special features of modal
notions.

In the context of counterpart theory, the paradoxical reasoning does
not even get off the ground, as I shall show by formalizing it in **QC**.

Rather than quantifying over counterparts, I shall assume they are
named: α_1 is a counterpart at w_1 of α_0 (which exists at the actual world, $w*$),
α_2 is a counterpart, at w_2, of α_1 (which exists at w_1), and so on.

7)   At $w*(F_0\alpha_0)$
At $w_1(F_1\alpha_1)$
...
At $w_n(F_n\alpha_n)$.

The first premise specifies α_0's actual constitution. The second premise
tells us that a counterpart of α_0 at w_1, α_1, is constructed out of slightly
different parts. To obtain anything paradoxical, we would need to assume
that α_n is a counterpart of α_0 at w_n. Then the conclusion would represent
the claim that α_0 could have been constructed out of completely different
parts. But, of course, we have every reason to suppose that α_n is not a
counterpart of α_0 at w_n, since it is completely unlike α_0 in its constitution.
In short, we have every reason to think that the counterpart relation is not
transitive, for it is based on the non-transitive relation of similarity.[†]

† See Exercise 159, page 367.

If there is an F_i such that it is neither definitely the case that α_0 could have been F_i, nor definitely the case that α_0 could not have been F_i, then the counterpart relation needs to be vague. This presents no immediate philosophical problem, since the similarity relation is vague.[32] If we are considering modal claims about the constitution of ships*, the counterpart relation will have to rule that there is a sharp step in the similarity curve: α_1 and α_2 are irrelevantly similar in that they differ by only one part, but relevantly dissimilar in that α_1 but not α_2 is a possible constitution for the ship* α_0. Hence an appropriate standard of similarity will draw a sharp line between α_1 and α_2. Where there is vagueness in the counterpart relation, its source is our ambition to be faithful to the relevant concepts, like ship as opposed to ship*. It does not appear that modality in and of itself is the source of the vagueness.

The counterpart relation will mirror the vagueness or precision of the concepts, but this does show that the *source* of the paradoxes of constitution lies with the expressly modal concepts. On the contrary, variations in the counterpart relation are owed to variations in the concepts invoked in the sentences it is used to interpret. This, in turn, is not to deny that, elsewhere, the concept of necessity imports vagueness. Counterfactual conditionals are an example.

11   Metaphysics

In the first four chapters, we managed mostly to steer clear of issues in metaphysics (the theory of being). We touched on the topic when, talking about formalizing empty names and descriptions, we pointed out that certain logical problems would not arise within a Meinongian ontology: one which includes non-existent things. We took no stand, but showed how the same logical problems are standardly treated within a non-Meinongian framework. If this treatment is adequate, logic does not at this point determine metaphysics; if it is inadequate, logical matters may well be used to settle metaphysical ones.

In the present chapter, metaphysical problems have arisen at every turn, although we have studiously avoided addressing them. The time has come to bring some of the most fundamental ones into the open. My aim is merely to display the connections between the (supposedly purely logical) problems we have been discussing, and metaphysical ones. I make no attempt to resolve the latter. I merely want to show that modality, if it is a branch of logic at all, is certainly not one which can be studied in isolation from the rest of philosophy.

First, *realism*. The picture is that you are a realist about a certain subject matter if you think that there really are facts belonging to the subject matter that are not mere artefacts of our thought or language. We are all instinctively realists about rocks and rivers: there they are, and they are what they are whether we think about them or not and irrespective of how we think about them if we do. To the facts about rocks and rivers, the mind makes no contribution. We are all instinctively non-realists about fiction. We do not think that there is a fact, independent of our thought or language, concerning whether the Red Duchess's pepper was black or white (in *Alice Through the Looking-Glass*). If there is such a fact, it was created by Carroll's thoughts. If Carroll's thoughts never turned to the matter, then there is no fact either way.

Modal realism (as I use this expression) is the view that there are modal facts, facts of the form "necessarily A" and "possibly A", which are what they are independently of our thought and language. Being a modal realist, as this term is properly used, involves no further commitment: in particular, it does not pronounce on whether these facts are properly formalized by **QN** or **QW** or **QC** or by some yet other language. Unfortunately, the phrase "modal realism" has been kidnapped by David Lewis to describe one particular brand of modal realism: that which holds that (i) modal facts are real and mind independent, (ii) they are best represented in Lewis's preferred form of quantifier treatment, viz. counterpart theory, and (iii) there "really" are non-actual individuals and worlds.[33]

Lewis's usage is unfortunate, as it places distinct views under a single umbrella. It may also speciously tempt someone who is a modal realist (in my sense) to side with Lewis on the other issues. I need other names for the other of Lewis's theses. We may call the second a species of "quantifierism". Quantifierism is the doctrine that ordinary modal idioms are best represented as quantifications over worlds, and counterpart theory is a special form of quantifierism. The third thesis embraces two distinguishable positions. First and foremost, it is "anti-actualist". An "actualist" is one who affirms, an "anti-actualist" one who denies, that everything is actual. If you are an anti-actualist, you face the question of whether you are a realist or not concerning non-actual things you claim exist. Are they "out there", or are they creatures of the mind? What Lewis calls "modal realism" thus consists in the conjunction of what I call modal realism with quantifierism (in particular, counterpart theory), anti-actualism and realism about non-actual entities. In what follows, I

Table 5.1

	Common sense	Lewis	Ersatzism
Modal realism	√	√	Neutral
Quantifierism	×	√	√
Actualism	√	×	√
Realism re non-actuals	—	√	—

shall keep to my narrower, and in my view more correct, usage of the term "modal realism".

It will be useful at the start to set out some of the positions in the form of table 5.1.

Common sense combines modal realism with rejection of quantifierism (for the natural expression of modality is by idioms which at least appear to be sentence connectives, and not quantifiers). In addition, common sense affirms that everything that exists actually exists, so it accepts actualism, and the question of realism or some other view with respect to non-actuals does not arise.

Lewis's view is as we listed it earlier, though thesis (iii) is now divided into the rejection of actualism together with the acceptance of realism about non-actuals.

The label for the third view comes from Lewis [1986a]. This is the view, available alike to modal realists and modal anti-realists, that possible worlds and their occupants, even non-actual ones, are relatively familiar actual objects.[34] I will discuss this shortly.

One could theoretically combine, without obvious inconsistency, being a modal anti-realist with being a non-actualist, but then one would be committed to being an anti-realist about non-actuals. If one were a realist about non-actuals, then one would believe things like "there really is a non-actual possible world such that A, and this is so independently of our thought and talk", and that would commit one to believing that some sentences of the form "possibly A" are really true, independently of our thought and talk.

How could one fail to be a modal realist? Kant said that experience can tell us how things are, but not that anything must be as it is.[35] This line of thought could ground modal anti-realism. All there really is in the world is open to observation, directly or indirectly. If observation cannot tell us how things must be or could be, but only how they actually are, then any necessity or possibility we believe in must have its source in us, in our language, our sensibility or our thought. A traditional favourite, though not Kant's, identifies the necessary with the "analytic" – that which is true in virtue of meanings. Supposedly, an example would be "All bachelors are unmarried". That this is *necessary* is no fact about bachelors (so the doctrine goes) but about our language. So this is an example of modal anti-realism. It has the disadvantage of disallowing the expression of de re modality, since it essentially involves seeing a modal expression as taking an entire sentence as its scope: this is the unit of language from which the necessity is said to derive.

Another form of modal anti-realism takes its cue from a species of moral anti-realism. Should one be realist or not about moral values? Should one suppose that they are genuine components of the world, or that they are artefacts of our responses to the world? If one takes the latter view, one might see the moral modalities – expressed by such sentences as "you must do this", "you ought not to do that" – as an attempt to objectify what is essentially a subjective response, and to warn that one's own response is something one is likely to act upon by approval, disapproval or whatever. One could extend this view to the modalities we have discussed in this chapter. The force of "$\Box A$", roughly speaking, is to indicate that one will not budge on the question of "A": it's non-negotiable.[36]

I mention these forms of anti-realism about modality not to recommend them, but only to help clarify what is involved in being a modal realist: rejecting these doctrines.

It might seem that the combination of modal realism and actualism points one towards **QN** rather than **QW** or **QC** as providing the proper forms for representing modal idioms. One might argue: "**QW** and **QC** involve quantification over non-actual objects, and I cannot accept this because I am an actualist. But I can accept **QN**, because that does not involve quantification over non-actual things. Modality is represented by sentence operators, and not by quantifiers over anything." This argument is flawed in both its parts. First, there is a case for saying that one can combine quantifierism with actualism: this combination is ersatzism. Secondly, it is not obvious that **QN** is really free of non-actuals, because the interpretation rules for **QN** are couched in terms involving non-actual worlds and non-actual denizens of them.

The ersatzist regards "non-actual" worlds as relatively familiar actual objects. For example, he might say that "worlds" are set-theoretic abstractions with the capacity for representing things as they are and also as they are not. What we call "non-actual" worlds are actual set-theoretic abstractions which represent things as being other than as they in fact are.

He might say: "I believe that a so-called non-actual world is an actual object, just as a picture is an actual object. It gets called 'non-actual' because the situation it purports to represent does not exist. But the anti-actualist thinks that when we say, of a fanciful picture, that it depicts some state of affairs, there really must exist such a state of affairs, if not in this world, then in another. That is a mistake." It is a *further* question for the ersatzist whether he is a realist or not about (ersatz) worlds. That will depend on whether he thinks that sets, for example, are mind-independent, or the creation of set-theorists. Notice that this question about realism is distinct both from modal realism and from realism about non-actuals. It is distinct from the latter, since it arises within a position according to which there are no non-actuals. It is distinct from the former, since the question of realism or anti-realism about sets (or whatever one takes worlds to be) is distinct from the question about the reality or otherwise of the facts that sets are used to represent: mind-independent sets could represent mind-dependent modal facts, and mind-dependent sets could represent mind-independent modal facts. If the inference from "there exist non-actual worlds" to "there exist non-actual things" is blocked, the ersatzist can consistently accept all the joys in the worlds.

I want to look more closely at the question of whether a modal realist who is also an actualist can simply dispense altogether with worlds, genuine or ersatz, as the common sense view tries to do. Such a theorist should reject any form of quantifierism, and nail his flag to **QN**. But the question arises: what is all this talk of non-actual worlds and non-actual individuals in the metalanguage in which the interpretation rules for **QN** are couched? How could one buy the language, without buying its semantics? If interpretation rules specify meanings, then the meaning of "□" according to **QN** is worlds-involving. I envisage two possible lines of reply. One is to take what one might call an "algebraic" view of the semantics for a language. **QN** was introduced for logical purposes, that is, for the purpose of formalizing arguments in order to enable them to be more readily assessed for validity. You might say that the so-called "semantics" for that language are a mere algebraic abstraction, designed to enable a calculation of whether or not an argument is valid. Worlds can be thought of as mere abstract objects, with no serious connection with thought, so one can take any view one likes about them, without any consequences for the realism one wishes to hold with respect to facts formalizable in **QN**. In short, as with ersatzism, you accept talk of non-actual worlds, but deny that this is a way of talking of non-actual objects.

There is an anaology with ersatzism, but it is only partial. The ersatzist

I have envisaged is a quantifierist. The combination would at least strongly suggest that the theorist thinks that what is actually going on when people use modal idioms is that they are quantifying over (ersatz) worlds: the sets (or whatever) are in some way before the minds of the users of the language. The common sense theorist who turns to an algebraic view of semantics to defend his position will deny this. The "worlds" that enter into the semantics are tools of the linguistic theorist, not objects of thought for the user of the language.

Here is an easily stated application of the algebraic view of semantics. Consider the talk of truth values in the semantics given in chapter 2 for the language **P**. One could say: the concept of *truth* does not enter into these semantics. The truth values are just a pair of arbitrary objects (many logicians choose the numbers 0 and 1, or even the numerals "0" and "1"). The calculation of validity simply requires the definition that an argument is valid iff any interpretation assigning a designated one of these objects to all the premises also assigns that object to the conclusion. The nature of these objects is irrelevant to ordinary speech, nor would any realist or non-realist thesis you might have about them have any special bearing on realism or non-realism with respect to the facts that are **P**-formalizable.

A quite different, non-algebraic, view of the role of semantics in logic has been adopted here. We have said that truth values do connect with the intuitive notion of truth: *having the truth value true* is simply *being true*. On this view, the semantics for **P** connect closely with other features of truth, for example: it is good to aim at believing something with truth value *true*, and it is good to reason **P** validly because then you can never go from truth to falsehood.

The other route available to the common sense combination of anti-quantifierism and actualism is to claim that the possible world quantifications in the metalanguage for **QN** owe their meaning to their sentence operator counterparts, and so are not "really" quantifiers at all. In justification of this view, one might, first, claim that there is an equivalence between sentence operator idioms like "□" and quantifications over possible worlds. An equivalence, however, is symmetric: it tells you either that the sentence operators are really quantifiers (the conclusion of quantifierism) or that the quantifiers are really sentence operators. Forbes urges us to take the latter course, on the ground that operators have a certain kind of primacy: they are closest to our mother tongue, our natural English modal idioms, and it is they that confer the meaning upon the quantifiers of the metalanguage, rather than the other way about.

A difficulty for this view is that quantifiers in general occur in all sorts of contexts: quantifications over books and rocks and numbers, as well as over worlds. The sense of a quantifier must be to some extent fixed by its occurrence in these other non-world contexts. This ensures that quantifiers do, quite generally, introduce objects, and thus do so when they quantify over worlds. To see the worlds-quantifiers of the metalanguage of **QN** as not really introducing entities would be to see them as quite separate idioms from our ordinary "all" and "some" (or "∀" and "∃"). It is open to the semantic theorist to *stipulate* that his metalanguage quantifiers, used in the semantics of **QN**, are just the ordinary, object-introducing ones. By contrast, it is of dubious intelligibility to claim, as Forbes does, that the metalanguage quantifiers are a special idiom, owing their sense to the operator counterparts in the object language.[37]

It should not be supposed that the only semantics for a language with the operator syntax of **QN** involve quantifiers. Just as one could specify the contribution of "not" by a sentence like

1) A sentence "not-*A*" is true iff it is not the case that "*A*" is true

so one could, arguably, specify the contribution of "□" or "necessarily" by a sentence like

2) A sentence "□*A*" is true iff it is necessarily the case that "*A*" is true.

Whether such semantics will serve the purpose of defining validity is another question, and a disputed one.[38]

Finally, I turn to a second more or less metaphysical issue: *analysis*. Those looking for a reductive explanation, or "analysis", of what modal concepts mean will not have found even the beginnings of one in this chapter. A reductive explanation of a concept is one any part of which you can fully understood without yet understanding the concept to be explained. However, to be told that "necessarily *A*" is true iff "*A*" is true in all possible worlds is circular, relative to reductive aims, because the account makes free use of the notion of "possible", and we all know that "necessary" and "possible" are interdefinable: if we already understood "possible" we would not *need* the account, and if we did not understand it the account would be *useless*.

However, it was no part of our aim at any point to provide a reductive explanation. Exactly the same objection of circularity could be levelled at the account of "all" and "∀", if that were supposed to be reductive. We start with English "all"-sentences. We formalize them using "∀" and explain the meaning of "∀" by giving an interpretation rule which crucially involves "all". Were this an exercise in analysis, it would be unrewardingly circular. It was not such an exercise, but rather an attempt to fashion an artificial language in which the notion of validity would be more accessible to theory than it is in ordinary English. If that is the modest aim of introducing a language like **QN**, the alleged "circularity" is beside the point.

The only candidate for a reductive account of modality of which I am aware is that offered by David Lewis [1986a]. He says that one can say what a possible world is without using any modal notions: a possible world is a sum of (not necessarily actual) objects linked by space-time relations. The explanation essentially involves anti-actualism. So if one found the explanation attractive, it might help one overcome the commonsensical appeal of actualism.

Notes

§1

For a comprehensive introduction to propositional modal logic see Chellas [1980]. For a philosophical account of modality, which includes both formal semantics and arguments for many substantive essentialist claims, see Forbes [1985]; and, for critical discussions of this, Edgington [1988] and P. Mackie [1987]. Lewis's most famous use of "the ways things could have been" as an introduction to possible worlds is in his [1973b], ch. 4.1, p. 84; for criticism, see McGinn [1981].

§2

The earliest presentation of Lewis's theory is Lewis [1973a]; the most comprehensive is his [1973b]; the most recent is his [1979]. The last contains discussions of criticisms, and would make a good introduction to his views. For Stalnaker's theory, and his response to Lewis's criticisms, see, for example, his [1984], ch. 7. See also bibliographical notes to ch. 3, above.

§3

A standard introduction to quantified modal logic is Hughes and Cresswell [1968]; see also their [1984]. The historical source of the accessibility relation is Kripke [1963]. See also Chellas [1980], and Bull and Segerberg [1983].

§4

In the empiricist tradition, necessity was held to be de dicto, and coextensive with analyticity. For an expression of this view, see Quinton [1963]. For distinctions between necessary, analytic and *a priori* see Kripke [1972]. These lectures were responsible for a considerable revival of interest in de re necessity. The present formulation of the contrast between de re and de dicto derives from Forbes [1985] p. 48. See also J. Mackie [1974] and Wiggins [1976].

§5

For Quine's arguments, see his [1953c], [1953d] and [1960] section 41. For criticism see Plantinga [1974], Appendix, pp. 222–51, and Linsky [1977], ch. 6.

§6

For Quine's version, see his [1960] pp. 148–9. Davidson uses the argument in a number of places, for example [1967c] p. 3. The account I follow most closely is Davies [1981] pp. 210–11.

§7

For discussions of the problems of trans-world identity, see: Plantinga [1974] pp. 88ff., Kaplan [1979], Forbes [1985], esp. chs 3 and 7, van Inwagen [1985], Fine [1985], and Lewis [1986a] pp. 210–20. For a discussion of the linguistic theory of necessity, see Pap [1958], esp. Part II, ch. 7 and van Fraassen [1977].

§8

See Lewis [1968] and [1986a]. In the latter, see pp. 5–20 for the introduction of the open-sentence formers, and pp. 199–202 for the case against treating intrinsic properties as relations. (NB although I use this to attack the extra argument place theory, in Lewis's book the argument has a different target.) Lewis, in both places, argues for counterpart theory (see §9). Note, however, that there are two separable questions: whether to adopt any quantifier treatment of modality, and whether to adopt the specific form of quantifier treatment embodied in counterpart theory. See also Hazen [1976], Davies [1981], ch. 9, esp. §1, and Forbes [1985] pp. 89–95.

§9

The main texts for counterpart theory are, again, Lewis [1968] and [1986a]. See also Forbes [1985], esp. chs 3.4 and 3.5, and Appendix 3; Mondadori [1983]; Ramachandran [1989]. For discussions of contingent identity see Forbes [1985], Wiggins [1980b], and, especially, Gibbard [1975].

§10

Lewis's remark about the non-transitivity of the counterpart relation shows *that* the paradox will not arise in his scheme, but it does not give any details of the semantic mechanisms. For such details, see Forbes [1985], ch. 7.

§11

For Lewis's views, see his [1986a]. See also Forbes [1985], esp. ch. 4, Kripke [1972] and esp. [1980]. Modal anti-realism is endorsed in, for example, Wittgenstein [1921], Ayer [1936]; and, in the form that necessary truths reflect merely human conventions, famously opposed by Quine [1936]. A new twist has been given by Blackburn in various writings, e.g. [1986b]. See also: Wright [1986] and [1989], Hale [1989], Craig [1985]. Ersatzism is argued for in Plantinga [1974], ch. 8, and Plantinga [1976] (though in this article Plantinga's target is the view that there are non-existent objects, rather than the view that there are non-actual objects). See also Plantinga [1987]. An argument against ersatzism by Lewis is discussed in van Inwagen [1986]. For a criticism of Lewis's [1973b] argument for the reality of non-actuals, see McGinn [1981].

---

1 The nomenclature "S5", "S4", derives from Lewis and Langford [1952].
2 For example, to establish that $\vdash_{PN} p \rightarrow \Diamond p$, first use (14) to establish that $\vdash_{PN} \neg\neg p \rightarrow \Diamond\neg\neg p$, and then apply $\vdash_{PN} X \leftrightarrow \neg\neg X$.
3 For Stalnaker's response to this objection see, for example, his [1984] pp. 133ff.
4 Cf. Bennett [1974], esp. pp. 387–8.
5 See Lewis [1979] for his account.
6 A world with an empty domain can be allowed for generality, but those of us who believe that there is at least one necessary existent will be less permissive.
7 (5) and (6) between them assume what Forbes calls the "Falsehood Principle": an atom is false with respect to a world if a name-letter in it is not assigned a referent with respect to that world (see Forbes [1985] pp. 29–31). There are other ways of proceeding, which would lead to different truth-upon-an-interpretation conditions. The motivation for my own choice is to keep as closely as possible to the semantics of **Q**, which required every name-letter to be assigned an object.
8 The semantics of **QN** are not fully determinate, since it was not specified whether there are any worlds with empty domains. If there are, then (10) will not be true upon an intended interpretation. Those who believe (as I do) that there is at least one necessary existent will require that no worlds have an empty domain (though this requirement is insufficient to embody the belief).
9 The interpretation rules for **QN** are designed to mirror a view of names proposed informally by Kripke [1972]: that they are rigid designators. This

view has been the subject of considerable debate: see e.g. McCulloch [1989] and bibliographical references associated with chapter **4.7** and **4.12**.

10 It is crucial to this point that "the number of the planets" be formalized by a name-letter.

11 See Quine [1960], section 41, the example of the mathematical cyclists. I have made trivial changes in the details, to avoid irrelevant objections.

12 Compare Quine [1960] pp. 148–9; Davidson [1967c] p. 3; Davies [1981] pp. 210–11. The source is supposed to be Frege [1892].

13 Cf. ch. **4.18**. An opaque position for a name is simply a non-extensional one.

14 For scepticism about some of these claims of non-extensionality in English, see K. Bach [1987] pp. 206–14.

15 Hence it is bad terminological practice to abbreviate "non-extensional" as "intensional".

16 Strictly speaking, the notion of an extension has not been defined for definite descriptions. This is no accident. From the perspective of languages like **Q** and **QN**, definite descriptions do not have extensions: the interpretation rules do not assign entities to them. However, for dialectical purposes it is obvious how the notion of the extension of a description should be understood.

17 Kaplan has suggested a context in which the inclination is strong: a museum has sent a philosopher to Greece to buy, crate and dispatch the ship of Theseus for reassembly in the museum. As the philosopher removes a plank, he replaces it with a brand-new one of exactly the same shape, so that when he has finished he has an assembled ship of new parts and a dismantled ship of old parts. He sends the latter to the museum. Should the museum director be seriously perturbed when he gets a phone call from the philosopher announcing that he has the *real* ship of Theseus, still in Greece?

18 What counts is the "at v" operator closest to the left of the first occurrence of the variable, say "x". The objects relevant to the quantification are those in the world which is assigned to "v", even if "x" subsequently occurs in the scope of some distinct operator "at v'".

19 Perhaps adding an "actually" operator is not the only way to express sentences like (**15**): see Teichmann [1990].

20 It may be that this approach requires the full resources of second order logic (see chapter **4.19**). There are many questions here whose complexity puts them beyond the scope of this book.

21 The example is due to Hazen [1976]. See also Forbes [1985] pp. 90–1.

22 For a more accurate specification, see Forbes [1985] p. 91, n. 28.

23 The "at y" operator appears to be redundant in this case, though it is not in (**6**) below. In what follows, some redundant occurrences are omitted.

24 Plantinga [1974] pp. 115–16; Kripke [1972] p. 344, n. For a good discussion, see Hazen [1979].

25 See Lewis [1968] p. 36. For a discussion, see Ramachandran [1989].

26 This proof goes back at least to Barcan [1947]. See Wiggins [1980b] pp. 109–11 and 214–17.

27 In this discussion, I wrench a strand from the complex and interconnected fabric of Lewis's thought. Such isolation does not do justice to the strength of his total position.

28 For a careful philosophical argument for contingent identity see Gibbard [1975].

29 Not that I agree with Lewis on this point, but I want to make plain that the paradox to be discussed is one that all views must confront.

30 $p \,\square\!\!\rightarrow q$ corresponds to "if it had been the case that p it would have been the case that q": see §2.

31 Some theorists treat iterated modus ponens as invalid, and so construe sorites reasoning as invalid: see Goguen [1969], or, for a discussion closer to present concerns, Forbes [1985] ch. 7, section 3.

32 There are some non-obvious philosophical problems, and, of course, technical ones. See Forbes [1985] ch. 7.

33 My characterization of Lewis's usage of "modal realism" in the text seems to be closer to his actual usage than his official claim in the preface to [1986a]. There he says that, as he uses the expression, modal realism claims simply the existence of certain objects (worlds and individuals) and "not the objectivity of a subject matter" (p. viii). However, he makes it plain that the objects in question do meet condition (i) of the text, if anything does; and he argues that (ii) is entailed by (his version of) (iii).

34 My usage departs from Lewis's. In his [1986a] he refers to what I call ersatzism as "ersatz modal realism" (see e.g. p. 136). This terminology would suggest a foreclosing of some of the possible combinations of views I wish to consider.

35 Kant [1787], A1. For a criticism, see Kripke [1980] pp. 158–60, and the text to which this relates, viz. Kripke [1972] pp. 35ff.

36 There are at least two importantly different versions of the approach I am calling anti-realist. One version does, the other does not, see an account of necessity, moral or logical, in terms of subjective responses as showing that our ordinary conceptions are in error. For the contrast, see Blackburn [1986b] and the suggested reading in the bibliographical notes to this chapter.

37 A similar point applies to an attempt to endorse **QW**- or **QC**-truths, yet reject the accompanying ontology.

38 One aspect of the dispute involves the relationship between truth theoretic semantics (as exemplified in (**1**) and (**2**)) and model theoretic semantics – the style of semantics in terms of interpretations provided throughout this book. Cf. Evans [1976].

6

The project of formalization

1 Logical versus grammatical form

In chapter 1 I gave some preliminary motivations for studying validity through the medium of artificial languages. In subsequent chapters I presented some of these languages, indicating how they could be used to formalize arguments expressed in English, and in many cases illustrating detailed limitations. We now have to raise our heads from the trees, and try to discern the overall character of the wood.

The main subject of my discussion is the view that formalizing a sentence or argument of a natural language in one of the artificial languages we have discussed reveals something about the nature of the natural language, something that would otherwise be apt to remain hidden. It is this revelation which justifies the efforts we have expended in formalizing. **Q** has been especially favoured by proponents of this view, so it will occupy centre stage.

The revealed must look at least superficially different from the concealed, or there could be no revelation. Let me list some apparent differences between natural language sentences and their formalizations.

1) Quantifiers: Some English universal (existential) quantifications not containing an occurrence of "if" ("and") are formalized by a **Q**-sentence containing "→" ("&").

2) Adjectives: English adjectival modification is formalized by **Q**-conjunction.

3) Descriptions: An English sentence of the form "The F is G" is subject–predicate but its **Q**-formalization is an existential quantification, containing constants like "→" and "=" whose correlates are not visible in the English.

4) "Exists" is a predicate in English but must often be matched by a quantifier in a **Q**-formalization.

5) Numeral adjectives: An English sentence like "Three men are at the door" contains neither an existential quantifier nor the identity sign, but its **Q**-formalization contains "∃" and " = ".

6) Verbs of action: an English sentence like "John walks" is subject–predicate but its **Q**-formalization (on Davidson's proposal) is an existential quantification.

7) Adverbs: Adverbial modification is (on Davidson's proposal) formalized by **Q**-conjunction.

8) Propositional attitudes: an English sentence like "John believes that the earth is flat" is a single complex sentence, but its **Q**-formalization (on Davidson's proposal) is two separate sentences.

9) Necessity: "Socrates is necessarily human" is in English a necessitated atom, whereas its **QN**-formalization is the necessitation of a conditional, containing the extraneous concept of existence.

10) Counterpart theory: "Socrates is necessarily human" is in English a necessitated atom, whereas its **QC**-formalization is a universally quantified conditional, containing the extraneous concept of a counterpart.

The confident classifications of English sentences (e.g. the assertion that an English sentence of the form "The F is G" is subject–predicate) are intended to reflect, not a theoretically grounded view, but the deliverances of our naive intuitive syntax. In the tradition I am describing, an important thesis is that formalization shows that many of our intuitive classifications of natural sentences are simply wrong. The tradition has it that universal quantifications in English are shown, by formalization, to be "really" quantified conditionals, definite description sentences are shown to be "really" existential quantifications; and so on. These facts are concealed from the naive and untrained eye, which sees only "grammatical form", but they are revealed by formalization, and these revelations are the main value of the project.

The thesis is sometimes expressed in the slogan

Grammatical form misleads as to logical form.[1]

The slogan may fail to capture the thesis. Using "logical form", as we are, to mean a sentence in some favoured artificial language, the truth of the slogan simply points up the existence of divergences like those noted in (1)–(10). The thesis, however, requires the further point that the divergences are only superficial, for *at bottom* the natural sentences have the features that are so readily visible in their logical forms. The slogan does more justice to the thesis upon a different interpretation of "logical form", according to which the phrase signifies not a formalization into some possibly alien language, but the intrinsic logical and semantic properties of the sentence. The thesis is that a natural sentence's logical form in my sense reveals its logical form in this other sense.

In the rest of the section, I shall elaborate some more precise versions of the traditional thesis about logical form. First, however, I want to emphasize how strange the thesis should initially appear. Nothing about our own language seems clearer than that "All men are happy" does not contain an expression for conditionality, or that "Shem kicked Shaum" does not contain an existential quantifier. Could one mitigate the strangeness of the thesis by likening logical theory to scientific theory? Let us see how this comparison would look.

Suppose we are concerned with the motion of real billiard balls on a real billiard table. We may find it convenient to make simplifying assumptions: the table is completely flat and frictionless, the balls are perfectly elastic, there is zero resistance from air and baize, and so on. We can with justice say that we are still studying the original concrete phenomena through the idealization. Laws statable in terms of the idealization can be applied to the real balls to yield fairly accurate (if not perfect) predictions of their motions. We could at any point achieve greater accuracy by removing some of the simplifying assumptions. No doubt we would do this until the cost of the additional time taken to solve the equations exceeded the benefit of extra accuracy, a balance that would be dependent on idiosyncratic needs and interests. There would be nothing sacrosanct about any one idealization, and there would be absolutely no temptation whatsoever, on finding, for example, that the assumption of perfect elasticity yielded adequately accurate predictions, to infer that the real balls are "really" perfectly elastic. The approach builds in the acknowledgement that the idealization differs from the phenomena.

The phenomena for logical theory are arguments in a natural language, say English, and the theory should pick out their validity-relevant features. An idealization can properly abstract from other features, for example from the actual mechanisms whereby specific truth-conditions are expressed. Though formalizations do not capture all the features of English, they enable one to give reasonably accurate predictions of the validity of English arguments. A divergence, for example a case in which an intuitively valid argument in English is formalized as invalid in some artificial language, may simply reflect, what we knew already, that the idealization does not exactly correspond to the phenomena. Alternatively, it may lead us to reconsider our intuitive judgement, though a divergence alone could never be a good reason to abandon such a judgement.

While one could not reasonably quarrel with this approach to validity in natural language, it is not one which yields the distinctive theses of the tradition I have in mind. The approach would not license the attribution to English sentences of all the features of the idealization. For example, to say that "All men are happy" is "really" a quantified conditional would be as ludicrous as saying that billiard balls are "really" perfectly elastic.

To understand the traditional conception of logical form, we need to distinguish variants. Let us start by making a very rough distinction, to be refined in §4, between *logical* features of a sentence or argument and *semantic* features. A prominent logical feature of a sentence would be its logical constants, and the pattern of occurrence of the non-logical expressions. Logical features would be those relevant to validity, or at least formal validity. Semantic features would include logical features, but would also include any other features pertaining to the meaning of the sentence and the words which compose it. We can distinguish two groups of theses about logical form: those which speak only to logic, and those which speak also to semantics. I shall begin by discussing a series of theses ((**11**), (**14**) and (**15**)) which fall into the latter category.

> 11) The logical form (i.e. adequate and deep formalization) of a sentence of a natural language gives a representation of its truth conditions.

We imposed the following condition of adequacy upon formalization: a formalization is adequate iff the recovered sentence or argument agrees with the original in truth conditions; iff the truth-upon-an-intended-interpretation conditions of the formalization agree with the truth conditions of the original. This ensures the truth of (**11**). Some of the most famous claims about the truth conditions of English sentences, like Russell's claim about the truth conditions of sentences containing definite descriptions, have arisen within the context of the project formalization.

(**11**) has no exotic consequences for the real logical nature of the natural language sentences. In general, two sentences may agree in truth conditions, but in other respects be quite unlike, for example the pair:

12)   Either snow is white or it is not

13)   If snow is white it is white.

The identity of the truth conditions gives us no reason to say that (**12**) is "really" a conditional, or that (**13**) is "really" a disjunction. The fact that a natural sentence's logical form matches the sentence in point of truth (-upon-an-interpretation) conditions does not in itself show that there is any other interesting relation between the two. Sameness of truth conditions is symmetric. It accordingly gives as much reason for thinking that the arrow in a **Q**-formalization of an English quantification is not "really" there as for thinking that the English "really" contains an (invisible) "if".[2]

The following strengthening of (**11**) would introduce an asymmetry:

14)   The logical form of a sentence of a natural language gives a *perspicuous representation* of its truth conditions.

One can make good sense of the notion of perspicuity. Let us say that a *syntactic category* is determined by the following test: expressions e_1 and e_2 belong to the same syntactic category iff, for every sentence s containing e_1, the result of replacing e_1 by e_2 is a (grammatical) sentence. Syntactic categories, on this account, are those of *naive* syntax, in the sense of chapter **1.12**. In **Q**, the syntactic categories include the following:

Unary sentence connectives: ¬
Binary sentence connectives: &, ∨, → and ↔
name-letters
1-place predicate-letters
2-place predicates (=) and predicate-letters
3-place etc
…
quantifiers: ∀, ∃
sentences (including sentence-letters)

Let us say that a *semantic category* is determined by the following test: expressions e_1 and e_2 belong to the same semantic category iff either they are assigned the same kind of entity by the interpretation rules, or else they are treated by the same kind of interpretation rule. In virtue of the first disjunct, all and only name-letters belong to one single semantic category, since they, and only they, are unrestrictedly assigned members of the domain; likewise all and only sentences belong to one single semantic category, since they, and only they, are invariably assigned truth values; likewise predicate-letters of any given degree, say, 3, belong to a single semantic category since they, and only they, are all assigned a set of triples of members of the domain. The second disjunct is intended to be read in such a way that all binary sentence connectives count as belonging to a single category, since they, and only they, are given rules of interpretation which fix a truth value for the resultant sentence on the basis of the truth values of two components. (We could have managed with just the first, more precise, disjunct in the definition had we required that an interpretation assign n-ary truth functions to each n-ary connective, and also made a suitable assignment of kinds of functional entity to the quantifiers.)

The perspicuity of **Q** can be defined as consisting in the fact that its syntactic categories coincide with its semantic categories.[3] By contrast, the syntactic and semantic categories of English do not coincide. Names like "Reagan" and quantifiers like "someone" arguably belong in the same syntactic category by the naive syntactic test; and names and definite descriptions are incontrovertibly in the same syntactic category by this test.[4] Yet the former pair must certainly be placed in different semantic categories; and so must the latter pair if one view considered in chapter 4 is correct: that names should be formalized by name-letters and definite descriptions should not be.

Perspicuity is connected with two features which are highly prized by logicians. First, a perspicuous language is one for which one can relatively easily formulate *rules of proof*, rules which, stated in terms of the physical make-up of sentences, specify derivations of sentences from others, in such a way as to ensure that a sentence derived from others is the conclusion of a valid argument if the others are its premises. Secondly, a perspicuous language is one for which one can relatively easily formulate *rules of interpretation*. For artificial languages, we actually possess rules of both kinds, whereas for English we possess rules of neither kind.

Thesis (14) injects an asymmetry in the relation of sameness of truth conditions, since a sentence of an artificial language like **Q** is a sentence of a perspicuous language, whereas its equivalent in a natural language is not. However, there is no inference from this to some more exotic thesis, expressible by such remarks as that English universal quantifications are really universally quantified conditionals. Since English is not perspicuous, there must be some vital difference between the way in which an English sentence and its formalization present their truth conditions. Indeed, there would be something paradoxical in the assertion that those features of the artificial languages which make the formulation of rules of proof and interpretation relatively easy are really present in the English sentences, for which the formulation of such rules is hard or impossible.

We get closer to what many people have had in mind by the idea of logical form in the following strengthening of (14):

15) The logical form of a sentence of a natural language gives a perspicuous and *systematic* representation of its truth conditions.

The notion of a systematic representation, which will be refined in §6 below, is linked to the notion of *compositional semantics*. The rough idea is as follows. The meaning of a sentence is determined by the meanings of the words of which it is composed, and by their manner of composition. Compositional semantics for a language will specify word meanings, and the semantic import of modes of composition, in such a way that, from these specifications, the meaning of any sentence in the language can be derived. As with meaning, so with truth conditions. Setting aside ambiguity, a word makes the same contribution to the truth conditions of any sentence in which it occurs, and a compositional semantics will specify this contribution. A *systematic* representation of truth conditions, alluded to in (15), is a representation within the perspective of a compositional semantics. On this view, a sentence's logical form will contribute to an understanding of how the words in the sentence contribute systematically to the truth conditions of the whole.

Thesis (15) offers an important aspiration. However, it is at first sight hard to see how its achievement would be consistent with some of the specific classical claims about logical form. For example, on the face of it, one who aspires to give a systematic representation of the truth conditions of universal quantifications in English should not pretend that such quantifications standardly contain an expression for conditionality.

I return to thesis (15) in §6. (15), in common with its weaker predecessors, (11) and (14), concerned semantic features. I now want to introduce two theses relating specifically to logical features:

16) If an argument in natural language is adequately formalizable as valid, then it is *formally* valid.

17) The formalization of a natural sentence renders the proposition it expresses accessible to formal deductive manipulations.

(**16**), which I discuss more fully in §2, does not entail that the nature of the features responsible for formal validity is intrinsically characterized by the logical form. For example, the **Q**-validity of the formalization of, say "All men are mortal, Socrates is a man, therefore Socrates is mortal" assures us, according to (**16**), that the argument is formally valid, but gives us no assurance that it is valid in virtue of having a quantified conditional as its first premise. The English will have a formal feature corresponding to being a universally quantified conditional, but (**16**) makes no commitment to the intrinsic nature of that feature. Such a position requires that one have a standard of formal validity independent of formalization. This question is discussed in §5. The notion of deductive manipulation, required by (**17**), is discussed in §3.

In the next section, I look briefly at a very ambitious version of the thesis of the revealingness of logical forms.

2   Analysis and the Tractarian vision

Russell toyed with the idea, and Wittgenstein embraced it in the *Tractatus*, that all validity is formal validity: indeed, is validity in virtue of **Q**-logical form. Contrary appearances are to be explained by the fact that our language is not fully analysed. Analysis would reveal all validity as formal. If we find a valid argument that apparently does not formalize as **Q**-valid, this shows that there is some defect in our formalization: we have not uncovered enough logical structure, we have not carried analysis far enough.

First, a historical correction. The language of logical forms that Russell had in mind was not **Q**, but the richer language of *Principia Mathematica*. The language of logical forms that Wittgenstein had in mind was also not **Q**, but a language that at least superficially seems less rich. The differences between them, and between them and **Q**, had no impact either on the basis of the vision or on its subsequent rejection by both philosophers.[5]

Let us see how this Tractarian vision would view a very simple case. A standard example of a valid argument which is not formally valid is:

1)   Tom is a bachelor, so Tom is unmarried.

One might suppose that the deepest **Q**-formalization is:

2) $F\alpha; \neg G\alpha$

with "α" corresponding to "Tom", "F" to "is a bachelor" and "G" to "is married". However, the Tractarian view has it that we have overlooked some hidden structure in the English. We can reveal it, and at the same time show (1) to be formally valid, by the formalization

3) $F\alpha \,\&\, \neg G\alpha; \neg G\alpha$

with correspondences as before except that "F" now corresponds to "is a man". The formalization depends upon the analysis of "is a bachelor" as "is a man and is not married".

(3) counts as an adequate formalization of (1) by the standards previously given, and nothing we have so far said gives grounds for objecting to it. These standards are consistent with intuitively more unacceptable proposals. In general, if an argument $A; C$ is valid, our current standard of adequacy counts as adequate a formalization "$p \,\&\, q$; q". For if A entails C, then "A" and "$A \,\&\, C$" have the same truth conditions. In this sense it is trivial that all validity can be represented as formal validity, and the maxim of formalizing in such a way as to maximize the amount of validity that can be represented would lead to this trivial result.

Russell always had doubts about the possibility of representing all validity as formal, and although Wittgenstein explicitly affirmed the position in the *Tractatus*, he explicitly retracted it later. For both of them, the decisive factor was that the vision was unrealizable because there are valid arguments which no amount of analysis can represent as formally valid. They were both influenced by the case of colour incompatibilities. Thus the valid argument

4)   This is red (all over), so it is not blue (all over)

cannot be represented as formally valid, however one analyses. One might try analysing "red" as "not yellow and not blue and not etc.". But first, it is doubtful whether the analysis can be brought to a conclusion, and secondly it would involve treating "yellow" etc. as primitive (not to be analysed) and a similar valid argument could be stated using just these primitives.[6]

Contemporary opinion would reject the Tractarian vision not just for this kind of reason, but also on the grounds that it fails to distinguish between a claim about logical form and a claim about analysis. Davidson, among others, has emphasized the distinction.[7] If it can be made good, it puts a curb on the Tractarian vision. Something like (3) would be said to be unsatisfactory as a logical form of (1) even if satisfactory as an analysis.

The distinction between analysis and form is quite intuitive. For example, a traditional "analysis of knowledge" might start off on lines such as these:

5) S knows that A iff
   (i) A
   (ii) S believes that A
   (iii) ... .

The rule of the game is that you do not re-use the word "knows" in the numbered clauses on the right of the biconditional. The clauses thus in some sense explain the meaning of "knows". This does not even begin to look like a claim about logical form. By contrast, Davidson's account of propositional attitudes rules that the logical form of "S knows that A" is

6) $F\alpha\beta, p$

(with "Fxy" corresponding to "x knows y", "α" to "S", "β" to "that" and "p" to "A"). This does not even begin to look like an analysis of knowledge. There is no attempt, in the logical form, to avoid the word "knows". Davidson's thought is that we must first get the logical form straight, leaving analysis as a separate issue.

It is not very difficult to see how Russell and Wittgenstein might have been unmoved by this distinction. Logical form reveals logical structure, even if that is hidden in the natural language. Revealing the premise of (1) as a conjunction is part and parcel of the very same enterprise as revealing "All men are happy" as a quantified conditional. Once you become hardened to, or even rejoice in, the marked differences between the logical and the natural language, one sort of difference is likely to seem much like another. The invisibility of the sign for conjunction in (1) will no more count against its formalization by (3) than the invisibility of the sign for conditionality in "All men are happy".

Davidson's own basis for the distinction depends upon considerations rather foreign to the discussion so far, considerations which will be introduced in §6 below. In the remainder of this section, I want to consider whether a well-grounded curb on the Tractarian vision can emerge from the kind of perspective we have adopted. In particular, the fact that (3) counts as adequate by our present standards may seem to constitute a criticism of these standards. The view for which I shall argue is that curbing the Tractarian vision on the basis of ideas we have already to hand is possible only by disqualifying, as inadequate, some standardly accepted **Q**-formalizations of English.

One way to motivate the distinction between logical form and analysis is to say that the logical form of a sentence should specify its logical constants, and the way in which they relate to the pattern of occurrence of the non-logical expressions. Analysis, by contrast, should concern the contribution to meaning of a specific expression, ideally providing a more complex equivalent expression built up out of unanalysable primitives. This simple-minded idea could be implemented by the following constraint upon formalizations:

> 7) A formalization is adequate only if each of its logical constants is matched by a single English expression making the same contribution to truth conditions.

On this view, formalization effects only notational changes, so far as the logical constants are concerned.

(7) corresponds to a deeply unambitious conception of logical form. The logical form of a sentence is given by how its logical constants occur, and the pattern of occurrence of the non-logical expressions. An artificial language would be just a convenient way of schematizing the non-logical expressions, rather as we did in chapter **1**, and providing a usefully standardized notation for the logical constants. The idea of an artificial language would not essentially enter into the account of the nature of logical form.

Although (7) would rule that formalizations like (3) of (1) are inadequate, since they introduce constants (in this case "&") having no corresponding expression in the English, it would play havoc with traditional practice in formalization. English universal quantifications could not be formalized by universally quantified conditionals since that would involve importing a constant ("→") to which there corresponded no English expression; similarly for English existential quantifications. Adjectival and adverbial modification could not be formalized by the conjunctive method. Necessitated atoms could not be formalized in **QN**. However, this is the narrow and unambitious view that one appears forced to take, if one is to hold to the thesis (**1.16**):

If an argument in natural language is adequately formalizable as valid, then it is *formally* valid.

If such a thesis is not to be trivial, it requires a conception of formal validity that applies directly to English without any detour through formalizations. The only available conception is that already offered in chapter **1.10**: an argument is formally valid iff its validity turns only upon the meanings of the logical constants it contains and upon the pattern of occurrence of the non-logical expressions. On this definition,

8)　This is a red house, so this is a house

is not formally valid. Hence, by (**1.16**), it should not be adequately formalizable as valid, and the only way to secure this result would appear to be the adoption of (7).

(7) is one way in which one could give a criterion for the difference between logical form and analysis, and thus limit the Tractarian dream. The motivation is coherent, but is far from doing justice to ambitious theses about logical form. Is there an alternative criterion?

We might prefer to think along the following lines. Logic, and thus formalization, must have no truck with non-logical expressions. These must be recognized as contributing only through the pattern of their occurrence, and not through their specific meaning. Hence a formalization should not, as (3) does, introduce a predicate-letter (in this case "F") which, according to the correspondence scheme, corresponds to an expression ("is a man") which does not occur in the sentence to be formalized. Let us call a generalization of this condition the "correspondence requirement". It rules that if the correspondence scheme associated with a formalization has it that, say, "F" corresponds to "...", then that actual expression, "...", must occur in the sentence of natural language which is formalized. This requirement is clearly failed by (3) relative to (1): the correspondence scheme mentions "is a man", which does not occur in (1). Yet the requirement would seem more liberal than (7), since it allows modes of composition in English to be represented as the application of **Q**-constants.

The correspondence requirement must be understood with some generosity if all its liberality is to be reaped. For example, the spirit of the requirement should allow the conjunctive formalization of adjectival modification. Yet in a standard **Q**-formalization of (8) as

9)　$F\alpha$ & $G\alpha$; $G\alpha$

we find ourselves saying that "*F*" corresponds to "is red", whereas that expression does not literally occur in the premise of (8). While the difference is important for a detailed understanding of the workings of natural language, let us agree to interpret the correspondence requirement in such a way that such differences will not count.

With this rather vague liberalization, the correspondence requirements comes closer than (7) to the traditional view, in that it allows the standard formalizations of sentences containing, for example, adjectival modification and quantifiers. It rules out a few of the **Q**-formalizations proposed in this book, but there is in addition at least one major category of formalizations which it brings under suspicion: Davidson's treatment of verbs of action and some of their adverbs. Consider, for example, the formalization of

10)   Shem kicked Shaum

as

11) $\exists x (Fx \ \& \ Gx\alpha \ \& \ Hx\beta)$

with "*F*" corresponding to "is a kick", "*Gxy*" to "*x* kicked *y*", "*Hxy*" to "*x* was received by *y*", "α" to "Shem" and "β" to "Shaum". Perhaps our vague liberalization of the correspondence requirement allows that (**10**) contains something close enough to "is a kick" (viz. "kicked"), and also contains something close enough to "*x* kicked *y*" where the position marked by *y* is supposed to be occupied by an expression standing not for a person or a rock but for a kick. But the liberalization cannot allow that (**10**) contains an expression close enough to "was received by". Either the tradition or else the correspondence requirement is at fault.

Moreover, the correspondence requirement is clearly violated by **QW** and **QC**.[8] On our current understanding of what is to count as a logical constant, furnished by the list in chapter **1.11**, the predicates "W" and "C" of these languages are not logical constants. Hence they are non-logical expressions, yet the English expressions to which they correspond ("is a world", "is a counterpart of ... in ...") do not occur in the natural language sentences.[†]

With the ideas currently to hand, I can find no way of curbing the Tractarian vision by providing a criterion for the distinction between logical form and analysis, except by stipulations which render incorrect

† See Exercise 160, page 367.

traditional logical form proposals. I believe this shows that the tradition is defective. It makes no room for a justification of the semantic theses concerning logical form which belong to the tradition. Such a justification must await the articulation of a concern with compositional semantics, discussed in §6, in a manner which goes beyond anything explicit in Russell and Frege, the founders of the tradition under discussion.

3 Proof

A comprehensive grasp of the activities of the logician requires understanding the notion of *proof*. In addition to its intrinsic importance, and its connection with certain traditional aspirations in logic, it has a role to play in giving an account of what it is to be a logical constant.

I mentioned in chapter 1 that a traditional logical aspiration has been the mechanization of inference. Where there is disagreement about what follows from what, it would be good to feed premises and conclusion into a calculating engine, and wait for the computation of a verdict. Such an engine has to be fed sentences. It cannot be fed propositions, for these are too abstract. It is important that one express the argument in a language in which the validity-relevant features correspond conveniently to the physical make-up of the sentences, for it is to this physical make-up that, in the first instance, we can expect the machine to be sensitive. The mechanical aspiration thus provides a motivation for creating perspicuous artificial languages.

We saw in chapter 4.14 that this aspiration cannot be satisfied in full with respect to a language as rich as **Q**, for there is demonstrably no decision procedure for **Q**. This fact has added to the importance of the notion of proof. For it is demonstrably the case for **Q** that there are systems of proof such that, if an argument is valid, a proof will be found for it, by following the system, in a finite number of steps. This does not add up to a decision procedure because if following the system has not resulted in a proof of a certain conclusion after a million or ten million steps, you do not know whether this is because a few or a million more steps are needed or because the argument under test is invalid.

There are various different kinds of rules of proof, but I shall give a sketch of a system of *natural deduction*. A reader wanting to become proficient in using natural deduction must look elsewhere (for example to Lemmon [1965]), and one already proficient should skip to §4.

Confining our attention just to propositional logic, a standard system of natural deduction associates with each sentence connective two kinds of *rules of proof*: *introduction rules*, which specify from what premises one can derive a **P**-sentence dominated by the connective, and *elimination* rules, which specify what conclusions one can derive from premises containing a **P**-sentence dominated by the connective. The intuitive idea behind the introduction rule for "&" is that, from a pair of sentences as premises you can infer their conjunction. For the elimination rule, the idea is that from a conjunction you can infer either conjunct. In setting out a full system, it is easiest to express these ideas in a slightly more complicated way. Using "Γ" and "Δ" to stand for any number of sentences (including zero), and a horizontal line to express that from what is above one can derive what is below, we write the rules as follows:

$$\text{\& E: if} \quad \frac{\Gamma}{X \,\&\, Y} \quad \text{then:} \quad \frac{\Gamma}{X} \quad \text{and} \quad \frac{\Gamma}{Y}$$

and

$$\text{\& I: if} \quad \frac{\Gamma}{X} \quad \text{and} \quad \frac{\Delta}{Y} \quad \text{then} \quad \frac{\Gamma \cup \Delta}{X \,\&\, Y}.$$

(Here "$\cup$" represents set-theoretic union: $\Gamma \cup \Delta$ is the set of all the sentences belonging to Γ or to Δ.) & E, the elimination rule, tells you that *if $X \,\&\, Y$ is derivable from Γ, then each conjunct is also derivable from Γ.* & I, the introduction rule, tells you that *if X is derivable from Γ and Y from Δ then the conjunction can be derived from the premises obtained by adding Γ to Δ.* "Derivable" in these contexts is not supposed to have any independent meaning. Rather, its meaning is to be fixed by the specification of the rules of proof for all the connectives.

For "$\neg$" the elimination rule is straightforward:

$$\neg \text{E: if} \quad \frac{\Gamma}{\neg \neg X} \quad \text{then} \quad \frac{\Gamma}{X}.$$

The introduction rule is based on *reductio ad absurdum* style argument:

$$\neg \text{I: if} \quad \frac{\Gamma, X}{Y \,\&\, \neg Y} \quad \text{then} \quad \frac{\Gamma}{\neg X}.$$

This rule tells you that if you can derive a contradiction from a sentence X

together with zero or more other premises Γ, then you can derive $\neg X$ from Γ.

In addition to the rules for each connective, the system also requires general structural rules. All the rules mentioned so far are hypothetical. They tell you that if such-and-such is derivable, then so is something else. But unless there is at least one categorical rule, there will be no categorical facts of derivability. The categorical rule is

A: $\dfrac{X}{X}$

sometimes called, perhaps misleadingly, the rule of assumptions. The only way in which derivations can get going is by one or more applications of A.

To illustrate an application of these rules, let us adopt a standard convention. We will write sentences on numbered lines, the numbers enclosed in parentheses. To the left of the line number, we will indicate, by numbers not enclosed in parentheses, the line number of "assumptions": sentences introduced into the proof by A and used as premises at some point, possibly remote, in the derivation of the sentence on the line. To the right we cite the rule to which appeal is made in writing the sentence.

$$1 \quad (1)\ p\ \&\ \neg p \qquad A$$
$$\quad (2)\ \neg(p\ \&\ \neg p) \qquad \neg I.$$

In the application of $\neg I$ on line (2), Γ has zero sentences, and the instance of $Y\ \&\ \neg Y$ is $p\ \&\ \neg p$. No number occurs to the left of "(2)", since no sentence is used as a premise in the derivation of $\neg(p\ \&\ \neg p)$: Γ has no members.

The system is completed by associating each connective with suitable introduction and elimination rules.[†]

The motivation for the rules is, of course, to reflect *valid* patterns of reasoning, but in the statement of the rules there is no mention of validity, truth or any other "semantic" notion. Whether or not something is a correct application of a rule can be determined just by inspecting the shapes of the sentences. The rules are amenable to mechanical testing. The rules fix the extension of a relation of derivability, which can be written "$\vdash_P$". Thus our little proof above established that:

1) $\vdash_P \neg(p\ \&\ \neg p)$,

† See Exercise 161, page 367.

in other words, that $\neg(p \mathbin{\&} \neg p)$ is derivable by the rules relating to **P** on the basis of no assumptions.

An obvious question is how "$\vdash_\mathbf{P}$" is related to "$\vDash_\mathbf{P}$". We shall say that for a language, L, any system of rules of proof, π, for L, and any semantics, σ, for L in terms of which validity is defined, π is *complete* with respect to σ iff

2)   if $\Gamma \vDash_\sigma X$ then $\Gamma \vdash_\pi X$;

and that π is *sound* with respect to σ iff

3)   if $\Gamma \vdash_\pi X$ then $\Gamma \vDash_\sigma X$.

There are standard systems of natural deduction which are both sound and complete with respect to the semantics we gave for **P**, so, using "$\vdash_\mathbf{P}$" to relate to such a system

4) $\Gamma \vdash_\mathbf{P} X$ iff $\Gamma \vDash_\mathbf{P} X$.

In short, "$\vdash_\mathbf{P}$" and "$\vDash_\mathbf{P}$" are equivalent, and there are rules for **Q** relative to which "$\vdash_\mathbf{Q}$" and "$\vDash_\mathbf{Q}$" are also equivalent.[9] For this result to have significance, it is obviously important that the two relations be defined in different ways. In the standard terminology, "$\vdash_\mathbf{P}$" is defined purely *syntactically* whereas "$\vDash_\mathbf{P}$" is defined *semantically*.[10]

A hypothesis is that there is no way of devising rules of proof for English that would come anywhere near being both sound and complete relative to our intuitive semantics. So if you want to be able to mechanize inference along the lines of rules of proof, it helps to transform English into some language for which there are rules of proof: **Q** would be an example of such a language. This hypothesis is, in effect, (**1.13**):

> The formalization of a natural sentence renders the proposition it expresses accessible to formal deductive manipulations.

It is beyond dispute that this claim can be understood in a way upon which it is true. However, it does nothing to establish an ambitious thesis according to which the semantic mechanisms of a formalization are the very ones at work in the natural sentence. By using a quantified conditional to express the proposition expressed in English by a quantification, you may be able to use rules of proof to prove some conclusion; but this does not begin to show that the English sentence is really itself a quantified conditional. It shows at most that it has the same truth (upon-an-intended-interpretation) conditions as such a conditional, but these truth conditions may be expressed by quite different mechanisms.

4   Formal and structural validity

A traditional view is that the logical form of a sentence shows how its primitive expressions are organized so as to engender its overall meaning: logical form displays *semantic structure*. One way in which this thought arises is as follows: formal validity is validity purely in virtue of structure, abstracted from content. However, Gareth Evans has argued, I think entirely convincingly, that it is possible to distinguish a coherent and important notion of *validity in virtue of semantic structure* (structural validity) which contrasts sharply with the traditional notion of formal validity. The main point of contrast is that formal validity turns on the specific contribution – the "content" – of the logical constants, whereas the conception Evans articulates is one in which validity depends upon no such specific contribution: the validity is *purely* structural.

Evans's starting point is the notion of a categorial semantics: a semantic theory having a number of semantic categories, each primitive expression being assigned to just one category. Members of the same category are assigned the same kind of entity by the semantics. Thus an n-place predicate might be assigned a set of n-tuples; an n-place sentence connective an n-ary truth function (i.e. a function from n-tuples of truth values to a truth value); and so on. Roughly speaking, a structurally valid inference is one whose validity is independent of the particular assignments that are made within the semantic categories, but which is wholly dependent on the pattern of the categories in the argument. For example, it is arguable that

1)   John is a large man, so John is a man

is structurally valid. Imagine a categorical semantics which treats "John" as belonging to the category of *names*, marked by the fact that to each member of the category the semantics assigns a member of the domain; which treats "man" as belonging to the category of *one-place predicates*, marked by the fact that to each member of the category the semantics assigns a set of members of the domain; and which treats "large" as belonging to the category of *extensional adjectives*, marked by the fact that to each member of the category the semantics assigns a function from a set (the set associated with the one-place predicate to which the adjective applies) to a subset of that set. Relative to this semantics, (1) is structurally valid, since any argument whose premise consists of a name followed by an extensional adjective followed by a one-place predicate, and whose conclusion consists of that name followed by that one-place predicate, is valid. The validity of (1) is thus due to its semantic structure: due only to the semantic categories to which its elements belong, and not to the special contribution the elements make within their categories.

If there were no restrictions on the categorical semantics, then any inference whatsoever would count as structurally valid (relative to some semantics or other). For example, we might subdivide the category of one-place predicates into three as follows: (a) those to each of which the set of all and only bachelors is assigned; (b) those to each of which the set of all and only unmarried persons is assigned; (c) those to which any other set of objects is assigned. In the light of this categorization

2)   Tom is a bachelor, so Tom is unmarried

is structurally valid, since any argument whose premise consists of a name followed by an (a)-type one-place predicate and whose conclusion consists of that name followed by a (b)-type one-place predicate is valid.

Evans places the followed constraint upon categorical semantics:

> We will construct a new category out of an older and more comprehensive category only when we can make an assignment to members of the new category which provides a *different* explanation for the behaviour which members of the new category had in common with the old, the provision of which explanation would show that the apparent unity in the behaviour of members of the old category was deceptive, concealing deep differences in functioning. ([1976], p. 64)

It is clear that a semantics which makes the envisaged threefold division among one-place predicates will not meet this constraint. It will be useful to compare this case with one where a division would be justified. Imagine a language containing both names and definite descriptions of the form "the so-and-so", and a semantic theory which lumps them into the single category of "singular terms". The theory assigns members of the domain to some (perhaps all) names and some (perhaps all) definite descriptions.[11] Some inferential behaviour is common to singular terms. For example, there is a valid pattern of inference from a premise applying to a singular term a predicate, F, to a conclusion which says that there is at least one F. Now imagine a new semantic theory which proposes a division of the category of terms, separating names from descriptions, treating names as before, and associating with "the" a function from pairs of sets to truth values: the function takes the value *true* iff the first set in the pair has a unique member and that member belongs to the second. The new semantic theory gives a new explanation of the inferential behaviour common to singular terms. In the case mentioned, there are two explanations, one for the case in which the premise contains a name, patterned on the explanation provided by the old theory, and another for the case in which the premise contains a definite description. In the second case, the explanation alludes to two assignments, that to "the" and that to "F", whereas the old theory alluded to a single assignment to "the F". This surely satisfies Evans's condition that the new theory's division of the category of singular terms shows "that the apparent unity in the behaviour of members of the old category was deceptive, concealing deep differences in functioning".

In our earlier listing of the semantic categories of $\mathbf{Q}$, the 2-place connectives were bundled into a single category, and this categorization is justified by Evans's test. The only inferential behaviour common to all of them is the substitution of equivalents, which can be represented in the following scheme:

3)   For any 2-place connective, ϕ, any interpretation, i, if $i(Y) = i(Z)$ then $i(X \phi Y) = i(X \phi Z)$.[12]

Specification of the different truth functions expressed by the three 2-place sentence connectives of $\mathbf{Q}$ gives no new explanation of this common behaviour. The explanation still resides simply in the fact that the connectives express some truth function or other. No further subdivision of the semantic category of two-place connectives is justified. Hence an argument of the form:

4) $p \ \& \ q; p$

while it may count as *formally* valid does not count as *structurally* valid. The relevant structure is:

5) | sentence$_1$ | sentence connective | sentence$_2$ | ; | sentence$_1$ |

But not all instances of this structure, permuting just within the specified categories, are valid; for example:

6) $p \vee q; p.$

The notion of a structurally valid argument corresponds to the idea that there are some arguments which are valid, not in virtue of the specific meanings of their expressions, but rather in virtue of the way in which the sentences are constructed. Though formal validity is sometimes thought to answer to this intuitive idea, it does not, for formal validity is validity partly in virtue of the specific meanings of certain favoured expressions, the logical constants, and if this idea is to have importance there has to be a deep reason for selecting some expressions rather than others as the logical constants.

On Evans's view, a sentence's semantic structure will be represented by a sequence of categories. A **Q**-formalization will not constitute such a representation. If you looked to **Q**-logical form to give answers about semantic structure you would, arguably, get wrong answers. You would infer that English quantifiers are unary, whereas the evidence is that they are binary. You would infer that English universal generalizations contain an expression belonging to the category of two-place sentence connectives, whereas they do not. You would infer that adjectives belong to the same category as predicates, being associated with a set of sequences, whereas the category they really belong to is of expressions assigned functions from sets to subsets.

It is instructive to compare an instance of adjectival detachment (for example (**2.8**)) with an instance of genuine (or, to speak more neutrally, explicit) conjunction elimination. In both cases, there is a generalization to be had. Evans's insight is that the former generalization can abstract from the specific meaning of any expression, whereas the latter cannot.

If the draconian standard of (**2.7**) is adhered to, there is a good sense in which a formalization reveals the logical structure of the natural sentence. It shows how the English sentence is structured around the logical constants it contains: what those constants are, and how the non-logical expressions contribute to truth conditions. The same is not true of the more relaxed standard provided by the correspondence requirement. (For example, the requirement permits the formalization of adjectival modification by conjunction, whereas there may be no expression for conjunction in such an English idiom.) None the less one can see how people might be led into thinking that formalizations meeting this requirement speak to the question of logical structure. Given, finally, that logical structure might well be confused with semantic structure one can see how the view that logical form reveals semantic structure found space within the traditional conception of logical form. Crudely put, the progression goes as follows: formalization singles out the logical constants of English and the way in which the non-logical material is organized by the constants; thus formalization identifies logical structure, the mechanisms whereby expressions contribute to truth conditions; thus logical form identifies semantic structure. The starting point is not secured by traditional practice in formalization, and the steps are unjustified.

5 Logical constancy

One goal we attributed to the logician is the characterization not merely of validity, but also of formal validity. A formally valid argument is one valid just in virtue of the meanings of the logical constants it contains and the pattern of occurrence of the other expressions. An account of the goal of logic thus essentially involves an account of what it is to be a logical constant. Once we have that, we will be able to specify the border between logic and other disciplines. A language fit for logic will have no constants other than logical constants. A sentence will be logically true iff it is true in virtue of the meanings of the logical constants it contains together with the pattern of occurrence of non-logical expressions. Logical necessity is necessity owed to the meanings of the logical constants, and to the meanings of no other expressions.

Within a formal language like **Q**, it is simply stipulated which are the logical constants: they are the expressions which receive a constant interpretation. However, there is no limit to what expressions could be accorded a constant interpretation. For example, one could introduce a **Q**-like language with the additional stipulation that some symbol was, upon every interpretation, to be assigned the set of all 1-tuples each of whose member is a cat. The present section asks what principles underlie our intuition that the symbol thus treated in that language as a logical constant is not really one.

In chapter 1 we gave a list of the logical constants, mentioning that this does not even hint at a rationale for grouping the listed expressions together. And I quite cavalierly ignored the original list in chapter 5, in order to investigate the logic of modality.

We will consider four accounts of what makes an expression a logical constant. We will find that there is a large measure of convergence between them. The first, briefly mentioned in chapter **1.11**, is that the logical constants are topic-neutral: they can be distinguished by the fact that they introduce no special subject matter. Thus "if" and "some" are intuitively not "about" anything at all, whereas a name like "Ronald Reagan" is about Reagan, and "happy" is about happiness.

The view is encouraged by the thought that logic concerns reasoning in general. It should therefore give special attention just to those expressions which can be used in reasoning on any subject matter whatsoever, and this may suggest that the logical constants should not introduce any specific subject matter.

If standard views about what expressions are logical constants are correct, the topic-neutral account as so far formulated does not give a sufficient condition for an expression to be a logical constant. Take a world like "therefore", which can certainly be used in connection with reasoning on any subject whatsoever, and is as topic-neutral as any expression one can think of. Yet orthodoxy does not include it among the logical constants.† One can see why. The validity of an argument cannot be mediated by the contribution made by the use of "therefore", so, while it may be used in any piece of reasoning, it contributes to the validity of no piece of reasoning.

The point itself suggests how the topic-neutral theory should be reformulated. A logical constant is an expression whose meaning contributes to the validity of arguments and which introduces no special subject matter. With this amendment, the main problem is the vagueness of the notion of topic neutrality, which leaves unsettled, concerning some expressions, whether they are logical constants or not. One example is "but", not normally accounted a logical constant. My earlier suggestion was that it makes the same contribution to validity as "and", since it expresses the same truth function, but that its meaning ensures that one who asserts that *A* but *B* represents himself as supposing that in the context it is surprising, poignant, or worthy of special note or emphasis, that *B* is true. Does this additional component in the meaning of "but" constitute the introduction of a specific subject matter (surprise, poignancy or special notice)? The topic-neutral theory delivers no answer, whereas traditionally, and intuitively, "but" does not count as a logical constant.

Though " = " is traditionally classified as a logical constant, the topic-neutral theory could be read so as to exclude this expression from the

---

† See Exercise 162, page 367.

logical constants. A natural view is that identity is a subject. However, it is also not unnatural to see the topic-neutral account as classifying " = " as a logical constant, since " = " introduces no specific objects. It is a relation that is quite neutral concerning what objects it relates. There are no actual or possible objects such that understanding " = " requires learning about them.

Thus understood, the criterion would arguably count "necessarily" as a logical constant, for even if a philosophical account of the notion introduces worlds, it would appear that learning about these is not essential to a grasp of the adverb. By contrast, the criterion would exclude the constants "W", "w∗" and "C" of **QW** and **QC** respectively, since there are objects, worlds and counterparts, which one must learn about in understanding these expressions.

The criterion would exclude the set-theoretic expression "∈" (for "is a member of"), as this is naively understood, since in order to understand this expression one has to learn about sets. The criterion in its original formulation would obviously exclude a putative sentence connective like "John knows that", since knowledge is a subject matter, and this is intuitively the appropriate result. It would therefore be best to disjoin the two tests we have been applying: an expression is a logical constant iff either it introduces no specific subject matter, or there are no objects one needs to learn about in coming to understand the expression.

Though this is still rather imprecise,[†] it does seem to correspond to an intelligible philosophical motivation, and also to justify the selection of the logical constants discussed in this book. It will be instructive to compare this criterion with others.

The second criterion starts off with a philosophical rationale close to that which animates the topic neutral account. Perhaps the essence of the matter is that a logical constant is an expression whose whole use is constituted by its role in reasoning. There is no question of it being used to speak of the world in the way that, say, names and predicates do. This thought could be implemented by the thesis that a logical constant is an expression whose meaning is fixed by what you can validly infer from a sentence dominated by it, and from what you can validly infer a sentence dominated by it. In short, a logical constant is an expression whose meaning is wholly determined by rules of proof. Let us call this the "rules" criterion for the logical constants.

The criterion requires further refinement. The envisaged rules are entirely schematic, except for the target expression. This will exclude

† See Exercise 163, page 367.

both rules like "From *A and B* infer *A or B*" and rules like "From *x is a bachelor* infer *x is unmarried*". The justification for the restriction is that the meaning of a constant should be inferrable from the rules alone, quite independently of the meaning of any other expression.

On plausible assumptions, the rules criterion determines that the English truth functional sentence connectives are logical constants. The assumptions are that rules for "and" just like those given for "&" enable the derivation only of valid arguments, and that the meaning of "and" is fixed by the fact that a conjunction is true iff both conjuncts are. The validity of the elimination rule ensures that the truth of a conjunction is sufficient for the truth of each conjunct, and the validity of the introduction rule ensures that the truth of each conjunct is sufficient for the truth of their conjunction.

Although exactly the same rules would be valid for the binary connective. "... and $2 + 2 = 4$ and ...", we need not fear that the latter will wrongly be counted a logical constant, for it is not plausible to say that its *meaning* is fixed by the rule that "*A* and $2 + 2 = 4$ and *B* is true iff *A* is true and *B* is true": a component of the meaning is left undetermined.

However, this sort of example poses another problem for the rules theory. Consider the connective "¢", where "*A¢B*" expresses the denial of the disjunction of "*not-A*" and "*not-B*".[13] The rules of "and"-introduction and elimination are valid for ¢, yet it is plausible to hold that although "¢" expresses the same truth function as "and", the two expressions differ in meaning. For example, someone could believe that *A and B* without believing that *A¢B*. Then one would have to confess that the rules criterion does not rule "and" a logical constant, since its meaning is not fixed by the validity of the rules, since the validity of the rules does not distinguish between the meaning of "and" and the meaning of "¢".

If we replace "meaning" by "truth conditions" in the rules criterion, we would be returned to the difficulty that "... and $2 + 2 = 4$ and ..." would counterintuitively be accounted a logical constant.

However, intuitively the meaning of "¢" is more detailed or specific than the meaning of "and". We can thus amend the rules criterion as follows: an expression is a logical constant iff its meaning is the least specific meaning which validates the introduction and elimination rules. The problem now is that "¢" cannot be a logical constant, either relative to the rules for "and" or to any others, since to assign it the less specific meaning of "and" will validate any rules which it would validate, yet will not assign it the correct meaning. This problem can be overcome by distinguishing between primitive and defined constants. The category of primitive logical constants is as determined by the amended rules criterion. The category of logical constants is the closure of this category under definition.[†]

Turning to "&" in **P**, the **P**-validity of the rules for "&" entails that every interpretation upon which a conjunction is true is one upon which each conjunct is true; and every interpretation upon which each conjunct is true is one upon which their conjunction is true. To recover the actual interpretation rules for "&", we need to add the assumption that every sentence not true upon an interpretation is false upon that interpretation.

Similar considerations, upon similar assumptions, substantiate the claims of the other truth functional sentence connectives, in English and in **P**, to be logical constants.

Quantifier elimination and introduction rules for **Q** would take the form:

$$\forall E: \quad \frac{\forall v X}{X_v^n},$$

where X_v^n results from X by replacing one or more occurrences of v by n.

$$\forall I: \quad \text{if } \frac{\Gamma}{X_v^n} \qquad \text{then}$$

$$\frac{\Gamma}{\forall v X} \qquad \text{provided that } n \text{ does not occur in } \Gamma.$$

The validity of $\forall E$ ensures that any interpretation upon which "$\forall v X$" is true is one upon which "X_v^n" is true, regardless of what the interpretation assigns to "n". Since each interpretation assigns just one thing to each name-letter, this enables us to infer the stronger: any interpretation upon which "$\forall v X$" is true is one such that "X_v^n" is true upon it, and upon any other interpretation agreeing with it in all assignments except to "n". The validity of $\forall I$ ensures that if "X_v^n" is true on some bunch of interpretations which include every possible assignment to "n" then "$\forall v X$" is true upon any member of the bunch. Thus the interpretation rules for $\forall$ are recoverable from the validity of the rules of proof.

It would not be easy to apply similar reasoning to English. For one thing, although one can see roughly what sort of rules should correspond to $\forall E$ and $\forall I$, it is unclear how they should be couched in detail. It would also be unclear whether the least specific meaning verifying these rules would be the meaning of "all" or "every". For example, an important

† See Exercise 164, page 367.

type of reasoning involving "all"-sentences, inductive reasoning from instances of a generalization to the generalization itself, might be argued to be partially constitutive of the meaning of "all", and it is simply not obvious whether this part would be captured by rules of "all" elimination and introduction. It might be better to argue indirectly: if "all" matches $\forall$ in meaning, and the latter is a logical constant, then so is the former.[†]

What does the rules criterion exclude from the category of the constants? Name-letters and predicate-letters are not even candidates for constancy, since they have no fixed interpretation. Likewise, names in English could not be constants, since there appear to be no valid rules special to any name.

The case of predicates is more delicate, since the classical view is that the English predicates "is the same as" and "exists" should count as logical constants, in view of their formalizability by the **Q**-constants "$=$" and "$\exists$". On anti-Meinongian assumptions, the rule

$$\text{EXISTS I:} \quad \text{if} \quad \frac{\Gamma}{\ldots n \ldots} \quad \text{then} \quad \frac{\Gamma}{n \text{ exists}}$$

is valid for extensional parts of English, and it would be plausible to argue that the least specific meaning assignable to "exists" which would validate it is its actual meaning. The significance of this is diminished by the fact that EXISTS I is not in general valid in English.[‡] There is a valid rule for **Q** corresponding to EXISTS I, so that one would expect an existential predicate to be a logical constant of **Q**, as indeed it is: "$\exists x(x = \ldots)$". Rather similar remarks apply with respect to identity. The non-extensionality of English means that there is no generally valid elimination rule, whereas the extensionality of **Q** means that there is such a rule.

This means that the rules criterion turns out to have an implicit relativity to a language. We must say either that "is the same as" and "$=$" do not mean the same, or else that of two expressions with the same meaning, it may be that one is a logical constant (in its language) while the other is not (in *its* language).

Though the topic-neutral criterion counts "$\square$" among the logical constants, this ruling has been disputed, so it would be useful if it could be confirmed by an alternative criterion. The elimination rule appropriate to **PN** is obvious:

$$\square\text{E:} \quad \text{if} \frac{\Gamma}{\square X} \quad \text{then} \quad \frac{\Gamma}{X}.$$

† See Exercise 165, page 368.   ‡ See Exercise 166, page 368.

Validating this rule (by the standards of **PN**-validity) requires that for all worlds w and interpretations i, if $\Box X$ is true at w upon i, then X is true at w upon i. By itself, this does not entail any significant feature of the interpretation rule for $\Box$. For example, it is consistent with $\Box$ having a much less specific meaning than it actually has, for example being a "null" operator, with the interpretation rule:

$\Box X$ is true at w upon i iff X is true at w upon i.

If the rules criterion is going to count "$\Box$" as a logical constant, everything must turn upon the introduction rules.

One possible such rule is:

$\Box$I(i)   if $\dfrac{\Gamma}{X}$, and every member of Γ has "$\Box$" as its main connective, then $\dfrac{\Gamma}{\Box X}$.

If this connection holds for **PN**-validity, then if for all interpretations i and all worlds w such that all the Γ are true at w upon i, X is true at w upon i, then for all interpretations i and all worlds w such that all the Γ are true at w upon i, $\Box X$ is true at w upon i. It is not hard to see that the standard interpretation rule will verify this conditional (given the assumption about the composition of Γ) but it is hard to see whether this is the least specific meaning that will do so. One cannot derive "$\Diamond X \to \Box \Diamond X$" from $\Box$I(i),[14] so this rule does not determine the meaning "$\Box$" possesses in **PN**.

The following rather similar rule, but with a less demanding restriction upon Γ,

$\Box$I(ii)   if $\dfrac{\Gamma}{X}$, and every sentence-letter in every member of Γ is within the scope of some occurrence of "$\Box$", then $\dfrac{\Gamma}{\Box X}$,

permits the derivation of just the arguments that are **PN**-valid. This is some evidence that it, together with the elimination rule, fixes the interpretation rules for "$\Box$"; but it is not decisive, since these interpretation rules may attribute more meaning than is needed for the validities.

Though not conclusive, the rules criterion adds some support to the case for the constanthood of "$\Box$". However, it would appear to give no encouragement to the constanthood of "W", "at", "w∗" or "C". So far as I know, no one has attempted to formulate elimination and introduction rules for these expressions, and I have no idea how such an attempt would begin. This is an odd result, given that what is expressible in **QN** is expressible in **QW** and **QC**. The criterion has some other counterintuitive features.

First, as it stands, the theory does not provide a sufficient condition for constancy. Let us stipulate that the meaning of the two-place sentence connective, "∗", is to be fixed by the two rules:

1) $\dfrac{X}{X * Y}$

2) $\dfrac{X * Y}{Y}$.

The rules together ensure that we can correctly infer any sentence from any sentence (by first using the introduction rule for "∗", then the elimination rule), and no expression can have a meaning which legitimates that.[†] Hence the fact that one can state introduction and elimination rules for an expression does not even ensure that the expression has a coherent meaning, let alone that it is a logical constant.

The argument shows that one who would introduce his constants by rules should place restrictions on the nature of the rules. A commonly suggested restriction, which would suffice to rule out the combination of ∗E and ∗I, is this.[15] Suppose we have a set, σ, of rules, one pair of which relates to a constant, ¢. Now imagine removing the ¢-rules from σ, and call the diminished result σ'. The proposed restriction is that σ' should permit all the derivations from assumptions not containing ¢ that σ permits. That a set containing just ∗ does not satisfy this restriction is shown by the fact that the derivation of Y from X (neither containing ∗), which is available in the presence of ∗, is unavailable in its absence. With this restriction, there is no reason to think that an expression could meet the rules criterion without being, intuitively, a logical constant. The doubt is not whether the account rules in too much, but whether it rules in enough.

This brings me to the second problem. It looks as if the meaning of "few" could not be fixed by elimination and introduction rules; and, depending upon the other expressive resources of the language, the same might go for "most". But "few" and "most" would appear, intuitively, to

† See Exercise 167, page 368.

have as much right to count as logical constants as "all" and "some", and this intuition is supported by the topic-neutral criterion. To exclude them is to show that one is willing to count as a logical constant only what yields the sorts of results in which one is interested, and that is to say that there is no objective borderline between the constants and other expressions.

The third criterion for constancy derives from Dummett [1981] and from Davidson [1973]. It is motivated by the thought that the logical constants are the expressions which act as "cement" in the construction of longer sentences out of the bricks of the symbols of the language; so I shall call it the "cement theory". Davidson has implemented this idea by saying that a logical constant is an expression which, in a semantic theory for a language containing it, receives a recursive rather than a basis clause. In the kind of semantic framework we have used, a basis clause is the specification of what an interpretation may or must assign to an expression (an object from the domain, a set of n-tuples, or whatever). The recursive clauses are the "interpretation rules". Their effect is to transform the question of the truth-upon-an-interpretation conditions for a sentence into a question about the truth-upon-an-interpretation conditions for one or more of its parts. Such a part may contain a further occurrence of the expression treated by the rule of interpretation, so that that rule, and also perhaps others, may have to be reapplied before the question of the truth-upon-an-interpretation conditions is resolved. The final resolution is purely in terms of the basis clauses. Looking at it from the bottom up, rather than the top down, once the assignments of the basis clauses are fixed, so is the truth upon the interpretation of every sentence in the language.

The criterion excludes " = ", since this receives a basis clause.[16] This is not a decisive refutation of the cement theory, since we cannot assume that our original list of logical constants is well motivated. More seriously, the cement theory will include as constants expressions which clearly are not, for example "large".[17] In terms of our semantic framework, we can find a rule which, rather than assigning something outright to "large", as a basis clause does, makes what is assigned depend systematically upon the assignments effected by the basis clauses. For example, one could introduce "large" into **Q** by the syntactic rule that attaching it to a 1-place predicate(-letter) forms a new 1-place predicate(-letter), and the semantic rule that an interpretation which assigns a given set, σ, to a 1-place predicate, ϕ, must assign the subset of large members of σ to "large ϕ". Just as assignments to the sentence-letters of **P** thereby determine the truth values of all **P**-sentences in virtue of the recursive rules for the sentence connectives, so the assignments to the "large"-free predicates and predicate-letters thereby determine the assignments to all the predicates and predicate-letters in virtue of the recursive rule for "large".

Though the cement theory must be rejected, one can see how holding it makes it hard to find use for a distinction between logical structure and semantic structure. Semantic structure might be thought of as the way a sentence is built up out of its parts. On the cement theory, this structure is fixed by the logical constants.

The fourth criterion is due to Peacocke.[18] I shall call it the "*a priori*" criterion. Consider this property of truth functional sentence connectives: if you know what truth values an interpretation has assigned to the components of a sentence dominated by a truth functional sentence connective, and you know the interpretation rule for the connective, then you can work out *a priori* what truth value the interpretation accords to the resultant. Perhaps some generalization of this property will constitute a distinctive feature of the logical constants.

In the semantic framework we have adopted, an interpretation assigns a truth value to every sentence of the language (relative to a domain, and perhaps also relative to a world – but we will omit these qualifications). It does so in the following way: it makes outright assignments of entities to the sentence-letters, name-letters and predicate-letters. The operators, that is, the sentence connectives and quantifiers, are not themselves assigned anything, but each is associated with a clause specifying a condition upon which a sentence dominated by the operator would be assigned the value true. The outright assignments are the basis clauses, the assignments upon a condition the recursive clauses. We can say that an outright assignment "treats" the expression to which the assignment is made, and that a recursive clause "treats" the expression which dominates the type of sentence which the clause addresses.

The version I shall consider of Peacocke's formulation of the generalization (adapted to our semantics and terminology) is:

3) α is a logical constant iff α is non-complex and, for any expression $\beta_1, \ldots \beta_n$ on which α operates to form the expression $\alpha(\beta_1, \ldots \beta_n)$, knowledge of what each interpretation assigns to each β_i, together with knowledge of how α is treated, enables one to infer *a priori* what each interpretation assigns to $\alpha(\beta_1, \ldots \beta_n)$.

This excludes the case in which some β_i or $\alpha(\beta_1, \ldots \beta_n)$ is an operator (and so not assigned anything upon an interpretation). This has no practical importance, and could be avoided without damaging the spirit of the proposal.

(3) clearly rules sentence connectives as constants. Also it debars, for example, predicate-letters. We can apply the test to "$F\gamma$", where we think of this as corresponding to "$\alpha(\beta)$" in (3) (so that "F" takes the place of "α" and "γ" of "β_1"). Suppose you know what each interpretation assigns to "γ", and so you know in particular that $i_1(\gamma)$ is Ronald Reagan. Suppose you also know how the unary predicate-letter "F" is treated (each interpretation assigns it a set of 1-tuples), and so you know in particular how i treats "F": for example, you know that $i_i(F)$ is the set of 1-membered sequences each having an honest man as member. You cannot infer *a priori* what truth value i_1 assigns to "$F\gamma$", because nothing in what you know about i_1's assignment to "γ" and treatment of "F" contains any information about Reagan's honesty. So a predicate letter fails, as it should, the test for constancy.

(3) does not count quantifiers among the constants. One reason is the trivial one that, given the semantics we have adopted, the truth-upon-an-interpretation condition of a quantification depends not upon how the interpretation treats any component, but rather upon how certain interpretations treat a sentence related in a certain way to the contained open sentence. I shall ignore this problem: let us assume, to assist the discussion, that we are dealing with a semantics in which interpretations assign objects to variables. Then it might seem as if (3) does count the quantifiers as constants. For suppose you know what each interpretation assigns to "Fx" (truth or falsehood, as the case may be), and you know the interpretation rule for "$\forall$", then it may seem that you can come to know *a priori* what each interpretation assigns "$\forall x Fx$". For you know that an interpretation, i, assigns truth to this sentence just on condition that all interpretations agreeing with i on their assignment to "F" assign truth to "Fx".

However, knowledge of what i assigns to "$\forall x Fx$" cannot be extracted *a priori* from the knowledge we have supposed. For one thing, one might not know that all the interpretations whose assignments to "Fx" one knew about are all the interpretations there are. For another thing, there is an unclarity about what it is to *know what* something, for example a predicate-letter, is assigned. Suppose that i_1 assigns the set of all 1-membered sequences whose member is a featherless biped to "F" and i_2 assigns to this letter the set of all 1-membered sequences whose member is a man. It is not an *a priori* matter whether they agree on their assignment to "F". You might know that "Fx" is true upon both i_1 and i_2, but because you do not know that these interpretations agree on their assignment to "F", you might not realize that i_2 is one of the interpretations relevant to the truth upon i_1 of the universal quantification. To deal with this problem, Peacocke in effect stipulates that included in the knowledge which is to be the basis for the *a priori* inference is this: for any letter, L, any pair of interpretations, if the interpretations agree on what they assign to L, you know that they do.[19]

As Peacocke stresses, this does not entail that, for any pair of letters, if an interpretation assigns the same thing to each, you know that it does. For this reason, identity does not, as the criterion stands, count as a logical constant. Suppose you know that an interpretation assigns Hesperus to "α" and Phosphorus to "β", and you also know that, like every interpretation, it assigns to "$=$" the set of ordered pairs such that the first member is the same as the second. You cannot work out *a priori* whether or not "$\alpha = \beta$" is true upon the interpretation. You need in addition the non-*a priori* information whether or not Hesperus is identical to Phosphorus.

This limitation, like the previous one relating to quantification, could be stipulated away, but then one would begin to wonder whether such stipulations have any rationale other than that of endorsing a predetermined list of constants, a list drawn up on the basis of some criterion other than (3). Why not add "is the same height as" to the list, by stipulating that the knower has access to the relevant information? It would seem that the way to argue for the rejection of such a suggestion is not from within the criterion (3), but from the independent intuition that information pertaining to heights must be irrelevant to logic. Again, "$\Box$" can be included among the constants, *if* we are prepared to add the following to the available information: with respect to which worlds the interpretations make the sentences of the language true, and, for each world, what each interpretation assigns to each letter with respect to that world. But the question arises whether or not we should be willing to suppose this information to be available.

It would, of course, be pleasing to have a way of giving our intuitions about constancy a precise technical embodiment, and this may well be provided by Peacocke's account, but it seems that it will not serve to underwrite these intuitions. Doubts about whether "$\Box$" or even "$=$" is a constant would not be resolved by the criterion.

To summarize this part of the discussion, it is not clear whether the first of the criteria – the topical-neutral one – is capable of any very marked improvement. It does not give definite rulings on certain cases, but perhaps this is simply because the concept of logical constancy is vague. The cement criterion seemed to be on the wrong lines, but the other three criteria showed some encouraging convergence. The indeterminacy in the rules theory related to the question whether the rules-determined meaning was really the least specific that would sustain the validities; and there were problems concerning whether it introduced a language-relative notion of constancy. The *a priori* criterion did not conflict with the topic-neutral criterion, but, rather, appeared to rely upon it when it came to crucial decisions about the formulation of the criterion.[20]

The boundary to the class of logical constants determines the boundary to the class of logical truths: truths true solely in virtue of the meaning of the logical constants, together with the pattern of occurrence of non-logical expressions. This in turn provides a boundary to the domain of logic: it could be identified with the class of logical truths.

Let us use "C" to refer to the claim that a logical truth is one true just in virtue of the meanings of the logical constants and the pattern of occurrence of other expressions. C has come under attack by Quine.[21] In Quine's work one finds attacks on at least the following five claims which might be held to be consequences of C:

4)   Logical truths are true by convention.

5)   Logical truths are true independently of how the world is.

6)   "The truths of logic have no content over and above the meanings they confer on the logical vocabulary". (p. 109)

7)   A truth of logic "is a sentence which, given the language, automatically becomes true". (p. 108)

8)   Logical truths are "true by virtue purely of the intended meanings ... of the logical words". (p. 110)[22]

The assertion of (4) is not part of the intention of the adherent of C, and I will ignore the question of whether it is a consequence of C. (It may seem to be, on the assumption that meanings are conventional.) C is certainly not intended to entail (5), and (5) is, on the face of it, false. As Quine remarks, the truth of "$\forall x\, x = x$" depends not just on words but on the world, on the fact that everything is self-identical. So we should understand C as saying that to the extent that meaning contributes to the truth of a logical truth it is only the meaning of the logical constants and the pattern of occurrence of the non-logical expressions that are relevant.[23] (6) is foreign to the approach of this book, in which no assumptions are made about how the logical vocabulary of English gets its meaning, and according to which if anything confers meaning on the **Q**-logical vocabulary it is the interpretation rules of the metalanguage and not the logical truths of the object language. (7) and (8), as recently qualified so as not to exclude a contribution to truth "from the world", are indeed consequences of criterion C.

Quine's argument against these last two theses, the only ones to which C is clearly committed, stems from a quite general scepticism about meaning. Certainly, if there is no such coherent concept as meaning there is no coherent doctrine of truth in virtue of meaning. It would take us too far afield to discuss this general scepticism, so I will simply turn aside from it.[24]

On one interpretation, to say that a sentence is logically necessary is just to say that it is logically true. This requires amplification. If truths like

9) Hesperus is Phosphorus

are *necessary*, then there is a kind of necessity that it is not logical necessity, in the sense just defined, while also not being merely physically, epistemically or morally necessary, or necessary in a way which is a restriction of logical necessity (in the sense of logical truth). (9) exemplifies the kind of necessity that Plantinga refers to as "broadly logical" necessity and which Kripke, for example, has called "metaphysical necessity". By contrast, the logical necessity that equates with logical truth is termed by Plantinga "narrowly logical" necessity. Not all broadly logical necessities, like (9), are knowable *a priori*. So if logic is to fulfil its promise of being an *a priori* subject, it should study and use only a narrower notion. This means that our original definition of validity in chapter **1.3** should be understood as employing a narrower notion.

6 Language, form and structure

The sentences of the artificial languages I have discussed have a strange feature upon which I have not explicitly remarked: for the most part they do not say anything, and cannot be characterized as true or false. The reason is that the letters, sentence-letters, name-letters and predicate-letters, have no fixed interpretation. They are empty vessels, waiting to be filled; the fillings are interpretations. Thus a **Q**-sentence like "$\exists x Fx$" does not say anything, nor is it true or false. Rather, it awaits an interpretation. *Relative to* an interpretation which, say, assigns the set of all 1-membered sequences which have a dog as member to "F", the sentence says that there are dogs, and is true. In **Q** there is no truth (or saying) *simpliciter*, but only truth (or saying) upon an interpretation.

We could have proceeded differently, introducing **Q**-symbols like "F" and "α" as abbreviations of specified expressions in a natural language, for example "is a dog" and "Fido". Let us call such a language **Q**∗. It will have the same definitions of interpretation and validity as **Q**, but its non-logical symbols will be thought of as (non-logical) constants for which just one interpretation is correct, and this interpretation, call it i∗, will be specified.[25] For **Q**∗-sentences, truth equates with truth upon i∗.

This alternative procedure makes **Q**∗ a real language, one whose sentences all say something, and are all true or false. For the purposes so far discussed, it makes little difference whether one formalizes in **Q** or in **Q**∗.[26] However, for this section we need to focus on **Q**∗. The suggestion I want to discuss is that we can use semantics for **Q**∗ to give a compositional semantics for natural language. In more detail, the suggestion is that in devising a semantics for English, we proceed in two stages:

1) First, English sentences will be translated into **Q**∗.

2) Secondly, the semantic resources discussed in connection with **Q** will be applied to the sentences of **Q**∗; the English expression "true" will be regarded as abbreviating "true upon i∗".

The second step assigns intuitively correct truth (that is, truth-upon-i∗) conditions to the **Q**∗-sentences, on the basis of the contributions systematically made by their parts. The first step ensures that none but correct truth conditions get assigned to English sentences which have **Q**∗-translations. Let us call a two-tiered theory of this kind a *proxy semantics*. In the version given above, English semantics are given by proxy of **Q**∗, which can be called the *proxy language*.

Something like this idea may well have been influential for some time, for example in Russell's early work, but it is only rather recently that a version of it has been explicitly formulated. Donald Davidson has proposed that English should be given proxy semantics. He uses **Q**∗ as an example of a suitable proxy language, while explicitly denying any commitment to the view that this is the only possible proxy language. He does not favour the type of semantic theory we have introduced, in which symbols are assigned entities of various kinds, preferring instead a "truth theory". I will not attempt to describe what this is, but will note that a truth theory recognizes exactly the same semantic categories as the semantics presented here, and, like our semantics, it provides truth conditions for every sentence on the basis of the elements from which it is constructed, and it is because of the perspicuity of $Q*$ that it is quite easy to produce a truth theory for it.

The essentially Davidsonian project of proxy semantics specified by (1) and (2) provides a definition of logical form in the spirit of (1.15):

3) the logical form of a sentence is the sentence of the proxy language into which it needs to be translated in order to provide its semantics.[27]

This gives the project of formalization an importance of a new kind. Formalization would not merely help to characterize validity, or formal validity, for a natural language, but, more widely, would help to give a general characterization of meaning, or at least truth conditions, for natural language. (3) would vindicate the view that logical forms provide insight into the semantic mechanisms of natural language.

On Davidson's view, logical form is essentially relative to a proxy language. Relative to a propositional language, the logical form of an English universal quantification will just be "p". This tells us very little about the semantics of the English sentence. For example, it would be absurd to infer that the English quantification is "really" unstructured. It is therefore important that some constraints be placed upon the choice of proxy language.

Intuitively, we want to say that an appropriate proxy language will mirror the structure of the natural language. However, it would be a pity to make this into a criterion of adequacy of a proxy language, for we wish to allow logical form proposals to provide a conduit for discoveries about the semantic structure of natural languages. For example, Davidson has proposed that a sentence like "Shem kicked Shaum" should be matched with an existential quantification in the proxy language, and it would defeat such a purpose to protest, on the basis of some pretheoretical "intuition" to the effect that the English sentence is atomic and not a generalization, that this proposal does violence to semantic structure.

The intuitive idea which, according to Davidson, should guide us is that semantics, including proxy semantics, should provide an answer to the question, asked with respect to the English: "What are these familiar words doing here?" An answer to this question, asked of "Shem kicked

Shaum" in the light of its $\mathbf{Q}_*$-translation, would be that "Shem" and "Shaum" serve to refer to objects, that "kicked" introduces kicks, and that the sentence as a whole introduces existential quantification. (I follow Davidson in ignoring tense.)

One more precise test of the adequacy of proxy semantics upon which Davidson insists is that if an English argument is, intuitively, formally valid, its translation in the proxy language should be formally valid.[28] For example, an important component in his argument for his logical form proposal for action sentences is that the $\mathbf{Q}_*$-translation (or $\mathbf{Q}$-formalization) of "Shem kicked Shaum in the face, so Shem kicked Shaum" is $\mathbf{Q}$-valid. This places important limitations upon what could be an adequate proxy language for English. For example, it rules out the adequacy of a propositional language.[†]

A further Davidsonian adequacy condition upon semantic theory is that the theory have only finitely many axioms. The effect of this condition is to force a semantic theory to recognize some structure in the object language. Languages typically have infinitely many sentences, so a finitely axiomatized semantic theory can deliver a correct assignment of truth conditions to every sentence only by seeing sentences as made up of parts drawn from a finite total stock, each part making a distinctive contribution to the truth conditions of whatever sentence may contain it. Applied to proxy semantics, this means that there must be a finite manual for translating English into the proxy language, and a finitely axiomatized semantic theory for the latter. Let us call this the finite axiomatization constraint.

The idea of proxy semantics enables one to make good sense of some extreme-sounding claims about logical form, for example, the claim that "Shem kicked Shaum" is "really" an existential quantification. The claim amounts to no more than that such a sentence will be translated into an existential quantification for the purposes of proxy semantics. Likewise the claim that "All men are happy" is "really" a quantified conditional amounts to no more than that such a sentence will be translated into a quantified conditional for the purposes of proxy semantics. This is more ambitious than the claim that, say, an English quantification and its $\mathbf{Q}_*$-rendering are alike in truth conditions, for their likeness in this respect is of a special kind, engendered by a single scheme of translation; yet it is suitably less ambitious than the absurd claim that an English quantification contains an invisible occurrence of "if" or "→".

† See Exercise 168, page 368.

Even so, some puzzlement may remain. Consider two logical form proposals for English quantifications: one made by standard **Q∗**-translations, the other by translations into a language containing binary quantifiers. Both proposals might meet all of Davidson's criteria, yet we are intuitively inclined to believe that they cannot both be right, cannot both be true to the semantic structure of English. Indeed, intuition clearly sides with treating English quantifiers as binary rather than unary. It seems perfectly possible that a natural language sentence should have a Davidsonian logical form, and yet exploit different mechanisms to achieve the same truth conditions from the same non-logical primitives. Hence it seems possible that a proxy semantics meeting Davidson's constraints should give an incorrect account of semantic structure.

The root of the puzzlement is that whereas Davidson's notion of logical form is relative to a proxy language, we tend to suppose that there is an absolute fact about the semantic mechanisms a sentence exploits. It remains to be seen whether there is any basis for this anti-Davidsonian supposition, and what impact it has upon his theory.

We saw that the choice of a suitable proxy language is crucial to the value of Davidson's conception of a proxy semantics. Intuitively, we want to say that the proxy language must mirror English in point of semantic structure. However, we should not make this a condition of correctness, or else the project of proxy semantics will be unable to contribute to our understanding of what the semantic structure of English is. Davidson's finite axiomatization constraint is an attempt to require proxy semantics to recognize semantic structure, without explicit allusion to this notion. However, it fails in this aim: even if it forces the recognition of some structure, it does not guarantee recognition of the *right* structure. For example, it does not prevent a proxy semantics, adequate by Davidson's standards, from regarding "Socrates is wise" as a conjunction, since the requirement of finite axiomatization would not be flouted by a finite translation manual with the rule that an English atom of the form "x is wise" be translated into **Q∗** by an atom of the form "Wise x & $(2+2=4)$".[29]

When we speak here of the "right" semantic structure, or when we spoke earlier of the "intuition" that English quantifiers are binary, to what standard are we appealing? An extreme Davidsonian might argue that there is no intelligible standard beyond what emerges from the project of proxy semantics. To object that, for example, it is "counterintuitive" to regard "kicked" as a three-place predicate is to invite the response that the intuition is worthless. Semantic classifications have no role other than to assist in the specification of truth conditions of all the sentences of a language on the basis of the contributions of subsentential expressions. If this is provided by a theory which regards "kicked" as three-place, there is no room for a standard relative to which this is the wrong semantic classification.

However, Davidson himself says that semantic structure should mirror the structure of our ability to speak and understand a language.[30] This suggests that an appropriate standard is psychological: semantic structure should reveal the features which one who understands a sentence exploits in coming to understand it. One cannot tell *a priori* what these features are, though certain "intuitions", for example, that English quantifications are not all quantified conditionals, constitute our unsubstantiated hunches. These need to be tested empirically, since we know quite well that introspection is not always a reliable guide to our mental processes.

Serious empirical data for such views could be obtained in a number of ways, of which the following is an extreme example. Suppose we found that severing a certain nerve in the brain had the effect of making a person incapable of understanding explicit conditionals, sentences like "if John is a man, then he is happy". Suppose that this same person continued to understand quantification like "All men are happy". This would be evidence that the semantic structure of English quantifications is not that suggested by their **Q**-formalizations.

The criterion of psychological correctness provides a basis for preferring one rather than another of two logical form proposals which meet all of Davidson's criteria.[31] It sets an absolute standard for semantic structure,[32] and harmonizes well with the tradition. For example, Russell on more than one occasion said, concerning a logical form proposal, that it gave a better picture of what was going on in the mind of a person using or understanding a sentence.[33] This would be assured, if logical form mirrors actual processing mechanisms. This makes logical form the form of thought as well as the form of language.

Suppose that we can find only one logical form proposal satisfying Davidson's criteria. Is it reasonable to believe that this correctly identifies the semantic structure of the English sentence? If one thinks of the psychological criterion as used to choose between two proposals, both of which are adequate by Davidson's criteria, an affirmative answer might seem appropriate. However, caution is required.

To illustrate one ground for caution, consider the claim that

4)   Tom is married

has the logical form of an existential quantification, "married" having the form of a *two*-place predicate. If we fix on languages like **Q**∗, there indeed appears to be no alternative way of explaining the truth conditions and inferential behaviour of (4). However, we here take for granted that *English* predicates have fixed degrees. This assumption stands in need of justification, and doubt is cast upon it by the existence in English of expressions like "are compatriots", which appear capable of taking as grammatical subject a conjunction of any number of names. If the assumption of fixed degrees is abandoned, one abandons any compelling reason for supposing that the truth conditions of (4) are correctly representable, for the purposes of compositional semantics, only by existential quantification into a position provided by a two-place predicate. The general moral is that, as with any science, a development which at one time could not be imagined (for example a formal language with predicates of no fixed degree) may soon become commonplace.

A further ground for caution is that the reasons that can be offered for a logical form proposal tend to be capable of different implementations in different proxy languages, even if this comes to light only later. Let me take two famous examples. The first is Russell's theory of descriptions. Every reason that Russell or anyone else has ever given for the theory is as much a reason for the view that "the" is a binary quantifier, forming sentences with the truth conditions proposed by Russell's theory, and so not involving, in the logical form, the extraneous machinery (two quantifiers, a conditional and an identity) attributed by Russell's theory. The second example concerns Davidson's proposal that many adverbs are predicates of events. Wiggins [1985] has shown how the underlying rationale is independent of the claim that the relevant sentences have **Q**∗-logical forms, and how the proposal can be transcribed, without loss, into a different setting. We should accordingly be wary of inferring that there is a unique way of providing proxy semantics from the fact that the only way we can at the moment envisage uses **Q**∗ as the proxy language. There is no sound inference from the premise that a natural sentence and its logical form are alike in truth conditions, in a way fit to serve the goal of proxy semantics, to the conclusion that they attain this likeness by exploiting the same semantic mechanisms.

Finally, we must not forget that the Davidsonian approach takes for granted that it must be possible to provide compositional semantics for English, directly or by proxy. However, this is debatable.[34] One source of doubt is as follows. The idea underlying compositional semantics is that the meanings of words determine the meanings of sentences. Correlatively, one who understands a sentence is regarded as computing its meaning on the basis of the meanings of the words it contains. This is supposed to be reflected in compositional semantics by the fact that the properties assigned to the words *entail* what the meanings of the sentences are. However, there are some grounds for supposing that interpreting sentences is not a matter of following the lines of entailment from word meaning to sentence meaning, but is, perhaps additionally, a matter of guesswork and probabilistic reasoning, so that there is no clear demarcation between our specifically linguistic competence and our general cognitive abilities. For example, it would appear that no correct assignment of semantic properties to "carpet", "vacuum", "sweeper" and "cleaner" *entails* that a carpet sweeper is something to sweep carpets with, whereas a vacuum cleaner is not for cleaning vacuums. Yet speakers of English have no trouble picking up the use of these compounds once they understand their constituents. Another example is the use of demonstratives, with respect to which general knowledge about what objects are likely to be conspicuous for speakers plays a crucial role in interpretation.

So a logical form proposal, even if it is the only one we can find capable of serving the purposes of proxy semantics, should be accepted only with caution. Having sounded this warning, it must be insisted that the Davidsonian theory of logical form articulates what the tradition of formalization, from Frege and Russell onwards, needed in order to make sense of its ambitious claims. The project of formalization may have started with concerns in which validity and proof played central roles, but the importance which it is rightly accorded today derives additionally from its role in theorizing about the structure of language and thought.

7 Conclusion

The original motivation for the introduction of artificial languages into logical studies was the lack of perspicuity in natural languages. If the only ambition is to use the perspicuity of an artificial language to attain some generalizations about validity in natural language, the connection between the artificial and the natural is unproblematic. The adequacy of a formalization ensures that if the formalization is valid, so is what it formalizes; and the notion that the formalization is an idealization diminishes any anxieties about validities in the natural language that are not reflected in adequate formalizations.

Once the aim is to segregate out *formal* validity in the natural language, a tension arises. This is because there are independent, even if vague, tests for whether or not a valid argument in natural language is formally valid. It turns out that arguments in natural language which, though valid, are not formally valid (for example, arguments involving adjectival detachment), can be adequately and validly formalized. On the one hand, it would seem a pity to pass up this opportunity; but doing so is inconsistent with concentrating upon only formal validity in the natural language.

The relationship between the artificial and the natural seems most strained in connection with exotic logical form proposals, for example the view that English universal quantifications are quantifications of conditionals. Making sense of such claims requires Davidson's conception of logical form, but this does not necessarily substantiate the claims. This is because resistant intuitions (for example, the intuition that English quantifiers are binary) are some evidence, albeit inconclusive, that the logical form proposals, even if they get the truth conditions right within the project of proxy semantics, do not satisfy the psychological criterion.

However, the semantics that can be provided for languages like Q_* offer a model of rigour and precision, and have provided an essential stimulus to research. What is overambitious is to suppose that these semantics are already, by proxy, what we seek for our natural language, since there is little evidence that they satisfy the psychological criterion. Indeed, it is overambitious to *assume* that compositional semantics of any form are possible for natural languages. But if they are not, that will be a most important fact, and one only statable against the background of what compositional semantics for artificial languages are like.

Davidson's conception offers the most ambitious prospect for the project of formalization, but let us not underestimate its other achievements. First, the very thought that specifying truth conditions is a useful contribution to specifying meaning (and perhaps sometimes exhausts what is required) grew up in the context of the project. Looking in his (somewhat distorting) rear-view mirror, Russell said that in his early years he thought that the logician's task in connection with "the" was to identify some weird abstract object to which it referred.[35] Specification of the way in which the truth of "the"-sentences depends upon the truth of their components is a vastly improved goal. Secondly, it is surprising how much room for disagreement there is about the truth conditions of the sentences of the language we actually speak. We have encountered cases in connection with apparently very simple sentences: universal and existential quantifications, and sentences containing definite descriptions. Such disagreements could in practice only be discovered and expressed in connection with attempts at formalization in artificial languages for which truth conditions could be precisely stated. Thirdly, it is hard to overestimate the importance of formalization in the resolution of ambiguity. For example, once you have been introduced to the quantifier shift fallacy in the context of formalization, your ability to spot instances of it in English is greatly enhanced.

A final reminder: my concern with formal logic has been guided by the limited aim of describing and understanding the project of formalization. It must not be forgotten that formal logic is a discipline – a branch of mathematics – which can be pursued for its own sake and with no eye to this project. Many of the results obtained in this discipline have – or have been claimed to have – a philosophical importance of a kind not envisaged within the project of formalization.[36]

Notes

The traditional view of logical form derives from Frege and Russell, though they never explicitly state it.

For the Tractarian vision, see Russell [1918] and Wittgenstein [1921]. For its repudiation, see Wittgenstein [1932].

For an introduction to the techniques of natural deduction see Lemmon [1965] or (for more advanced texts) Prawitz [1965] and Tennant [1978].

For discussions of the logical constants see Peacocke [1973] and [1987], Prawitz [1977] and [1979].

For compositional semantics see Davidson [1965], [1967a], [1970b], [1973]; Davies [1981]; Schiffer [1987].

For Davidson's conception of logical form see esp. his [1970a] and [1973]. For a gentle shift of perspective, see Wiggins [1985]. For a quite different interpretation from mine of Davidson's view, see Lycan [1984] pp. 31–2.

Further reading in philosophical logic could well take the form of following up works mentioned in connection with specific discussions. For something both wide-ranging and introductory I recommend Haack [1978]. For a much more advanced collection, I recommend Gabbay and Guenthner [1983], [1984], [1986] and [1989]. For a splendid and relatively recent example of how a philosophical outlook can be informed by something like the project of formalization, see Quine [1961].

1 Cf. Strawson [1952] p. 51.
2 I use "truth conditions" in the way explained in ch. **1.9**: the truth conditions of a sentence are the actual or possible circumstances in which it is or would be true. There is, however, another usage, according to which the truth condition (singular) of a sentence is its meaning.

3 On the presentation of **Q** in this book, the truth of this remark requires that we effect syntactic classifications in the light of "constructional history": cf. ch. **4.2**.

4 I stress that it is on the *naive* syntactic criterion that names and descriptions fall into a common category. Current syntactic theory is not naive, and typically differentiates names and descriptions.

5 The Tractarian dream was dreamed before the notion of completeness for logical systems was available. No doubt the incompleteness of second order logic would have affected Russell's version of the dream. (A consequence of incompleteness is that some valid sentences would be formalized by unprovable ones.)

6 It is presupposed that an expression has its analysis once and for all. Otherwise, the second point could be evaded by proposing different sets of primitives for different cases.

7 See Davidson [1967a] pp. 31ff., [1970a].

8 Even one who accepted the correspondence requirement should not be worried by this fact. Those who have done so much fruitful work on languages like **QW** and **QC** should not be alarmed if some of the results get classified as contributing to analysis rather than as contributing to a specification of logical form.

9 This is not quite correct. The definitions of $\vDash_P$ and $\vDash_Q$ allow for the case in which infinitely many sentences stand to the left of the turnstile, whereas the specification of $\vdash_P$ does not. We need to augment the rules of proof to allow for infinite sets of premises. This can be achieved by the stipulation that if $\Gamma \vdash_P X$ and $\Gamma' \supseteq \Gamma$, then $\Gamma' \vdash_P X$.

10 The contrast is not as clear as it might appear. Using the word "truth" in a definition does not automatically render it semantic, and the "semantic" features ascribed by interpretation rules are identified on the basis of purely syntactic features, the shapes of the sentences. The algebraic view of semantics mentioned in ch. **5.11** detaches semantics from the ordinary notions of truth etc., so could be regarded as belonging to syntax, yet it mentions no rules of proof. Equally, if in stating rules of proof you read the horizontal line as showing that what is below *follows from* what is above, you are viewing the rules in a semantic light.

11 The theory must specify, for each description, whether it assigns it an object and if so which. To make the case more plausible, imagine that some syntactic restriction requires that "the" immediately attach just to a simple predicate.

12 Strictly, we also need a further clause to show that the substitution is available with respect to the first sentence position in a ¢-sentence.

13 The example derives from Peacocke [1987].

14 Cf. Prawitz [1965]. $\Box$I(i) determines the logic known as S4. (In this system of classification, deriving from Lewis and Langford [1932], **PN** is known as S5.) Arguably, the English word "necessarily" is ambiguous, and one can see some modal logics, including S4, as specifying various disambiguations of it.

15 Cf. Belnap [1961].

16 Some logics involve a sentence constant (as opposed to a sentence letter) "f" or " ∧ ", with the constant interpretation of the truth value false; this would receive a basis clause, and so would not be counted a logical constant.

17 See Evans [1976] p. 69.

18 Peacocke [1973]. He expresses the criterion in terms of a different semantic framework, but claims that it can be adapted at least to an intuitionistic framework (p. 238). Though my adaptation is much more minor, the possibility remains that, along with deliberate oversimplifications, there are also distortions.

19 Cf. Peacocke [1973] pp. 226–7.

20 It is worth noting that all criteria for constancy include predicate quantifiers as well as name quantifiers among the logical constants.

21 Quine [1963].

22 In this account, I am guided by Peacocke [1987].

23 Not that Quine would allow this bifurcation between the contribution to truth made by meaning and the contribution made by the world.

24 See Quine [1951], [1960], ch. 2, and [1968]. One fruitful argument against Quine's position is in Evans [1975].

25 The stipulation that $\vDash_Q = \vDash_{Q*}$ aims to use **Q**-validity to define formal validity in **Q**∗. There will, of course, be **Q**∗-arguments which are valid in the sense of $\vDash$, but which are not **Q**∗-valid.

26 **Q**∗ has the practical disadvantages that it is hard to remember what the abbreviations are, and if we have many examples we soon have to use letters made unsightly by clusters of primes. However, **Q**∗ has, in effect, been regarded as theoretically superior: see Smiley [1982], who may well regard the use made of **Q** in this book as an instance of the "schematic fallacy".

27 See Davidson [1977], esp. p. 203.

28 Davidson would say "provable". We would get an equivalent condition, where the proxy language is **Q**∗, by saying "**Q**∗-valid".

29 For essentially the same point, applied more directly to Davidson's position, see Foster [1976].

30 [1967a] p. 25.

31 Davidson accepts this (personal communication, February 1988). However, he holds that it is in practice virtually impossible to find even one, let alone two, proposals meeting his criteria, so that the practical importance of a further criterion is small.

32 The truth is a good deal more complicated: see Davies [1983]. By "absolute" here I mean "independent of a language of formalization". There remains a relativization to speakers.

33 See, for example, Russell [1912] p. 29.

34 See Schiffer [1987], ch. 7.

35 Russell [1959] p. 150.

36 To choose two examples from dozens: John Lucas [1961] claimed that Gödel's incompleteness theorem shows that men are not mechanisms; and Hilary Putnam [1980] claimed that the Löwenheim–Skolem theorem has far-reaching consequences for metaphysics.

Exercises

Chapter 1

1 Suppose that Henry is indeed poor. How would you show that (3) is not a good argument, that is, that its premise does not constitute a good reason for its conclusion? Would you appeal to any facts about the actual income of playwrights, or to any further facts about Henry?

2 What is the statistical truth, of which the Monte Carlo fallacy is a distortion? Can the truth be used to formulate a rational betting policy for roulette?

3 What is likely to be at issue between one who does and one who does not think that the evidence shows that smoking increases the risk of heart disease?

4 (a) Why should this not be a physical, rather than a logical, possibility?

    (b) Give three examples, not in the text, of states of affairs which are logically possible but physically impossible.

    (c) Give three examples, not in the text, of states of affairs which are logically impossible.

    (d) Are any states of affairs logically impossible but physically possible?

    (e) Are the following logically impossible, physically impossible, or neither? ("Don't know" may be the best response in some cases!)

  (i) Mr Stamina has at last perfected his perpetual motion machine.
  (ii) Mary has precisely twice as many children as Jane: she has the twin girls and little William.
  (iii) Jock ("the Flash") McVite ran the mile in under 3.5 minutes.

(iv) The notorious swindler, Siva Malgavany, who died in 1880 in the arms of one of his many mistresses, was reincarnated in 1987 as a small Irish terrier, owned by Mrs Fortescue-Brown of Egham, Surrey.

(v) On 15 March 1988, Dr Chronowski stepped into his time machine, and stepped out again to find himself present at the battle of Waterloo.

5 (a) Is it nevertheless inductively strong? Compare its inductive strength with that of (1), justifying your comparison.

(b) Would it make a difference to the validity of the argument if the first premise were "This creature *is* a finch"?

6 (a) In each of the following cases, say whether the argument is deductively valid, inductively strong, or neither. Justify your answers:

(i) I told you to add more pepper if you thought the stew was too bland. You added more pepper even though you didn't think the stew was too bland. So you disobeyed me.

(ii) You must know how to make a sling, if you are really a qualified nurse. But evidently you don't know how to make a sling. So you are not a qualified nurse.

(iii) He has coughed up arterial blood, so his lungs must be extensively damaged.

(iv) My father was born in London, yours in Paris. So we can't be brothers.

(v) This plant grew from a seed produced by a tomato plant. So it, too, is a tomato plant.

(vi) Frank Whittle died in poverty. Therefore, the inventor of the jet engine died in poverty. [Historical note: Frank Whittle invented the jet engine.]

(vii) The satellite picture shows a cold front sweeping in from the Atlantic, and it should reach our shores by morning. So it will probably rain tomorrow.

(viii) The explanation of any fact is in terms of other facts. So if any facts can be explained, there are facts which cannot be explained.

(ix) If you know anything, you can't be mistaken. So any proposition that could be false is one which you do not know.

(x) The proposition "All propositions are false" is false. For if it were true it would be false. [NB: In ordinary English, the conclusion of an argument may be presented before the premises.]

(xi) Geraniums are not frost hardy. There are frequent frosts in Iceland. So geraniums are not native to Iceland.

(xii) If either the CIA agent or the KGB agent killed the President, he used a gun. But the President died of cyanide poisoning. So neither the CIA agent nor the KGB agent killed him.

(b) Think of a stack of four cards with numbers on one side and letters on the other. Someone tells you that all cards which have an even number on one side have an "E" on the other. Say which of the following cards you need to turn over in order to tell whether what he said is true:

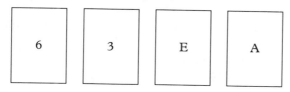

Explain how this example relates to the validity of various arguments involving what the person said as one of its premises.

(c) We can use cards in the manner of the previous example to specify situations. Thus each card might represent a person in a bar, one side indicating whether he is an adult or a minor, the other side indicating whether he is drinking alcohol or a soft drink. Suppose there are again four cards having visible sides as follows:

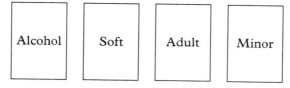

Which cards do you need to turn over in order to check whether the bar is violating the rule that only adults may drink alcohol? Explain how your answer relates to the validity of arguments having as a premise "Only adults drink alcohol".

7   Without using "not" (or any equivalent word or phrase) form a contradictory of each of the following:

(i) Tom has no children.
(ii) Richard Nixon died in 1981.
(iii) London is north of Paris or south of Paris.

8 Which of the following pairs consists of a proposition and its negation?

   (i) I hope you will come.
       I hope you will not come.
  (ii) I hope you will come.
       I do not hope you will come.
 (iii) I am pleased with your progress.
       I am not at all pleased with your progress.
  (iv) Everyone who goes to see Mick the Fix is satisfied with his handling of their problem.
       Everyone who goes to see Mick the Fix is not at all satisfied with his handling of their problem.
   (v) You must not walk on the grass.
       You must walk on the grass.

9 Give three examples (not in the text) of collections of propositions which are inconsistent, but where no member of the collection is a contradictory of any other member.

10 Is the argument valid? Is the conclusion true?

11 Give it.

12 State the definition.

13 (a) What is the relation between ";", as used here in setting out arguments, and "hence" or "therefore"?

   (b) For each of the following pieces of discourse, state what argument, if any, you think the speaker is intending to propound. If you think he is intending to propound one, say what its premises and conclusion are.

   (i) My opinion, for what it's worth, is that you should take to robbing banks. This is a demanding career, requiring a high degree of responsibility, but offering greater than average financial rewards. Moreover, it gives you the opportunity to be your own boss and to develop your talents of leadership and initiative. If all else fails, it provides you with free board, lodging and protection at the expense of Her Majesty's Government.

(ii) If John and Mary are godparents of the same child, as indeed they are, they cannot be married according to Christian rites.

(iii) Since John and Mary are godparents of the same child, they cannot be married according to Christian rites.

(iv) John and Mary are godparents of the same child. Therefore, they cannot be married according to Christian rites.

(v) Treason doth never prosper. What's the reason?
Why, if it doth, then none dare call it treason.

(vi) No contemporary politician understands the French as well as de Gaulle did. Mitterrand is at home with the intellectual Left, and d'Estaing knows the aristocracy and the *haute bourgeoisie*. But de Gaulle knew something about the traditional peasant class, which is what sets him apart from our contemporaries.

(vii) Authors of sentimental novels were not being so ridiculous as they appeared to later generations when they described so many abductions, ladders, musketeers and escapes from convents. All these exciting events actually took place, and frequently, too. It was part of the current vogue for aggressive Spanish manners.

(viii) It was late and I was tired. So I took a taxi.

14 (a) Are there any valid arguments with false premises and a true conclusion? If so, give an example. (Here and elsewhere, select examples of propositions which are well known to be true, or false, as the case may be.)

(b) Are there any valid arguments with true premises and a false conclusion? If so, give an example. If not, say why not.

(c) If an argument is invalid and at least one of its premises is false, what can be inferred about whether or not its conclusion is false?

(d) If an argument is valid and has a false conclusion, what can be inferred about whether or not its premises are true?

15 Using the displayed definition of validity near the end of §3, show that (**6**) is true.

16 Using the displayed definition of validity near the end of §3, show that (**7**) is true.

17 Use this notation to express the claim: "If a collection of propositions is inconsistent, it remains inconsistent whatever proposition is added to it." Now argue for the truth of this claim.

18   (a) Show that:

If [⊨ B], then:

[A_1, ... A_n ⊨ C] if and only if [A_1, ... A_n, B ⊨ C].

This shows that necessary truths are redundant as premises. You may now wish to think back to your answer to Exercise 6 (avi).
(b) Show that:

If [(B) ⊨], then [A_1, ... A_n, B ⊨ C], whatever A_1, ... A_n, C may be.

Could "if" here be strengthened to "if and only if"? Justify your answer.

19   (a) Say whether or not you think that:

[A_1, ... A_n ⊨ C] if and only if [⊨ If A_1 and ... and A_n then C].

Briefly justify your view.
(b) Show that:

[(A) ⊨] if and only if [⊨ It is not the case that A].

(c) Is it true that

[(A) ⊨] if and only if [⊭ A]?

Justify your answer.

20   (a) Could there be invalid arguments which are:

(i) sound?
(ii) relevant?
(iii) persuasive?

If so, give examples. If not, say why not.
(b) An argument can be sound and relevant, yet fail to be persuasive through being too elliptical. Give an example.

21   (a) Suppose Tom utters the following sentence, directed at you: "I like kissing you". Give an example of a sentence you could utter and thereby express the same proposition as Tom expressed.

(b) Suppose Tom yesterday in London uttered the sentence "It is raining here". Give an example of a sentence you could utter today in Paris and thereby express the same proposition as Tom expressed.

(c) Suppose Tom, walking in the woods yesterday, dimly perceived what was in fact a boa constrictor (though he did not know it was), and, referring to that object, uttered: "That is the strangest looking rabbit I ever saw". Give an example of a sentence that you could utter today, in the comfort and security of your own home, and thereby express the same proposition as Tom expressed.

22   Show that this is so.

23   Give two further instances (not in the text) of (**4**). Can you find any invalid instances?

24   Why is this qualification – "necessarily" – needed?

25   Give a further instance (not in the text) of (**7**). Can you find any invalid instances?

26   (a) Give a further instance (not in the text) of (**9**). Can you find any invalid instances?

(b) Specify valid argument-forms for any of the following which are formally valid:

(i) Some politicians are liars and all liars are charming. So some politicians are charming.

(ii) Some politicians are liars. Veredici is a politician. So Veredici is a liar.

(iii) If you are going to die by drowning, then there is no point in learning to swim. If you are not going to die by drowning, then there is no point in learning to swim. Therefore, there is no point in learning to swim.

(iv) Everyone who has studied logic is able to spot an invalid argument when he sees one. You haven't studied logic. So you are not able to spot an invalid argument when you see one.

(v) 7 is prime. $7 = 5 + 2$. So $5 + 2$ is prime.

(vi) The battle of Marengo occurred before the French invasion of

Moscow, and the battle of Waterloo came after that. So the battle of Marengo occurred before the battle of Waterloo.

(vii) He will die unless he is given a blood transfusion. But he will not be given a blood transfusion. So he will die.

27 (a) Justify the claim that (**11**) is invalid by finding an invalid instance of it. (To make the invalidity plain, pick an example with an obviously true premise and obviously false conclusion.)

(b) Show by example that each of (**1**), (**2**) and (**6**) are instances of invalid argument-forms.

(c) Show that *every* argument is an instance of at least one invalid argument-form.

28 Consider the following objection to the test mentioned in the text:

Despite the fact that "cut" is ambiguous, "I cut the string with my pocket knife" is not. Hence the fact that a word can occur in unambiguous sentences does not show that it is unambiguous.

29 If any of the following is ambiguous, give unambiguous paraphrases of the alternative interpretations:

(i) I am going to buy a book.
(ii) Everyone has a problem.
(iii) "In the whole wide beautiful world, Aldo Cassidy was the only person who knew where he was." (Le Carré [1971], p. 8)
(iv) "Most of all I would like to thank my students, who have taught me more than they know." (E. Bach [1974], p. vi)

30 What, if anything, is the reading of (**15**) upon which it is valid?

31 Give two further invalid instances of (**14**).

Chapter 2

32 Show that this remark needs qualification. (Compare (**1.12.20**).)

33 (a) Does the same hold for substituting one *false* component for another in the context "Napoleon knew that ..."?

(b) On the assumption that "necessarily", in, for example, "necessarily no even prime number is greater than 2", is a sentence connective, use the substitution test to determine that it is not a truth functional one.

34 Why would it be incorrect to speak of *the* intended interpretation?

35 Show this, either by informal reasoning using the definitions of the connectives and **P**-validity given in §1, or, if you know one, by a formal method.

36 Specify an interpretation which demonstrates this, by saying what truth values the interpretation assigns to the **P**-letters of (**17**).

37 Give assignments to the letters of (**26**) to establish its **P**-invalidity. Show (by informal reasoning, or by a formal method) that (**27**) is **P**-valid.

38 Give English versions of (**26**) and (**27**) which make plain the contrast between them.

39 (a) Demonstrate the **P**-validity of (**29**).
(b) Formalize each of the following arguments in **P**, showing your correspondence scheme. Determine the **P**-validity of the formalizations, and say what you think can be inferred about the validity of the English:

(i) Peter or Quentin killed Richard. If it was Peter, then the motive was jealousy. If it was Quentin, the motive was greed. But in fact the motive was not greed. So Peter killed Richard.

(ii) Peter will win the election unless Quentin does. Quentin won't win unless he buys all the electors drinks, and that is something he won't do. So Peter will win.

(iii) Neither Peter nor Richard will win the election unless Quentin doesn't stand. But Quentin's not standing would ensure Richard's success. So whatever happens, Peter will not win.

(iv) Although protective legislation has been enacted in most communities, the number of sperm whales is continuing to decline. The cause is either illegal whaling by nationals of countries participating in the protective legislation, or whaling by nationals of non-participating countries, which, under the circumstances, cannot be regarded as illegal. If the first alternative is ruled out, then the decline in numbers will be halted by bringing pressure on the non-participating governments. No doubt they will yield in exchange for subsidized loans. So if the decline in numbers of the sperm whale is

not caused by illegal whaling, halting the decline requires making subsidized loans to non-participating governments.
(v) We can infer that the local plonk is good. For if it is not good, some people will not be drinking it. But some people *are* drinking it.
(vi) Either, if Tokyo is the capital of Japan, the EEC will collapse before 1992, or else, if the EEC will collapse before 1992, then Tokyo is the capital of Japan.

40   Show that there is a case for thinking that "He will die unless he is given penicillin" is equivalent to "He will be given penicillin or die".

41   (a) Suppose "*p*" corresponds to "Some cows eat grass". Why does "¬*p*" not formalize "Some cows do not eat grass"? (Check the definition of "¬" in §2.)

(b) Suppose "¬*q*" formalizes "Some cows do not eat grass". To what does "*q*" correspond?

(c) Are there any difficulties in interpreting "not" as it occurs in the following sentences as expressing the same truth function as "¬"? If there are none, provide a **P**-formalization (showing your correspondence scheme). If there are any difficulties, explain.

(i) I am not very optimistic about the upshot of the talks.
(ii) The world will end not with a bang but with a whimper.
(iii) You ought not to smoke.
(iv) All the passengers who have not got tickets will wait in line.

42   There is an analysis of (**8**) according to which "and" functions in (**8**) the same way as it does in (**10**). Can you discover it?

43   Are there any difficulties in interpreting "and" as it occurs in the following sentences as expressing the same truth function as "&"? If there are none, provide a **P**-formalization (showing your correspondence scheme). If there are any difficulties, explain.

(i) John and Mary bought a boat.
(ii) All elephants have short ears and long tails.
(iii) Some elephants have short ears and long tails.
(iv) You and I are the only people who matter.

44   However, (**21**) does not seem to entail that Tom and Mary could both help you (simultaneously), for it could be true in a situation making it impossible for there to be more than one helper. Give an example to show this. Does this suggest a way of avoiding the unattractive suggestion that "or" in (21) really means "and"? Provide details.

45 Show how the exclusive disjunction of any **P**-sentences X and Y can be expressed using just the **P**-connectives already defined (in §2).

46 A restaurateur who puts on the menu "dessert or fruit" commits himself only to allowing you one of these, but he does not falsify his menu, or violate any undertaking to which his menu commits him, if he gives you both. Does this support the inclusive or the exclusive interpretation of "or"? Reflection on this case should, in my view, remove any appearance of exclusive disjunction in (28). Can you explain how this line of thought runs?

47 Why does this follow? (Refer back to the definition of "→" at (1.4).)

48 It might be suggested that the following example shows that "if" sometimes expresses a truth function, but not that of "→":

If you need bandages, there are some in the first aid box.

Here, according to the suggestion, the conditional is true iff its consequent is. Discuss the suggestion.

49 Give a context (e.g. a narrative) within which "Oswald hadn't shot Kennedy" straightforwardly has a truth value.

50 What, if anything, would be wrong with the translation:

(someone is in debt)→(someone should curb his expenditure)?

51 Provide the remaining three examples of "if ... then ..." sentences, being sure to select sentences whose truth values are well known.

52 Explain why.

53 (a) To serve as an example, (9) needs to be false. Put the case for the view that it is true (even if you don't accept this view!).

(b) Can you think of a convincing example to show the falsehood of (11)?

(c) Suppose you believe that there will be a third world war whether or not there is a summit meeting next spring. Suppose you also firmly believe that there will be no summit meeting next spring. Should you accept or reject the conditional "If there is a summit meeting next spring, there will be a third world war"? What bearing does your answer have upon whether (11) is true?

54 Which answer to Exercise 53c undermines the claim that (14) is invalid? Explain your answer.

55 (a) Explain how one who thinks that (25) is invalid could best justify his position.

(b) Explain how the following argument could be used to strengthen the case against the truth functional interpretation of "if".

If John is in Paris, then he is in France.
If he is in Istanbul, then he is in Turkey.
So if he is in Paris, he is in Turkey, or if he is in Istanbul, he is in France.

56 Say which of the following could have been used by S to convey his unflattering opinion:

(i) Either he is a good philosopher or he isn't.
(ii) He has a beautiful wife.
(iii) He has appalling handwriting.
(iv) His handwriting is average.

In general, what are the necessary and sufficient conditions for an utterance to be usable by S, in this context, to convey his low opinion of Jones?

57 Can one say that A and thereby convey that $\neg A$? If so, give an example. If not, why not?

58 Why is the parenthetical qualification needed?

59 Can you suggest an example of something which is rather uninformative but highly assertible? If so, could (3) be amended to allow for your example? If not, is there a way of showing that there could be no such example?

60 Evaluate the following argument in favour of this claim:

The truth of the antecedent would make the truth of the consequent impossible, and it is hard to see how more than this could be required for the conditional as a whole to be false.

61 Jackson has argued that one should sometimes assert something weaker in the interests of robustness. Suppose "*A* & *B*" is something of which you are sure and is relevant to the conversational needs in question, but that all you really care about is that your audience believe *A*. Show how the demands of robustness will conflict with the requirement of being maximally informative.

62 Apply (**9**) to (**5.20**).

63 What is the best response a defender of the truth functional interpretation can make to the claims related to (**5.17**), (**5.23**), (**5.25**), (**5.28**) and (**5.30**)?

64 Can you think of any basis for this denial?

65 Explain how a true conditional with true antecedent and false consequent would refute (**7**).

66 Explain how.

67 Which step, if any, do you regard as suspect, and why?

68 Explain how, contrary to what is said in the text, a Frege-style concealed component truth functional analysis might be applied to "*A* therefore *B*".

69 Could it be persuasively argued that, provided the scope-indicating brackets are held firmly in place, the truth of (**4**) does not require the truth of "Mary quarrelled with John"?

70 Give examples.

71 Explain why.

72 Give an example of a sentence in which "when" occurs as a genuine sentence connective. Is "when" truth functional, as it occurs in your example?

73 For each of the following expressions, say whether it always, sometimes, or never occurs as a sentence connective. If always or sometimes, say whether it expresses a truth function on these occurrences, and, if so, which. Use examples to justify your answers.

   (i) Iff;
  (ii) It is surprising that;
 (iii) It is true that;
 (iv) It is probable that;
  (v) Unfortunately;
 (vi) Fortunately;
(vii) Science shows that;
(viii) Although;
 (ix) Unless;
  (x) Hopefully;
 (xi) Even;
(xii) Yet;
(xiii) Not only ... but also ...;
(xiv) Hastily; (compare chapter **4.6**)
 (xv) In the park;
(xvi) Before; (enthusiasts may wish to compare the discussion in Davidson [1970a] pp. 138–9)
(xvii) Possibly,
(xviii) Despite the fact that.

74 Say what is wrong with the following reflections on some English argument, call it ϕ:

The deepest **P**-formalization of ϕ I can find is **P**-invalid.
Hence ϕ is not valid in virtue of its **P**-logical form.
Hence, in virtue of its **P**-logical form, ϕ is invalid.

75 Using the definition of a **P**-interpretation, establish the following:

   (a) If $[X_1, \ldots X_n \vDash_P Z]$ then $[X_1, \ldots X_n, Y \vDash_P Z]$ whatever Y may be. (Compare (**1.6.5**).)
   (b) If $[X_1, \ldots X_n \vDash_P Z]$ and $[Y_1, \ldots Y_k, Z \vDash_P W]$, then $[X_1, \ldots X_n, Y_1, \ldots Y_k \vDash_P W]$. (Compare (**1.6.6**).)
   (c) If Z is among the $X_1, \ldots X_n$ then $[X_1, \ldots X_n \vDash_P Z]$. (Compare (**1.6.7**).)
   (d) If $[(X_1, \ldots X_n) \vDash_P]$ then $[X_1, \ldots X_n \vDash_P Y]$, whatever Y may be. (Compare (**1.6.8**).)
   (e) If $[\vDash_P X]$ then $[Y_1, \ldots Y_n \vDash_P X]$, whatever $Y_1, \ldots Y_n$ may be. (Compare (**1.6.9**).)

Chapter 3

76   Supply a diagram to demonstrate the falsehood of (**6**).

77   If any of the following argument-claims is false, use a diagram to demonstrate the fact:

  (i) [It is not the case that if A then B] ⊢ A.
 (ii) [It is not the case that if A then B] ⊢ it is not the case that B.
(iii) [If A then not-B, B] ⊢ not-A.
 (iv) [If (A and B) then C, A] ⊢ [if B then C].
  (v) [If A then B] ⊢ [if (A and C) then B] whatever C may be.

78   (a) is (**16**) acceptable? Compare (**2.3.28**), which might be abbreviated to: "If common sense is correct, then common sense is incorrect. Therefore common sense is incorrect."

  (b) Stalnaker also gives an argument (pp. 124–5) against contraposition. The crucial assumption is:

   if [B ⊨ C] then [if A then B ⊨ if A then C].

Can you complete the argument? What would be the best response that a defender of the material implication interpretation could make?

79   Give an apparent example of this equivalence, and then suggest the best account for one who wants to deny it.

80   Why does (**6**) not constitute such a constraint?

81   Show how to use (**4**) to establish

   $Pr_i((B \text{ and } A)|C) = Pr_i(A|C) \times Pr_i(B|(A \text{ and } C))$.

(Hint: you have to use (**4**) more than once; and a strict setting out of the proof would also invoke (**2**).)

82   For example, they agree in finding "either (if A then B) or (if A then not-B)" valid. Stalnaker calls this "conditional excluded middle". Can you explain why this follows from his semantics?

83   Fill in a column representing your own views on the issues on the rows of the table, and say what you think the best justification is for each choice.

Chapter 4

84 What is the degree of "$=$"?

85 For not very interesting technical reasons, that definition cannot simply be transposed to **Q**, with "operator" replacing "connective". Enthusiasts are invited to work out an appropriate definition of scope for **Q**.

86 (a) The following arguments present problems for the standard **Q**-formalizations of English universal quantifications. Formalize the arguments, showing your correspondence scheme, state the problems, and briefly indicate how, if at all, you think they might be resolved:

(i)   It is not the case that all bodies acted on by no forces undergo random changes of velocity. Therefore there are no bodies acted on by no forces.
(ii) If anyone plays cricket, he does not also play squash. So anyone who plays both cricket and squash does not play cricket.
(iii) Every parent who loves all of his children is saintly. So every person who loves all of his children is saintly. (Cf. McCawley [1981] p. 163.)

(b) Formalize the following in **Q**, noting any problems or deficiencies. The only quantifier to be used is "∀".

(i)   If Pedro owns a donkey, he beats it.
(ii)   If Pedro owns a donkey, he is lucky.
(iii) Old soldiers never die.
(iv)   It is not invariably the case that love is requited. [α's love for β is requited iff, in addition, β loves α.]
(v)   Anyone who needs something should have it.
(vi)   Jones never leaves his desk.
(vii) If a is greater than 0, 0 is less than a.
(viii) None but the brave deserve the fair.
(ix)   Someone who ever lies is someone you should never trust.
(x)   If no one telephones her Jane will be miserable.
(xi)   No one runs faster than John.
(xii) No one runs faster than himself.
(xiii) Not everyone runs faster than himself.
(xiv) If all the people in the world were stretched end to end, they would circle the globe.

87 Consider the following pair:

(i) Some Buddhists are vegetarians.
(ii) Some vegetarians are Buddhists.

Is the apparent difference between them one which could affect the validity of some argument? Can the apparent difference be reflected in different **Q**-formalizations? If so how? If not, does this give a new reason for thinking that the meaning of "some" differs from that of "∃"? (Cf. McCawley [1981] p. 123.)

88 (a) Assuming "∀" is primitive in **Q**, show how you could introduce "∃" by a definition.

(b) Assuming "∃" is primitive in **Q**, show how you could introduce "∀" by a definition.

(c) Formalize the following in **Q**, noting any problems or deficiencies.

  (i) If Pedro owns a donkey he is lucky. [Do not use "∀".]
 (ii) There is a town between Oxford and London.
(iii) I met a man.
 (iv) A puppy is a young dog.
  (v) Some men are touchy and vain.
 (vi) Some people have everything.
(vii) Every cloud has a silver lining.
(viii) When beggars die there are no comets seen.
 (ix) There is a skeleton in every cupboard.
  (x) Nothing lasts for ever.
 (xi) Jones never feeds any of his dogs.
(xii) Who laughs last laughs longest.

89 Show why not.

90 Explain why not. Is there any other standard of formal validity which would be more appropriate?

91 Can you find expressions T_1 and T_2 such that there is no inconsistency in supposing that someone could be happy for a T_1 but not happy for a T_2?

92 (a) Formalize:

This is a gold ingot. Therefore it is made of gold.

(b) Give three examples each (not in the text) of predicative and non-predicative adjectives.

(c) Provide **Q**-formalizations of the following noting any inadequacies:

(i) Some greedy men are vain.
(ii) Some vain men are greedy.
(iii) John loves a beautiful actress.
(iv) John loves a former actress.

93 What considerations, if any, should incline one to classify "necessarily", "probably" and "rarely" as sentence adverbs?

94 (a) Formalize the following, stating your correspondence scheme, and noting any inadequacies:

(i) John ambled down the hill.
(ii) Quickening his pace, John strode up the hill.
(iii) John worked for the whole night.
(iv) John worked in Boston for four years.

(b) Davidson has argued that pairs of sentences like

(i) Shem kicked Shaum. It happened in the gymnasium.

constitute evidence in favour of the view that sentences like the first in the pair introduce events. For the "it" of the second sentence appears to be anaphorically linked to the first sentence in the way that "he" is linked to "a man" in the pair

(ii) I met a man. He was bald.

(The "he" here is called an "anaphoric" pronoun.) Formalize (i), showing your correspondence scheme.

95 Formalize (**6**) and (**8**), showing your correspondence scheme in each case.

96 Formalize the following, using " = " where appropriate:

(i) Mr Hyde is the same person as Dr Jekyll.
(ii) Everything is what it is and not another thing.

  (iii) John is wiser than anyone else.

  (iv) Apart from John no one brought a present.

  (v) Mary loves only John.

  (vi) If Mary loves anyone John is that person.

 (vii) Mary loves John but John loves someone else.

(viii) Mary loves John and Sally loves someone else.

  (ix) John loves himself and so does Mary.

  (x) Who laughs last laughs longest.

97 (a) Following the pattern of (**15**) and (**16**), provide inductive definitions of the "at least" and "at most" quantifiers.

    (b) Explain what, if anything, would be wrong with the following in place of (**16**):

$$\exists^n Fx =_{df} \exists^1 x(Fx \ \& \ \exists^{n-1}y \ (Fy \ \& \ x \neq y)).$$

    (c) Can "There are finitely many foxes" be adequately **Q**-formalized as:

$$\exists x(Gx \ \& \ \exists^x y(Fy))$$

with "G" corresponding to "is a natural number" and "F" to "is a fox"?

98 (a) Formalize the following sentences, showing your correspondence scheme. (You may if you wish use the "exactly" quantifier defined by (**15**) and (**16**), and the quantifiers you defined in your answer to Exercise 97a.)

  (i) Mary kissed two people at once.

 (ii) Three kings came to Bethlehem.

(iii) If two persons are present, there is a quorum. There are three of us present. So there is a quorum.

(iv) John is happy and so is Mary. So at least two people are happy.

 (v) Bananas and apples are fruits. You have eaten two bananas and three apples. So you have eaten five fruits.

(vi) $2 + 3 = 5$.

    (b) Is the following consistent?

    I don't have two racquets, I have three.

What does your answer show about the correct interpretation of numeral adjectives in English? See K. Bach [1987] p. 78.

(c) Formalize the following, showing your correspondence scheme:

Shem kicked Shaum twice.

(d) Formalize the following, noting any problems.

Shem kicked Shaum twice. It happened in the gymnasium.

(e) Does the displayed sentence in (d) above pose any problem for the alleged evidence, presented in Exercise 94b, in favour of the event analysis of action sentences? In your discussion, consider also the fact that the following seems ungrammatical, yet arguably should be acceptable on the event analysis:

Three times the red flag was unfurled. *They* happened in the main square.

99 One argument Russell gave in favour of his theory of descriptions goes roughly as follows:

"The King of France is bald" is obviously false, since France is a republic. But then it would seem that "The King of France is not bald" ought to be true. My theory can explain how this is possible.

The explanation involved discerning a scope ambiguity in "The King of France is not bald". Use distinct **Q**-formalizations of this sentence (stating the intended correspondences) to bring out the ambiguity. Cf. Russell [1905] p. 53.

100 (a) Use the "exactly 1" quantifier of (**9.15**) to formalize (**1**).

(b) Formalize the following, giving alternatives in any cases of ambiguity:

(i) The cat sat on the mat.
(ii) The cat did not sit on the mat.
(iii) The cat did not sit on the mat, but on the table.
(iv) The average family has 2.4 children.
(v) Someone was the unique author of *Principia Mathematica*.
(vi) Sally ate some of the cakes.

101 Assess the following argument, which Russell used against Meinong's theory:

Since the golden mountain does not exist, the existent golden mountain also does not exist. But all instances of "The *FG* is *F*" – for example, "The red book is red" – are true. Hence in particular "The existent golden mountain exists" is true. But this contradicts the fact that the existent golden mountain does not exist.

102 Explain why (**13**) is **Q**-valid.

103 Specify the formalization, stating the correspondences.

104 (a) Show how this specification of Russell's account of names makes it immune to the following objection.

Suppose the description associated with "Reagan" is "The President of the United States (in 1987)". Then the proposition expressed by "Reagan is the President of the United States" is the same as that expressed by "The President of the United States is the President of the United States". But the second of these is trivial and the first is not, so they cannot express the same proposition.

(b) Comment on the following argument, preferably in the light of reading Russell [1912] pp. 29–31:

The claim that names are descriptions is ambiguous between an ∀∃∀ claim (that for every name, there is a description such that whenever the name is used it is equivalent to the description) and an ∀∀∃ claim (that for every name, on every occasion of its use, there is a description to which it is equivalent). The former is wildly implausible; Russell's is the latter.

105 Formalize the argument to show that it is an instance of the quantifier shift fallacy. (Note: The answer to Exercise 104b depends upon spotting an example of the fallacy.)

106 (a) Why cannot the "he" in (**9**) be regarded as a stylistic variant of the reuse of "someone"?

(b) Why cannot the "himself" in the following be regarded as a stylistic variant of the reuse of "every American"?

Every American admires himself.

(Cf. McCawley [1981], p. 126.)

107   (a) Use distinct **Q**-formalizations to bring out any ambiguities in the following:

(i)   Winston is always smoking a big cigar.
(ii) He always carries a large stick.
(iii) Only non-smoking males are eligible for this job.
(iv) I watched the tennis-match in bed.
(v)   Only sensible dogs are taken by somebody kind for all their walks.
(vi) Everyone has a problem.

(b) Use mixed English and **Q**-formalization (in the manner of (**21**), (**22**) etc.) to bring out any ambiguities in the following:

(i)   "In the whole wide beautiful world, Aldo Cassidy was the only person who knew where he was." (Le Carré [1971] p. 8)
(ii) "Most of all I would like to thank my students, who have taught me more than they know." (E. Bach, [1974] p. vi)
(iii) Gerry means everything he says ironically.
(iv) If John is to enter a university, he must pass his examinations.
(v)   I thought you were someone else.
(vi) Mary knows a man taller than Michael. (See K. Bach [1987] p. 74.)

See Quine [1956] for some relevant examples. For doubts about the adequacy of the view that all these are examples of scope ambiguities see K. Bach [1987] pp. 206–14.

108   Establish the truth of (**10**) and (**11**), in the latter case by providing a counterexample.

109   (a) Argue for the truth of the following:

(i) $\forall x(Fx \rightarrow Gx) \vDash_{\mathbf{Q}} \exists x(Fx \rightarrow Gx)$.
(ii) $\vDash_{\mathbf{Q}} \exists x(Fx \vee \neg Fx)$.
(iii) $\vDash_{\mathbf{Q}} \forall x((Fx \rightarrow Gx) \vee (Gx \rightarrow Fx))$.

(b) Devise counterexamples to the following (a counterexample to an argument is an interpretation upon which the premises are true and the conclusion false; a counterexample to a sentence is an interpretation upon which it is false):

(i) $\forall x(Fx \rightarrow Gx)$; $\exists xFx$.

(ii) $\exists x \exists y \ x \neq y$.

(iii) $\forall x(Fx \rightarrow Gx)$, $\exists x \neg Fx$; $\exists x \neg Gx$).

110 (a) Explain how someone who agreed that (**16**) is not valid might argue that this fact is consistent with the substitutivity of identicals: (**8.5**).

(b) Could the argument for the invalidity of (**16**) be undermined by adopting Russell's strategy of formalizing names like "Phosphorus" and "Hesperus" by definite descriptions? Explain your reasoning.

111 Are there occurrences of "Only F" which are more plausibly interpreted as lacking this entailment? (Cf. McCawley [1981] pp. 180–2.)

112 Provide a counterexample to (**20**).

113 (a) Formalize the following arguments and determine the validity of the formalizations.

(i) Harry loves anyone who loves him. Mary loves Harry. So Harry loves Mary.

(ii) Only Harry loves himself. So only Harry loves Harry.

(iii) No one but Harry lives in the house and Harry is not mad. So the inhabitant of the house is not mad.

(b) This book is silent on problems of time and tense. Some appreciation of the problems will become apparent by considering the following:

(i) George is marrying Mary, Mary is an orphan; so John is marrying an orphan.

(ii) George married Mary, Mary is a widow; so John married a widow.

Can you formalize them in such a way that their **Q**-validity corresponds to the judgements you make about their validity? See Lacey [1971] for an excellent introduction to problems of formalization connected with tense.

114 Davidson's own account makes the pronoun refer to the sentence-token which follows (see his [1969] pp. 105–6). (For the distinction between *token* and *type*: if you are asked to write a sentence out twice, you are being asked to write two sentence-tokens of one sentence-type.)

(a) Evaluate the suggestion that the existence of beliefs that never have been and never will be expressed argues for regarding "that", in sentences like (2), as referring forward to a sentence type rather than a sentence token.

(b) Evaluate the suggestion that the truth of a sentence like

John believes that she is happy,

uttered in the circumstances which make plain the referent of "she", argues for regarding "that", in sentences like (2), as referring forward to a sentence token rather than a sentence type.

115 (a) Using Davidson's paratactic proposal, formalize (making explicit any assumptions you need to make):

John believes that Hesperus is a planet, "Hesperus is a planet" is true; therefore John believes something true.

(b) Referring back to Exercise 114, does the argument in Exercise 115a support the type or the token construal of the referent of the demonstrative "that", or is it neutral?

116 (a) Using Davidson's paratactic proposal formalize:
Gallileo said that the earth moves, and Newton said the same. So there is something that they both said.

(b) Can Davidson's proposal be modified so as to provide an adequate formalization of, for example,

Referring to Jones' murderer, John said that he was innocent?

See Platts [1979], ch. 5, §5, Hornsby [1977].

(c) Say whether Davidson's proposal gives an adequate account of the truth conditions of the following:

John believes something which no one has ever or will ever express in an utterance.

See Schiffer [1987] pp. 122–38.

117 One has to proceed by an enumeration of cases. The ones considered in the text show that the falsity of the antecedent of the supposed connective must not be sufficient for the truth of the compound, and that the truth of both antecedent and consequent must not be sufficient for the truth of the compound. Establish the result for the remaining cases.

118 (a) Would it make any difference to replace "most" in (**11**) by "more than half"?

(b) Give an account of "Most natural numbers are non-prime" upon which the sentence is false.

(c) Give an account of "Most natural numbers are non-prime" upon which the sentence is true. Would the account also give intuitively the right result when applied to finite cases? For the claim that "most" and "few" are not formalizable in "first order logic" (approximately: are not **Q**-formalizable) see van Benthem and Doets [1983] p. 277.

119 Another view of quantifiers in English is that they take a predicate or open sentence to form a "subject" expression: one which forms a (closed) sentence out of a predicate or open sentence. Is this view better, worse, or as good as the view that English quantifiers are binary, as in **QB**? Could you give suitable interpretation rules for quantifiers treated as formers of subject expressions?

120 (a) Define a binary quantifier (by giving the syntax and an appropriate rule of interpretation) suitable for formalizing "the". Illustrate by examples of formalizations. Do the truth conditions your formalizations attribute differ from those attributed by Russell's Theory of Descriptions?

(b) Define a binary quantifier (by giving syntax and an appropriate rule of interpretation) suitable for formalizing the quantifier "no". Formalize the following pair, using binary quantifiers in both cases:

(i) Every student will pass if he works.
(ii) No student will pass if he works.

Compare this treatment of (i) and (ii) with that awarded by unary quantifiers (i.e. their **Q**-formalizations). (For discussion, see Higginbotham [1986].)

121 Show how imposing this requirement would lead to unwanted truth-upon-an-interpretation conditions for sentences like "$Ex\ EyFxy$".

122   "Snow is white" is true iff snow is white.

This appears to hold not just for the sentence "snow is white", but quite generally. Attempt to formalize an appropriate generalization both in **Q** and in **QS**, and comment on the success of your efforts.

123   Is there a **QP**-valid **QP**-formalization of (**9**)?

124   Using the style of formalization exemplified by (**12**), provide a **Q**-formalization of (**6**) and a counterexample.

125   How does Russell try to make room for the intuitions regarding (**1**), (**2**), (**3**) and (**4**)?

126   (a) How would the guiding motivations be thwarted if the **Q**-rules for the quantifiers were adopted as the **QF**-rules?

   (b) Establish the **QF**-invalidity of (**1**) and (**4**) by giving detailed counterexamples.

127   Show why.

128   Provide details of the interpretation rules for **QF** atoms and quantifiers, including rules governing the use of square brackets.

129   Show that revising the quantifier rules by omitting the words "assigns something to n and" would not affect the validity of (**6**), and would be unsatisfactory for independent reasons.

130   Explain how this is so.

131   Using the definition of a **Q**-interpretation, establish the following:

   (a) If $[X_1, \dots X_n \vDash_Q Z]$ then $[X_1, \dots X_n, Y \vDash_Q Z]$ whatever Y may be. (Compare (**1.6.5**).)
   (b) If $[X_1, \dots X_n \vDash_Q Z]$ and $[Y_1, \dots Y_k, Z \vDash_Q W]$, then $[X_1, \dots X_n, Y_1, \dots Y_k \vDash_Q W]$. (Compare (**1.6.6**).)
   (c) If Z is among the $X_1, \dots X_n$, then $[X_1, \dots X_n \vDash_Q Z]$. (Compare (**1.6.7**).)
   (d) If $[(X_1, \dots X_n) \vDash_Q]$ then $[X_1, \dots X_n \vDash_Q Y]$, whatever Y may be. (Compare (**1.6.8**).)
   (e) If $[\vDash_Q X]$ then $[Y_1, \dots Y_n \vDash_Q X]$, whatever $Y_1, \dots Y_n$ may be. (Compare (**1.6.9**).)

Chapter 5

132 Assuming that the sentences to which "□" can be applied are as rich as the sentences of English, show that this involves treating "□" as a non-truth functional sentence connective.

133 Show how this rule is derived from the interpretation rule for "□" together with the definition of "◇".

134 (a) Draw a diagram illustrating a set of worlds and an interpretation which establish:

$$\Diamond p, \Diamond(p \to q) \nvDash_{\mathbf{PN}} \Diamond q.$$

(b) Say whether the following is true, and argue informally for your view:

$$\Box p, \Diamond(p \to q) \vDash_{\mathbf{PN}} \Diamond q.$$

135 Give an argument of this form in English in which the premises are plainly true and the conclusion plainly false.

136 (a) What further equivalences are required for the property of modal collapse to obtain? Argue informally for the **PN**-validity of these equivalences.

(b) Formalize the following argument in **PN**. If the formalization is **PN**-invalid draw a diagram to establish this. If it is **PN**-valid, comment on whether you think the English argument is sound.

It is possible that God should necessarily exist. Therefore, God necessarily exists.

137 (a) For each of the argument patterns in (**12**), (**15**), (**17**) and (**18**) draw a diagram of interpretations, sets of worlds, and accessibility relations, according to which the premises are true and the conclusion false.

(b) Give an example of an accessibility relation which would invalidate (**21**).

138 Formalize the following in **PN**. If the formalization is **PN**-valid, show that this is so by informal reasoning; if it is not, provide a diagrammatic counterexample.

(i) Necessarily, if there is a first moment in time, the history of the universe up to now is finite; therefore if there had to be a first moment in time, the history of the universe up to now has to be finite.

(ii) Whatever will be must be.

(iii) It is possible to run a mile in 4 minutes. Inevitably, if a mile is run in exactly four minutes, a mile will be run in three and a half minutes. So it is possible to run a mile in three and a half minutes.

139 Assess the following argument:

The antecedent of (**6**) *is* false, taken quite literally, though what is intended is true, and could be more properly expressed along the lines: "If I had added sugar to this cup of coffee and everything else had remained as much the same as possible, then it would have tasted good".

Cf. Urbach [1988] p. 197.

140 What is the truth value upon an intended interpretation of an adequate **PNS**-formalization of the following (applying (**11**))?

If Bizet and Verdi had been compatriots, they would have either both been Italian or else both French.

What impact does your answer have upon the correctness of Lewis's proposal?

141 Rewrite (**11**) in such a way that this is explicit.

142 Give an example of a subjunctive conditional and contexts, such that we are inclined to assign different truth values to the conditional in the different contexts.

143 What happens if the two premises of (**12**) are conjoined into a single premise?

For a defence of transitivity for $\square\!\!\rightarrow$ (the principle of which (**13**) is an instance), see Urbach [1988] p. 198.

144 (a) Say which of the following are in your opinion true:

If Julius Caesar had been in command of the US troops in Vietnam, he would have used nuclear weapons.

If Julius Caesar had been in command of the US troops in Vietnam, he would have used catapults.

What account of counterfactuals is best in accordance with your verdict?

(b) Discuss the claim that the account at the end of §2 of what similarities are relevant should be modified to take account of the following example:

Suppose that just as he was about to fire, Oswald had a heart attack, but that on a sudden impulse a moment later, someone else (a Dallas cop run berserk) shot Kennedy. There are two worlds just like ours up to the moment which in fact verifies "Oswald fired": one in which he didn't fire and Kennedy was not shot, and one in which he didn't fire and Kennedy was shot by the berserk cop. Yet, on the historical assumptions of this book, we still want to affirm "If Oswald had not fired, Kennedy would not have been shot" and deny, "If Oswald had not fired, Kennedy would still have been shot".

Compare Bennett [1974] esp. pp. 394–7; Lewis [1973b] pp. 75–7 and 91–5; and Lewis [1979] pp. 43ff.

145 Assess the following argument:

True, valid sentences should correspond to necessary truths. The problem raised by (11), however, is a problem with **Q**, not with strong necessity. For **Q** invites us to formalize the doubtful sentence "Homer exists" by the **Q**-valid $\exists x\, x = \alpha$. Modify **Q** so that it is a free logic, and (11) will be, as it should be, true.

146 Formalize the following in **QN**, commenting on the validity of your formalization of (ii) and (iii):

(i) Causation is a necessary relation.
(ii) A necessary condition for the possibility of experience is that there

are causally related events. So the existence of our experience
establishes the necessity of the causal relation.

(iii) It is possible for me to exist yet my body not to exist. Therefore I am
not my body.

(iv) What is known must be true.

 (v) A married man must be married *to* someone.

147   Give a pair of examples to contrast "negation de re" with "negation
de dicto".

148   Formalize the following:

Not every belief can be justified, so there are beliefs which cannot be
justified.

Is the argument as formalized **QN**-valid? What does your answer show
about the relation between de dicto and de re necessities? (Cf. Anscombe
[1959] pp. 138–9.)

149   Justify this remark.

150   Provide a **QW**-formalization of (**4**), showing your correspondence
scheme.

151   Provide a **QW**-formalization of (**5**), showing your correspondence
scheme. Assuming that **QW** uses the standard notion of validity
(paralleling **PN**- and **QN**-validity), is your formalization **QW**-valid?

152   Evaluate the following response on behalf of the extra-argument-
place account:

On my view, to say that rectangularity is a property this page has "in
itself" is simply to say that it is a *two-place* property of this page: a
property the possession of which relates this page to only one other
subject, a world. A relational property of this page is one which is
more than two-place.

153   Show why.

154   Explain why not, by showing how the truth-upon-an-intended-
interpretation conditions of (**17**) and (**18**) differ from the truth conditions
of (**15**). (See Davies [1981] pp. 220–1.)

155 Give the best formalization of the following in (a) **QN** and (b) **QW**:

My car could have been the same colour as yours actually is.

See Forbes [1985] p. 92.

156 Is there a response to the objection that if an English sentence formalizable as "$\Box F\alpha\beta$" is true, there is no world containing a counterpart of α but not one of β?

157 How else might one give a **QN**-representation of the thought that Caesar is essentially related to his death? Could someone who is only contingently mortal be essentially related to his death?

158 Can you find an example in English, not involving the constitution of artefacts, to establish the falsehood of

$$\Diamond A, A \Box\!\!\rightarrow \Diamond B \vDash \Diamond B?$$

159 Give an example, not involving modality, which shows that similarity is not a transitive relation.

Chapter 6

160 Does the same point apply to **QN**?

161 State suitable introduction and elimination rules for v, $\rightarrow$ and $\leftrightarrow$.

162 Discuss the following suggestion:

"$\vdash$" is a logical symbol and represents "therefore", so the latter is, in effect, counted as among the logical constants.

163 Consider a sentence operator: "It was the case that ...". Does this count as a logical constant by the topic-neutral criterion? Are there any other reasons for judging the issue either way?

164 (a) How does the rules criterion treat "but"?
(b) Could there be a logical constant that was primitive in the sense

of being typically learned directly, and not via a definition, and yet which had the meaning of "¢"?

165 Propose suitable introduction and elimination rules for " = ". (See Tennant [1978] p. 77.)

166 Give a counterexample.

167 Show why no expression can have a meaning such that (**1**) and (**2**) are both valid. Cf. Prior [1960].

168 Could the same effect be achieved by requiring that the translation procedure envisaged in (**1**) be statable in a finite number of rules? Could this requirement rule out the claim that an English quantification has a sentence-letter as its logical form?

Glossary

Chapter 1

1.1   *Proposition*: Something which can be asserted, believed, denied, supposed, etc. p. 7. See also G 1.18.

1.2   *Argument*: A collection of propositions, one of which is designated as the *conclusion*, the remainder being the *premises*. Subsequently amended so as to apply also to a collection of sentences, of which one is the conclusion, the remainder (if any) the premises. p. 7.

1.3   *Deductive Logic*: The study of what it is for conclusions to *follow from* premises. p. 8.

1.4   *A priori*: A proposition can be known *a priori* just on condition that it can be known without recourse to experience or experiment. p. 8.

1.5   *Monotonic*: Deductive validity is *monotonic* in that, if you start with a deductively valid argument, then, no matter what you *add* to the premises, you will end up with a deductively valid argument. p. 11. Cf. (**1.6.5**).

1.6   *Invalid*: Not valid. p. 10.

1.7   *Valid*: An argument is valid if and only if it is logically impossible for all the premises to be true yet the conclusion false. Compare also **P**-valid and **Q**-valid. p. 9, 15.

1.8   *Inconsistent*: A collection of propositions is *inconsistent* if and only if it is logically impossible for all of them to be true. p. 15.

1.9   *Contradictories*: Two propositions are contradictories if and only if it is logically impossible for both to be true and logically impossible for both to be false. p. 16.

1.10 *Negation*: The negation of a proposition is what results from inserting "not" (or some equivalent expression) into it in such a way as to form a contradictory of it. p. 16.

1.11 *Turnstile*: ⊢ (see entry in list of symbols). p. 19.

1.12 *Argument-claim*: An argument claim refers to an argument, and says of it either that it is valid (a positive argument-claim), or that it is invalid (a negative argument-claim). p. 19.

1.13 *Transivity*: To say that validity is transitive is to say that: If $[A_1, ... A_n \vDash C]$, and $[B_1, ... B_k, C \vDash D]$ then $[A_1, ... A_n, B_1, ... B_k \vDash D]$. p. 21.

1.14 *Reflexivity*: To say that validity is reflexive is to say that: If C is among the $A_1, ... A_n$, then $[A_1, ... A_n \vDash C]$. p. 22.

1.15 *Sound*: An argument is sound if and only if it is valid and has true premises. p. 23.

1.16 *Reductio ad absurdum*: Reasoning which uses the validity of an argument, together with the falsehood of the conclusion, to infer the falsehood of one of the premises. p. 24.

1.17 *Persuasive*: An argument is persuasive for someone only if that person is willing to accept each of the premises but, before encountering the argument, is unwilling to accept the conclusion. p. 25.

1.18 *Proposition*: What is expressed, in a given context, by a meaningful, declarative, indicative sentence. p. 25.

1.19 *Declarative sentence*: One that could be used to make an assertion, to affirm that something is or is not the case. p. 26.

1.20 *Indicative sentence*: Contrasted with *subjunctive sentence*, and introduced by examples. p. 26.

1.21 *Truth conditions*: Circumstances under which a sentence is or would be true. p. 29.

1.22 *Formal validity*: An argument is formally valid if and only if it is valid in virtue of its form or pattern. p. 29.

1.23 *Argument-form*: Pattern from which an argument results by replacing letters or variables by expressions from a natural language. p. 30.

1.24 *Predicates*: Expressions which take one or more names to make a sentence. They include verbs, like "runs", adjective phrases like "is red" and noun phrases like "is a man". p. 31.

1.25 *Logical constant*: See list at (**11.1**) p. 33.

1.26 *Structural features*: Facts about the recurrence of non-logical elements in a sentence. p. 34.

1.27 *Language of propositional calculus, of predicate calculus*: These are the languages **P** and **Q** introduced in chapters **2** and **4** respectively. p. 34.

1.28 *Lexical ambiguity*: Ambiguity in a sentence due to some "lexical item", i.e., roughly, word, which it contains. p. 35.

1.29 *Structural ambiguity*: Non-lexical ambiguity in a sentence. p. 36.

1.30 *Syntax* or *grammar*: A set of rules which determines how sentences are constructed from the language's vocabulary. p. 38.

1.31 *Naive syntactic test*: Two expressions belong in the same syntactic category by the naive syntactic test if and only if you can replace one by the other, wherever it occurs, without turning sense into nonsense. p. 39.

1.32 *Syntactic irregularity*: Failure of match between logical properties and syntactic properties as determined by the naive syntactic test. p. 41.

Chapter 2

2.1 *Sentence connective*: Expression which forms a sentence out of one or more sentences. p. 44.

2.2 *Tilde*: Name for "¬". p. 44.

2.3 *Ampersand*: Name for "&". p. 44.

2.4 *Vel*: Name for "∨". p. 44.

2.5 *Semantics*: Rules which (in some sense) specify meanings; e.g. the interpretation rules for **P**. p. 45.

2.6 *Truth value*: A sentence possesses the truth value *true* just on condition that it is true; the truth value *false* just on condition that it is false. p. 45.

2.7 *Iff*: If and only if. p. 46.

2.8 *Truth tables*: Tabular representation of the way in which truth values of complex sentences depend upon the truth values of their compounds. p. 46.

2.9 **P**-*validity*: An argument in **P**, $X_1, \ldots X_n$; Y, is **P**-*valid* iff every interpretation upon which all the premises are true is one upon which the conclusion is true. Abbreviation: $X_1, \ldots X_n \vDash_P Y$. p. 48.

2.10 *Component*: Sentence which a sentence connective takes to make a complex sentence. p. 48.

2.11 *Resultant*: Sentence formed by a sentence connective from one or more components. p. 48.

2.12 *Scope (of an occurrence of a sentence connective)*: The shortest sentence in which it occurs. p. 49.

2.13 *Dominates*: An occurrence of a sentence connective *dominates* the sentence which is its scope. p. 49.

2.14 *Truth functional*: A sentence connective is *truth functional* iff whether or not any resultant sentence it forms is true or false is determined completely by whether its components are true or false. p. 49.

2.15 *Substitution test*: Replacing a component by one with the same truth value invariably leaves the truth value of the resultant unchanged. The test is for the truth functionality of the connective dominating the component to which the test is applied. p. 50.

2.16 *Correspondence scheme (for a formalization)*: Determines which English sentences are replaced by which **P**-letters. p. 51.

2.17 *Recovered argument*: The result of replacing the **P**-letters in a formalization by the corresponding English sentences (as determined by the associated correspondence scheme), and then replacing the **P**-connectives by the corresponding English connectives. p. 53.

2.18 *Adequate*: A formalization is *adequate* iff the recovered argument (sentence) says the same as the original English. p. 53.

2.19 *Intended interpretation*: One which assigns to all the **P**-letters of a formalization the same truth value as the corresponding English sentences actually possess. p. 53.

2.20 *Deep*: One adequate formalization is *deeper than* another iff it reveals more of the structure of what it formalizes. p. 55.

2.21 *Entails*: A entails B iff it is logically impossible for A to be true without B being true. p. 56.

2.22 *Stronger than*: A is *stronger than* B iff A entails B but B does not entail A. p. 56.

2.23 *Weaker than*: A is *weaker than* B iff B entails A but A does not entail B. p. 56.

2.24 *Valid in virtue of **P**-logical form*: Has an adequate and **P**-valid **P**-formalization. p. 59.

2.25 *Wide scope*: An occurrence of a sentence connective has *wide scope* relative to another occurrence of a connective iff the the scope of the latter is a proper part of the scope of the former. p. 60.

2.26 *Equivalent*: There are many different standards of equivalence. The standard at this point of the discussion is: X and Y are equivalent iff every **P**-interpretation assigns both the same truth value. p. 65.

2.27 *Inclusive disjunction*: The truth function expressed by "∨". p. 67.

2.28 *Exclusive disjunction*: The truth function which associates a pair of different truth values with true and a pair of the same truth values with false. p. 67.

2.29 *Conditionals*: Sentences of the form "if ... then ...". p. 68.

2.30 *Antecedent*: First component of a conditional. p. 68.

2.31 *Consequent*: Second component of a conditional. p. 68.

2.32 *Indicative conditionals*: Exemplified by (**4.34**): "If Oswald didn't shoot Kennedy, someone else did." p. 69.

2.33 *Subjunctive conditionals*: Exemplified by (**4.35**) "If Oswald hadn't shot Kennedy, someone else would have." p. 69.

2.34 *Transitivity*: Of "→": $[X \rightarrow Y, Y \rightarrow Z] \vDash_P [X \rightarrow Z]$. Of "if": [if A then B, if B then C] $\vDash$ [if A then C]. p. 75.

2.35 *Implicature*: What is conveyed but not said. See §6.

2.36 *Cancellable*: If A implicates but does not entail B, then "A but not B" is consistent. Grice called this the cancellability of the implicature. p. 79.

2.37 *Assertible*: An utterance's degree of assertibility varies inversely with the extent to which it is open to criticism, on any grounds at all. Being misleading or uninformative will lower an utterance's degree of assertibility. p. 81.

2.38 *Robust with respect to the antecedent*: Evidence for the antecedent will not undermine evidence for the conditional. p. 85.

2.39 *Modus ponens*: The principle licensing inference from a conditional, together with its antecedent, to its consequent. It can be expressed by the assertion: $[A$, if A then $B] \vDash [B]$. p. 85.

2.40 *Conditional Proof*: If $A_1, \ldots A_n, B \vDash C$ then $A_1, \ldots A_n \vDash$ if B then C. p. 88.

2.41 *Binary sentence connective*: One which takes two sentences to make a sentence. p. 92.

Chapter 3

3.1 *Conditional probability*: The probability of A, given B, written: $\Pr(A|B)$. p. 104.

3.2 $\Pr(A)$: The probability of A. p. 104.

3.3 $\Pr(B|A)$: $\Pr(B \& A) \div \Pr(A)$, if $\Pr(A) > 0$. p. 104.

3.4 *Probabilistically valid*: $[A_1, \ldots , A_n \vdash C]$ iff it is logically impossible for all of $A_1, \ldots , A_n$ to have high probability and C to have low probability. p. 105.

3.5 *Contraposition*: For "if", this obtains only if: [if A then not-B] $\vDash$ [if B then not-A]. p. 108.

Chapter 4

4.1 *Name-letters*: α, β, γ, α', ... etc. These will be used to correspond to ordinary English names like "John". p. 134.

4.2 *Predicate-letters*: F, G, H, F', ... etc. These will be used to correspond to English verbs, like "runs", some adjectives, like "hungry", and some nouns, like "man". p. 134.

4.3 *Variables*: x, y, z, x', ... etc. p. 134.

4.4 *Operators*: The sentence-connectives of **P**, together with "$\forall$" (the universal quantifier, corresponding to "all") and "$\exists$" (the existential quantifier, corresponding to "some" or "a"). p. 134.

4.5 *Predicate*: An expression which takes one or more names to form a sentence. p. 134.

4.6 *Degree*: The degree of a predicate is the number of names it takes to form a sentence. p. 134.

4.7 *Atom of* **Q**: A predicate or predicate-letter of degree n combined with n names. p. 135.

4.8 *Existential quantifier*: In English, "there is a" or "some" or equivalent expressions. In **Q**, "$\exists$". p. 135.

4.9 *Universal quantifier*: In English "all" or "every" or equivalent expressions. In **Q**, "$\forall$". p. 135.

4.10 *Existential (universal) quantification*: Sentence dominated by an existential (universal) quantifier. p. 136.

4.11 *Dominant operator*: Expression taking the whole of the sentence in question as its scope. p. 136.

4.12 *Sequence*: A series of objects in a fixed order. Thus one could write the series consisting just of Reagan and the number seven as $\langle$Reagan, 7$\rangle$. The importance of the order is brought out by the fact that $\langle$Reagan, 7$\rangle \neq \langle$7, Reagan$\rangle$. p. 138.

4.13 *Ordered pair*: Sequence of two objects. p. 138.

4.14 *N-tuple*: A sequence of n objects. Thus a 3-tuple (or triple) is a sequence of three objects. p. 138.

4.15 X^n_v: The result of replacing every occurrence of the variable v in X by the first name-letter n not occurring in X. p. 138.

4.16 **Q**-*validity*: An argument is **Q**-*valid* iff every interpretation, with respect to any domain, upon which all the premises are true is one upon which the conclusion is true. p. 140.

4.17 *Constructional history*: Metaphorically, the steps taken in building up a sentence from the symbols of the language. Literally, the sequence of sentences that have to be cited in using the definition of *sentence* to establish that a given expression is a sentence. p. 140.

4.18 *Quantifier shift fallacy*: Argument from premise of the form ∀∃ to conclusion of the form ∃∀. p. 177.

4.19 *Predicative adjective*: The following condition is a rough test for the predicativity of any adjective, *A*: where *n* is any name, *C* any noun *A* qualifies: *n* is a(n) *AC* ⊧ *n* is *A* and *n* is a(n) *C*. p. 149.

4.20 *Substitutivity of identicals*: An interpretation upon which "α = β" is true is one upon which: "... α ..." is true iff "... β ..." is true. p. 158.

4.21 *Law of identity*: The validity of every instance of "α = α". p. 158.

4.22 *Definite description*: Expression of the form "The so-and-so". p. 156, 162.

4.23 *Counterexample* (to a **Q**-argument): An interpretation upon which the premise(s) are true and the conclusion false. p. 182.

4.24 *Decision procedure for* **Q**: A mechanical method for determining, with respect to an arbitrary **Q**-sentence, and in a finite number of steps, whether or not it is valid. (NB: there is no decision procedure for **Q**.) p. 184.

4.25 *Verbs of propositional attitude*: Verbs expressing the *attitude* (e.g. belief, saying, wondering) adopted by a subject to a proposition. p. 191.

4.26 *Intersection*: The intersection of two sets, Γ and Γ′, is the set of things belonging to both Γ and Γ′.

4.27 *Open sentence*: Expression resulting from a **Q**-sentence by replacing some name by a variable not already contained in the sentence. p. 194.

4.28 *Unary quantifier*: One which takes one open sentence to form a sentence. p. 195.

4.29 *Binary quantifier*: One which takes two open sentences to form a sentence. p. 195.

4.30 *Objectual quantifier*: A quantifier (for example, the **Q**-quantifiers) interpreted in terms of how things are with some or all objects. p. 197.

4.31 *Substitutional quantifier*: A quantifier interpreted in terms of the truth or otherwise of sentences resulting by replacing the quantified variables by constants. p. 197.

4.32 *Opaque context with respect to names*: One in which there is no guarantee that two co-referring names can be substituted for one another without affecting the truth value of the whole. p. 199.

4.33 *Name quantifier*: Quantifier whose variables occupy a position in a sentence appropriate to a name (name-position). p. 201.

4.34 *Predicate quantifier*: Quantifier whose variables occupy a position in a sentence appropriate to a predicate (predicate position). p. 201.

4.35 *Proper subset*: Γ is a proper subset of Γ' ($\Gamma \supset \Gamma'$) iff every member of Γ is a member of Γ' and some member of Γ' is not a member of Γ.

4.36 *One–one mapping*: A one–one mapping between the sets Γ and Γ' is a relation, R, such that each member of Γ' bears R to exactly one member of Γ' and each member of Γ' bears R to exactly one member of Γ. Since we can think of a relation as a set of n-tuples, we can think of a one–one mapping between Γ and Γ' as a set Σ of ordered pairs meeting the following conditions: for any x in Γ, there is exactly one y in Γ' such that $\langle x, y \rangle$ is in Σ; and for any any x in Γ', there is exactly one y in Γ such that $\langle y, x \rangle$ is in Σ.

4.37 *Singleton*: A one-membered sequence.

4.38 *Extensional*: A sentence is extensional iff its truth or falsity turns on nothing except what objects the names it contains refer to, of what things the predicates it contains are true, and the truth values of any unstructured sentences it contains. See also below (G5.15) for a more precise definition.

Chapter 5

5.1 *Modal idioms*: Examples are: "possibly", "necessarily", "must", "may", "might", "has to", "contingent". p. 220.

5.2 *Contingent*: A proposition is contingent (or contingently true) iff it is true but is not necessarily true.

5.3 *Box*: Name of "$\Box$" (corresponding to "necessarily"). p. 221.

5.4 *Diamond*: Name of "$\Diamond$" (corresponding to "possibly"). p. 222.

5.5 *Box arrow*: Name of "$\Box\!\rightarrow$", where $p \Box\!\rightarrow q$ corresponds to "if it had been the case that p it would have been the case that q". p. 230.

5.6 **PN**-*validity*: $X_1, \ldots X_n \vDash_{\mathbf{PN}} Y$ iff for all **PN**-interpretations i and all sets of worlds, W, if, for any world w in W, all of $X_1 \ldots X_n$ are true at w upon i, Y is true at w upon i. p. 222.

5.7 *Modal collapse*: Given a non-modal sentence, X, there are only two non-equivalent fresh sentences, $\Box X$ and $\Diamond X$, that can be formed from it just by adding modal operators. p. 228.

5.8 *Accessibility*: Relation between worlds used to restrict which worlds are relevant to truth upon an interpretation. p. 228.

5.9 *Weak necessity*: $\Box X$ is true on an interpretation i iff X is true on i at every world at which all the objects i assigns to any name-letters in X exist. p. 238.

5.10 *Strong necessity*: $\Box X$ is true on an interpretation i iff X is true on i at every world. p. 238.

5.11 **QN**-*validity*: $X_1, \ldots X_n \vDash_{PN} Y$ iff for all **QN**-interpretations i and all sets of worlds, W, if, for any world w in W, all of $X_1, \ldots X_n$ are true at w upon i, Y is true at w upon i. p. 237.

5.12 *De re necessity*: **QN**-formalizable by a sentence in which either there is a name within the scope of some modal operator or a modal operator within the scope of a quantifier. p. 239.

5.13 *De dicto necessity*: Necessity which is not de re. p. 240.

5.14 *Extension*: Of a sentence, its truth value; of a name, its bearer; of a predicate of degree n, the set of n-tuples of which it is true. p. 244.

5.15 *Extensional*: A sentence is *extensional with respect to a position for an expression (sentence, name or predicate)* iff replacing an expression in that position with any other expression having the same extension leaves the truth value of the whole sentence unchanged. A sentence is *extensional* iff it is extensional with respect to all its positions for sentences, names and predicates. p. 244.

5.16 *Intension*: The set of all ordered pairs whose first member is a possible world and whose second member is the extension of the expression with respect to that world. p. 246.

5.17 *Modal realism*: Modal facts are real and mind-independent. p. 280.

5.18 *Quantifierism*: Ordinary modal idioms are best represented as quantifications over worlds. p. 280.

5.19 *Actualism*: Everything is actual. p. 280.

5.20 *Analytic*: True in virtue of meanings. p. 282.

5.21 *Ersatzism*: Quantifierism plus actualism. p. 282.

Chapter 6

6.1 *Syntactic category*: e_1 and e_2 belong to the same syntactic category iff, for every sentence s containing e_1, the result of replacing e_1 by e_2 is a (grammatical) sentence. p. 295.

6.2 *Semantic category*: e_1 and e_2 belong to the same semantic category iff either they are assigned the same kind of entity by the interpretation rules, or else they are treated by the same kind of interpretation rule. p. 296.

6.3 *Perspicuous*: Language whose syntactic and semantic categories coincide. p. 295–6.

6.4 *Rules of proof*: Rules stated in terms of the physical make-up of sentences, specifying derivations of sentences from others, in such a way as to ensure that a sentence derived from others is the conclusion of a valid argument if the others are its premises. p. 296.

6.5 *Compositional semantic theory*: Theory assigning properties to words and modes of composition, on the basis of which the meanings or truth conditions of all the sentences of the language can be derived. p. 297.

6.6 *Tractarian vision*: View, found in Wittgenstein's *Tractatus*, that all validity is formal validity. p. 298.

6.7 *Correspondence requirement*: If the correspondence scheme associated with a formalization has it that, say, "*F*" corresponds to " … ", then that actual expression, " … ", must occur in the sentence of natural language which is formalized. p. 302.

6.8 *Introduction rule for e*: Specifies from what premises one can derive a sentence dominated by *e*. p. 305.

6.9 *Elimination rule for e*: Specifies what conclusions one can derive from premises containing a sentence dominated by *e*. p. 305.

6.10 *Complete*: π (a system of proofs) is *complete* with respect to σ (a semantics) iff: if $\Gamma \vDash_\sigma X$ then $\Gamma \vdash_\pi X$. p. 307.

6.11 *Sound*: π is *sound* with respect to σ iff: if $\Gamma \vdash_\pi X$ then $\Gamma \vDash_\sigma X$. p. 307.

6.12 *Topic-neutral criterion*: Logical constants are expressions which introduce no special subject matter, or understanding which does not require learning about any actual or possible objects. p. 313.

6.13 *Rules criterion*: Logical constants are expressions whose meaning is exhaustively determined by introduction and elimination rules. p. 314.

6.14 *Cement criterion*: Logical constants are expressions which serve to combine the primitive expressions of the language into more complex ones. p. 320.

6.15 *C*: A sentence is a logical truth iff true in virtue of the meanings of the logical constants it contains. p. 324.

6.16 *Proxy semantics*: The semantics for a language given by translating it into another, and giving semantics for the latter. p. 326.

6.17 *Proxy language*: Language given semantics directly, and thus giving proxy semantics for any language translatable into it. p. 326.

6.18 *Finite axiomatization constraint*: Applied to proxy semantics, the constraint is that there be a finite manual for translating English into the proxy language, and a finitely axiomatized semantic theory for the latter. p. 328.

6.19 *The psychological constraint* is met by a logical form proposal only if the logical form reflects the way in which the corresponding sentence of natural language is processed in an act of understanding. p. 330.

List of symbols

Chapter 1

A, B, C — etc. are used as variables ranging over English sentences or propositions.

$A_1, \dots A_n; C\,D$ — gives the general form of an argument (in English) with n premises, $A_1, \dots A_n$, and conclusion C.

$\vDash$ — "$A_1, \dots A_n \vDash C$" abbreviates: "$A_1, \dots A_n; C$" is valid. As a special case, "$(A_1, \dots A_n) \vDash$" abbreviates: "$(A_1, \dots A_n)$" is inconsistent. As a further special case, "$\vDash A$" abbreviates: "it is logically impossible for A to be false".

$\nvDash$ — "$A_1 \dots A_n \nvDash C$" abbreviates: "$A_1, \dots A_n; C$" is not valid.

F, G, H — etc. are used in chapter **1** to mark places which can be occupied by predicates, and in chapter **4** as predicate-letters (more or less the same role).

α, β, γ — etc. are used in chapter **1** to mark places which can be occupied by names, and in chapter **4** as name-letters (more or less the same role).

Chapter 2

P — the propositional language.

$p, q, r, \dots$
$\neg, \&, \vee, \rightarrow, \leftrightarrow$ Symbols of **P**: see §2.

X, Y, Z — etc. are used as variables ranging over **P**-sentences (later, over sentences in whatever artificial language is under discussion).

$\vDash_{\mathbf{P}}$ — abbreviates "**P**-valid" (see G2.9).

Chapter 3

|	Pr$(A\|B)$ abbreviates: the probability of A, given B.
⊢	$A_1, \ldots, A_n$⊢C iff it is logically impossible for the premises all to have high probability and the conclusion to have low probability.

Chapter 4

Q	the quantificational language (sometimes called the language of first order (predicate) logic, or the language of quantification theory).
∃, ∀	quantifiers of **Q**.
x, y, z, x'	etc. variables of **Q**.
⟨ … ⟩	denotes a sequence. The names of its members fill the dots, distinct names separated by commas.
⊨$_Q$	abbreviates "**Q**-valid" (see G4.16).
$\alpha = \beta$, $\alpha \neq \beta$	abbreviate, respectively "$= \alpha\beta$" and "$\neg = \alpha\beta$".
T, W	unary quantifiers supposed to correspond to "most" and "few"
Q+	**Q** augmented by "T" and "W".
μ, ϕ	binary quantifiers corresponding to "most" and "few".
QB	**Q** augmented by "μ" and "ϕ".
λ	binary quantifer corresponding to "every".
A, E	substitutional universal and existential quantifiers.
QS	**Q** augmented by names and "A" and "E".
△, ▽	unary universal and existential predicate quantifiers.
$f, g \ldots$	variables for "△" and "▽".
⊇	set theoretic extension: $\Gamma \supseteq \Gamma'$ iff every member of Γ' is a member of Γ.
QP	**Q** augmented by "△", "▽" and "f", "g", ….
QF	a language with the syntax of **Q** but with interpretation rules not requiring every interpretation to assign every name-letter an object.

Chapter 5

□	box, corresponding to "necessarily".
◇	diamond, corresponding to "possibly".
QN	**Q** augmented by "□" and "◇".
PN	**P** augmented by "□" and "◇".
□→	box arrow: $p\square\to q$ corresponds to "if it had been the case that p it would have been the case that q".
PNS	**PN** augmented by "□→".
$\delta\sigma$	expression denoting 1 or 0, depending as σ is true or false.
W	predicate constant of **QW** assigned, on every interpretation, the set of all 1-membered sequences each of which has a world as member.
QW	**Q** augmented by "W", and, later, by "at", and, later, by "w∗".
at v	such operators, for some variable v, form open sentences from closed sentences. "At v, p" is to be true upon an interpretation assigning a world, w, to the name-letter which replaces "v" iff p is true upon the interpretation with respect to w.
w∗	name in **QW** of the actual world.
Ⓐ	corresponds to "actually".
QC	**QW** augmented by "C".
C	$Cxyz$ corresponds, in **QC**, to "x is a counterpart in z of y".

Chapter 6

∪	set theoretic union: something belongs to $\Gamma\cup\Gamma'$ iff either it belongs to Γ or it belongs to Γ'.
⊇	set theoretic extension: $\Gamma\supseteq\Gamma'$ iff every member of Γ' is a member of Γ.
$\vdash_P$	the derivability relation in **P**.

Bibliography

Adams, E. W. [1970] "Subjunctive and indicative conditionals" *Foundations of Language* 6, pp. 89–94.

—— [1975] *The Logic of Conditionals* Reidel, Dordrecht.

Anderson, A. R., and Belnap, E. [1975] *Entailment* Princeton University Press.

Anscombe, G. Elizabeth M. [1959] *An Introduction to Wittgenstein's Tractatus* Hutchinson, London.

Appiah, Anthony [1985] *Assertion and Conditionals* Cambridge University Press, Cambridge.

Armstrong, David [1978] *Nominalism and Realism: Universals and Scientific Realism*, vol. 1, Cambridge University Press, Cambridge.

Ayer, Alfred J. [1936] *Language, Truth and Logic* Gollancz, London; 2nd edn, 1946.

Bach, E. [1974] *Syntactic Theory* Holt, Rinehart and Winston, London.

Bach, Kent [1987] *Thought and Reference* Clarendon Press, Oxford.

Barcan Marcus, Ruth [1947] "The identity of individuals in a strict functional calculus of second order" *Journal of Symbolic Logic* 12, pp. 12–15.

Barwise, Jon, and Cooper, R. [1981] "Generalized quantifiers and natural language" *Linguistics and Philosophy* 4, pp. 159–219.

Belnap, N. [1961] "Tonk, plonk and plink" *Analysis* 22, pp. 130–4.

Bencivenga, Ermanno [1985] "Free logics" in Gabbay and Guenthner [1986] pp. 373–426.

Bennett, Jonathan [1974] "Counterfactuals and possible worlds" *Canadian Journal of Philosophy* 4, pp. 381–402.

Blackburn, Simon [1986a] "How can we tell whether a commitment has a truth condition?" in Charles Travis, ed., *Meaning and Interpretation*, Basil Blackwell, Oxford, pp. 201–32.

—— [1986b] "Morals and Modals" in Graham Macdonald and Crispin Wright, eds, *Fact, Science and Morality*, Basil Blackwell, Oxford.

Boolos, George [1975] "On second-order logic" *Journal of Philosophy* 72, pp. 509–27.

—— and Jeffrey, Richard C. [1974] *Computability and Logic* (3rd edn, 1989), Cambridge University Press, Cambridge.

Bull, Robert A., and Segerberg, Krister [1983] "Basic modal logic" in Gabbay and Guenthner [1983] pp. 1–88.

Burge, Tyler [1986] "On Davidson's 'Saying that'" in LePore [1986] pp. 190–208.

Carroll, Lewis [1872] *Through the Looking Glass and What Alice Found There* Macmillan, London.

Chellas, B. F. [1980] *Modal Logic* Cambridge University Press, Cambridge.

Clark, Romane [1970] "Concerning the logic of predicate modifiers" *Noûs* 4, pp. 311–35.

Cohen, L. Jonathan [1962] *The Diversity of Meaning* Methuen, London.

Craig, Edward [1985] "Arithmetic and fact" in I. Hacking, ed., *Exercises in Analysis* Cambridge University Press, Cambridge.

Davidson, Donald [1965] "Theories of meaning and learnable languages" in Yehoshua Bar-Hillel, ed., *Proceedings of the 1964 International Congress for Logic, Methodology and Philosophy of Science* North-Holland, Amsterdam; reprinted in Davidson [1984] pp. 3–15.

—— [1967a] "Truth and meaning" *Synthese* 7, pp. 304–23; reprinted in Davidson [1984] pp. 17–36.

—— [1967b] "The logical form of action sentences" in N. Rescher, ed., *The Logic of Decision and Action*, University of Pittsburgh Press, Pa., pp. 115–20; reprinted in Davidson [1980] pp. 105–21.

—— [1967c] "Causal relations" *Journal of Philosophy* 64, pp. 691–703; reprinted in E. Sosa, ed., *Causation and Conditionals*, Oxford University Press, 1975, pp. 82–94.

—— [1969] "On saying that" *Synthese* 19 (1968/9), pp. 130–46; reprinted in D. Davidson and J. Hintikka, eds, *Words and Objections* Reidel, Dordrecht, 1969, pp. 158–74; and in Davidson [1984] pp. 93–108.

—— [1970a] "Action and reaction" *Inquiry* 13, pp. 140–8; reprinted as "Reply to Cargile" in Davidson [1984] pp. 137–46.

—— [1970b] "Semantics for natural languages" in *Linguaggi nella società e nella tecnica* Milan; reprinted in Davidson [1984] pp. 55–64.

—— [1973] "In defense of convention T" in H. Leblanc, ed., *Truth, Syntax and Modality* North-Holland, Amsterdam; reprinted in Davidson [1984] pp. 65–75.

—— [1977] "The method of truth in metaphysics" in P. French, T. Uehling and H. Wettstein, eds, *Midwest Studies in Philosophy, II: Studies in the Philosophy of Language*, University of Minnesota Press, Minneapolis, pp. 244–54; reprinted in Davidson [1984] pp. 199–214.

—— [1980] *Essays on Actions and Events* Clarendon Press, Oxford.

—— [1984] *Inquiries into Truth and Interpretation* Clarendon Press, Oxford.

Davies, Martin K. [1981] *Meaning, Quantification, Necessity: Themes in Philosophical Logic* Routledge & Kegan Paul, London.

—— [1983] "Meaning and structure" *Philosophia* 13, pp. 13–33.

Delong, Howard [1970] *A Profile of Mathematical Logic* Addison-Wesley Publishing Co., Reading, Mass.

Donnellan, Keith [1966] "Reference and definite descriptions" *Philosophical Review* 77, pp. 203–15.

Dudman, V. H. [1984a] "Conditional interpretations of 'if'-sentences" *Australian Journal of Linguistics* 4, pp. 143–204.

—— [1984b] "Parsing 'if'-sentences" *Analysis* 44, pp. 145–53.

—— [1987] "Appiah on 'if'" *Analysis* 47, pp. 74–9.

Dummett, Michael [1973] *Frege: Philosophy of Language* Duckworth, London.

—— [1981] *The Interpretation of Frege's Philosophy* Duckworth, London.

Edgington, Dorothy [1988] Review of Graeme Forbes, *Metaphysics of Modality*, *Philosophical Quarterly* 38, pp. 365–70.

Evans, Gareth [1975] "Identity and predication" *Journal of Philosophy* 72, pp. 343–63; reprinted in Evans [1985] pp. 25–48.

—— [1976] "Semantic structure and logical form" in Evans and McDowell, eds [1976] pp. 199–222; reprinted in Evans [1985] pp. 49–75.

—— [1977a] "Pronouns, quantifiers and relative clauses (I)" *Canadian Journal of Philosophy* 7, pp. 467–536; reprinted in Evans [1985] pp. 76–152.

—— [1977b] "Pronouns, quantifiers and relative clauses (II)" *Canadian Journal of Philosophy* 7, pp. 777–97; reprinted in Evans [1985] pp. 153–75.

—— [1979] "Reference and contingency" *The Monist* 62, pp. 161–89; reprinted in Evans [1985] pp. 178–213.

—— [1982] *The Varieties of Reference* Clarendon Press, Oxford.

—— [1985] *Collected Papers* Clarendon Press, Oxford.

—— and McDowell, John, eds [1976] *Truth and Meaning*, Oxford University Press, Oxford.

Fine, Kit [1985] "Plantinga on the reduction of possibilist discourse" in James Tomberlin and Peter van Inwagen, eds, *Alvin Plantinga: A Profile* Reidel, Dordrecht, pp. 145–86.

Forbes, Graeme [1985] *The Metaphysics of Modality* Clarendon Press, Oxford.

Foster, John [1976] "Meaning and truth theory" in Evans and McDowell [1976] pp. 1–32.

Frege, Gottlob [1879] *Begriffsschrift, eine der arithmetischen nachgebildete Formelsprache des reinen Denkens* Nebert, Halle; English translation in T. W. Bynum, ed. and trans. *G. Frege, Conceptual Notation and Related Articles*, Clarendon Press, Oxford, 1972, pp. 101–23.

—— [1884] *Die Grundlagen der Arithmetik. Eine logisch-mathematische Untersuching über den Begriff der Zahl* Koebner, Breslau; reprinted with English translation by John L. Austin as *The Foundations of Arithmetic* Basil Blackwell, Oxford, 1950.

—— [1892a] "Über Begriff und Gegenstand" *Vierteljahrsschrift für wissenschaftliche Philosophie* 16, pp. 192–205; translated as "On concept and object" in P. Geach and M. Black, eds, *Translations from the Philosophical Writings of Gottlob Frege*, Basil Blackwell, Oxford, 1952, pp. 42–55.

—— [1892b] "Über Sinn und Bedeutung" *Zeitschrift für Philosophie und philosophische Kritik* 100, pp. 25–50; translated as "On sense and reference" in P. Geach and M. Black, eds, *Translations from the Philosophical Writings of Gottlob Frege* Basil Blackwell, Oxford, 1952, pp. 56–78.

Gabbay, Dov M., and Guenthner, Franz [1983] *Handbook of Philosophical Logic,* vol. 1 Reidel, Dordrecht.

—— [1984] *Handbook of Philosophical Logic,* vol. 2 Reidel, Dordrecht.

—— [1986] *Handbook of Philosophical Logic,* vol. 3 Reidel, Dordrecht.

—— [1989] *Handbook of Philosophical Logic,* vol. 4 Reidel, Dordrecht.

Geach, Peter [1956] "Good and evil" *Analysis* 17, pp. 33–4.

—— [1972] "A program for syntax" in D. Davidson and G. Harman, eds, *Semantics of Natural Languages* Reidel, Dordrecht, pp. 483–97.

Gibbard, Allan [1975] "Contingent identity" *Journal of Philosophical Logic* 4, pp. 187–221.

—— [1980] "Two recent theories of conditionals" in Harper, Stalnaker and Pearce [1980] pp. 221–47.

Goguen, J. A. [1969] "The logic of inexact concepts" *Synthese* 19, pp. 325–73.

Goodman, Nelson [1955] *Fact, Fiction and Forecast* Harvard University Press, Cambridge, Mass; 2nd edn, Bobbs-Merrill, Indianapolis, 1965.

Grice, H. P. [1961] "The causal theory of perception" *Proceedings of the Aristotelian Society,* Supplementary Volume 35, pp. 121–54; reprinted in G. Warnock, ed., *The Philosophy of Perception* Oxford University Press, 1967, pp. 85–112.

—— [1975] "Logic and conversation" in Peter Cole and Jerry L. Morgan, eds, *Syntax and Semantics,* vol. 3: *Speech Acts* Academic Press, New York, pp. 41–58.

—— [1978] "Further notes on logic and conversation" in Peter Cole, ed., *Syntax and Semantics,* vol. 9: *Pragmatics* Academic Press, New York, pp. 113–17.

Haack, Susan [1974] *Deviant Logic* Cambridge University Press, Cambridge.

—— [1978] *Philosophy of Logics* Cambridge University Press, Cambridge.

Hale, Bob [1989] "Necessity, caution and scepticism" *Proceedings of the Aristotelian Society,* Supplementary Volume 63, pp. 175–202.

Hanson, William H. [1991] "Indicative conditionals are truth-functional" *Mind* 100, (forthcoming).

Harper, W. L., Stalnaker, R., and Pearce, G. eds [1980] *Ifs: Conditionals, Belief, Decision, Chance, Time* Reidel, Dordrecht.

Hazen, Allen [1976] "Expressive incompleteness in modal logic" *Journal of Philosophical Logic* 5, pp. 25–46.

—— [1979] "Counterpart-theoretic semantics for modal logic" *Journal of Philosophy* 76, pp. 319–38.

Higginbotham, James [1986] "Linguistic theory and Davidson's program in semantics" in LePore [1986] pp. 29–48.

—— [1988] "Is semantics necessary?" *Proceedings of the Aristotelian Society* 88 (1987/8) pp. 219–41.

Hintikka, Jaakko [1973] *Logic, language-games and information: Kantian themes in the philosophy of logic,* Clarendon Press, Oxford.

Hochberg, Herbert [1957] "On Pegasizing" *Philosophy and Phenomenological Research* 17; reprinted in Hochberg [1984] pp. 101–4.

—— [1984] *Logic, Ontology and Language: Essays on Truth and Reality* Philosophia Verlag, Munich.

Hodges, Wilfrid [1977] *Logic* Penguin, London.

—— [1983] "Elementary predicate logic" in Gabbay and Guenthner, pp. 1–131.

Hornsby, J. [1977] "Saying of" *Analysis* 37, pp. 177–85.

Hughes, G. E., and Cresswell, Max J. [1968] *An Introduction to Modal Logic* Methuen, London.

—— [1984] *Companion to Modal Logic* Methuen, London.

Jackson, Frank [1979] "On assertion and indicative conditionals" *Philosophical Review* 87, pp. 567–89.

—— [1981] "Conditionals and possibilia" *Proceedings of the Aristotelian Society* 81 (1980/1), pp. 125–37.

—— [1987] *Conditionals* Cambridge University Press, Cambridge.

Kamp, J. A. W. [1975] "Two theories of adjectives" in Edward L. Keenan, ed., *Formal Semantics of Natural Language* Cambridge University Press, Cambridge, pp. 123–55.

Kant, Immanuel [1787] *Critique of Pure Reason*, English translation by Norman Kemp Smith, Methuen, London, 1929.

Kaplan, David [1979] "Transworld heir lines" in Loux [1979] pp. 88–109.

Kirwan, Christopher [1978] *Logic and Argument* Duckworth, London.

Kneale, William, and Kneale, Martha [1962] *The Development of Logic* Oxford University Press.

Kripke, Saul [1963] "Semantical considerations on modal logic" *Acta Philosophica Fennica* 16, pp. 83–94; reprinted in Linsky, ed. [1971] pp. 63–72.

—— [1972] "Naming and necessity" in D. Davidson and G. Harman, eds, *Semantics of Natural Language* Reidel, Dordrecht; reprinted with added preface (referred to as Kripke [1980]) as *Naming and Necessity* Basil Blackwell, Oxford, 1980.

—— [1976] "Is there a problem about substitutional quantification?" in Evans and McDowell [1976] pp. 325–419.

Lacey, Hugh M. [1971] "Quine on the logic and ontology of time" *Australasian Journal of Philosophy* 49, pp. 47–67.

Lambert, Karel [1965] "On logic and existence" *Notre Dame Journal of Formal Logic* 6, pp. 135–41.

—— [1983] *Meinong and the Principle of Independence* Cambridge University Press, Cambridge.

Le Carré, John [1971] *The Naive and Sentimental Lover* Hodder and Stoughton, London.

Lemmon, Edward J. [1965] *Beginning Logic* Nelson, London.

Leonard, H. S. [1956] "The logic of existence" *Philosophical Studies* 5, pp. 49–64.

LePore, Ernest [1966] *Truth and Interpretation: Perspectives on the Philosophy of Donald Davidson* Blackwell, Oxford.

Lewis, C. I. and Langford, C. H. [1932] *Symbolic Logic* Century Co., New York.

Lewis, David [1968] "Counterpart theory and quantified modal logic" *Journal of Philosophy* 65, pp. 17–25; reprinted with added postscripts in Lewis [1983] pp. 26–46.

—— [1970] "General semantics", *Synthese* 22, pp. 18–67; reprinted with an added postscript in Lewis [1983] pp. 189–232.

—— [1973a] "Counterfactuals and comparative possibility" *Journal of Philosophical Logic* 2, pp. 418–46; reprinted in Harper, Stalnaker and Pearce [1981] pp. 57–85.

—— [1973b] *Counterfactuals* Basil Blackwell, Oxford.

—— [1976] "Probabilities of conditionals and conditional probabilities" *Philosophical Review* 85, pp. 297–315; reprinted in Harper, Stalnaker and Pearce [1981] pp. 129–47; and with a postscript, in Lewis [1986b] pp. 133–56.

—— [1979] "Counterfactual dependence and time's arrow" *Noûs* 13, pp. 455–76; reprinted with a postcript in Lewis [1986b] pp. 32–66.

—— [1983] *Philosophical Papers* Oxford University Press, Oxford.

—— [1986a] *On The Plurality Of Worlds* Basil Blackwell, Oxford and New York.

—— [1986b] *Philosophical Papers II* Oxford University Press, Oxford.

—— [1986c] "Probabilities of Conditionals and Conditional Probabilities II" *The Philosophical Review* 95, pp. 581–9.

Linsky, Leonard ed. [1971] *Reference and Modality* Oxford University Press, Oxford.

—— [1977] *Names and Descriptions* University of Chicago Press, Chicago.

Loux, Michael J., ed. [1979] *The Possible and the Actual: Readings in the Metaphysics of Modality* Cornell University Press, Ithaca.

Lucas, John [1961] "Minds, machines and Gödel" *Philosophy* 36, pp. 112–27.

Lycan, William G. [1984] *Logical Form in Natural Language* MIT Press, Cambridge, Mass.

McCawley, James D. [1981] *Everything that Linguists have Always Wanted to Know about Logic but were Ashamed to Ask* University of Chicago Press, Chicago.

McCulloch, Gregory [1989] *The Game of the Name: Introducing Logic, Language and Mind* Clarendon Press, Oxford.

McDowell, John [1977] "On the sense and reference of a proper name" *Mind* 86, pp. 159–85; reprinted in M. de B. Platts, ed., *Reference, Truth and Reality* Routledge and Kegan Paul, London, 1980, pp. 141–66.

McGinn, Colin [1981] "Modal reality" in R. Healey, ed., *Reduction, Time and Reality* Cambridge University Press, Cambridge, pp. 143–205.

Mackie, John L. [1974] "*De* what *re* is *de re* modality?" *Journal of Philosophy* 71, pp. 551–61.

Mackie, Penelope [1987] "Essence, origin and bare identity" *Mind* 96, pp. 173–201.

Marsh, R. C., ed. [1956] *Logic and Knowledge*, Allen and Unwin, London (a collection of essays by Russell).

Meinong, Alexius [1904] "The theory of objects", in Meinong, ed., *Untersuchungen zur Gegenstandtheorie und Psychologie*, Barth, Leipzig; English translation by Isaac Levi, D. B. Terrell and Roderick M. Chisholm, in Roderick M. Chisholm, ed., *Realism and the Background of Phenomenology* The Free Press, New York, 1960.

Mill, John Stuart [1979] *System of Logic* Longmans, London.

Mondadori, Fabrizio [1983] "Counterpartese, counterpartese∗, counterpartese_D" *Histoire, Epistémologie, Langage* 5, pp. 69–94.

Pap, Arthur [1958] *Semantics and Necessary Truth* Yale University Press, New Haven.

Parsons, Terence [1972] "Some problems concerning the logic of grammatical modifiers" in Donald Davidson and Gilbert Harman, eds, *Semantics of Natural Languages* Reidel, Dordrecht.

—— [1974]: "Prolegomena to Meinongian semantics" *Journal of Philosophy* 71, pp. 561–80.

Peacocke, Christopher A. B. [1973] "What is a logical constant?" *Journal of Philosophy* 73, pp. 221–40.

—— [1975] "Proper names, reference and rigid designation" in S. Blackburn, ed., *Meaning, Reference and Necessity* Cambridge University Press, Cambridge, pp. 109–32.

—— [1987] "Understanding logical constants: a realist's account" *Proceedings of the British Academy* Oxford University Press, Oxford, 1988, pp. 153–200.

Plantinga, Alvin [1974] *The Nature of Necessity* Oxford University Press, Oxford.

—— [1976] "Actualism and possible worlds" *Theoria* 42, pp. 139–60; reprinted in Loux [1979] pp. 253–73.

—— [1987] "Two concepts of modality: modal realism and modal reductionism" in James E. Tomberlin, ed. *Philosophical Perspectives*, I: *Metaphysics* Ridgeview, Atascadero, Calif., pp. 189–231.

Platts, Mark de Breton [1979] *Ways of Meaning* Routledge and Kegan Paul, London.

Pollock, John L. [1982] *Language and Thought* Princeton University Press, Princeton NJ.

Popper, Karl [1959] *The Logic of Scientific Discovery* Hutchinson, London.

Prawitz, Dag [1965] *Natural Deduction* Almqvist and Wiksell, Stockholm.

—— [1977] "Meaning and proofs: on the conflict between classical and intuitionistic logic" *Theoria* 43, pp. 2–40.

—— [1979] "Proofs and the meaning and completeness of the logical constants" in Jaakko Hintikka, Ilkka Niiniluoto and Esa Saarinen, eds, *Essays on Mathematical and Philosophical Logic*, Reidel, Dordrecht, pp. 25–40.

Prior, Arthur [1960] "The runabout inference ticket" *Analysis* 21, pp. 38–9.

Putnam, Hilary [1980] "Models and reality" *Journal of Symbolic Logic* 45, pp. 464–82; reprinted in his *Realism and Reason*, Cambridge University Press, Cambridge, 1983, pp. 1–25.

Quine, Willard van O. [1936] "Truth by convention" in O. H. Lee, ed., *Philosophical Essays for A. N. Whitehead* Longmans, New York, 1936; reprinted in Quine [1966] pp. 70–99.

—— [1940] *Mathematical Logic* Harvard University Press, Cambridge, Mass.

—— [1948] "On what there is" *Review of Metaphysics* 2, pp. 21–38; reprinted in Quine [1953a] pp. 1–19.

—— [1951] "Two dogmas of empricism" *Philosophical Review* 60, pp. 20–43; reprinted in Quine [1953a].

—— [1952] *Methods of Logic* Routledge and Kegan Paul, London.

—— [1953a] *From A Logic Point of View* Harvard University Press, Cambridge, Mass. 2nd edn, 1961.

—— [1953b] "Logic and the reification of universals" in Quine [1953a] pp. 102–29.

—— [1953c] "Reference and Modality" in Quine [1953a] pp. 139–59.

—— [1953d] "Three grades of modal involvement" *Proceedings of the Eleventh International Congress of Philosophy*, North-Holland, Amsterdam, pp. 65–81; reprinted in Quine [1966] pp. 156–74.

—— [1956] "Quantifiers and propositional attitudes" *Journal of Philosophy* 53, pp. 177–87; reprinted in Quine [1966] pp. 183–94.

—— [1960] *Word and Object* MIT Press, Cambridge, Mass.

—— [1963] "Carnap and logical truth" in P. A. Schilpp, ed., *The Philosophy of Rudolph Carnap*, Open Court, La Salle; reprinted in Quine [1966] pp. 100–25.

—— [1966] *Ways of Paradox and Other Essays* Random House, New York.

—— [1968] "Ontological relativity" *Journal of Philosophy* 65, pp. 185–212; reprinted in Quine [1969b] pp. 26–68.

—— [1969a] "Existence and quantification" in Quine [1969b] pp. 91–113.

—— [1969b] *Ontological Relativity and Other Essays* Columbia University Press, New York.

—— [1970] *Philosophy of Logic* Prentice-Hall, Englewood Cliffs, NJ.

—— [1976] "Worlds away" *Journal of Philosophy* 73, pp. 859–63.

Quinton, Anthony [1963] "The a priori and the analytic" *Proceedings of the Aristotelian Society* 64 (1963/4), pp. 331–54; reprinted in P. F. Strawson, ed., *Philosophical Logic*, Oxford University Press, 1967.

Ramachandran, Murali [1989] "An alternative translation scheme for counterpart theory" *Analysis* 49, pp. 131–41.

Ramsey, Frank P. [1929] "General propositions and causality" in *Foundations: Essays in Philosophy, Logic, Mathematics and Economics*, edited by D. H. Mellor, Routledge and Kegan Paul, London, 1978.

Read, Stephen [1988] *Relevant Logic* Basil Blackwell, Oxford.

Routley, R. [1970] "Non-existence does not exist" *Notre Dame Journal of Formal Logic* 11, pp. 289–320.

Russell, Bertrand A. W. [1905] "On denoting" *Mind* 14, pp. 479–93; reprinted in Marsh, ed. [1956] pp. 442–54.

—— [1908] "Mathematical logic as based on the theory of types" *American Journal of Mathematics* 30, pp. 222–62; reprinted in Marsh, ed. [1956] pp. 59–102.

—— [1912] *Problems of Philosophy*, Williams and Norgate, London, reprinted by Oxford University Press, Oxford, 1967.

—— [1914] *Our Knowledge of the External World as a Field for Scientific Method in Philosophy*, Allen and Unwin, London.

—— [1918–19] "Lectures on the philosophy of logical atomism" *Monist* 28, pp. 495–527, and *Monist* 29, pp. 32–63, 190–222, 345–80; reprinted in Marsh, ed. [1956] pp. 59–102.

—— [1919] *Introduction to Mathematical Philosophy* Allen and Unwin, London and New York.

—— [1959] *My Philosophical Development* Allen and Unwin, London.

—— and Whitehead, Alfred N. [1910–13] *Principia Mathematica* Cambridge University Press, Cambridge.

Sainsbury, R. M. [1979] *Russell* Routledge and Kegan Paul, London.

Salmon, Nathan [1989] "Reference and information content: names and descriptions" in Gabbay and Guenthner, pp. 409–61.

Schiffer, Stephen [1987] *Remnants of Meaning* MIT Press, Cambridge, Mass.

Schock, R. [1968] *Logic Without Existence Assumptions*, Almqvist and Wiksell, Stockholm.

Searle, John [1958] "Proper names" *Mind* 67, pp. 166–73; reprinted in P. F. Strawson, ed., *Philosophical Logic*, Oxford University Press, Oxford, 1967, pp. 89–96.

Segal, Gabriel [1989] "A preference for sense and reference" *Journal of Philosophy* 86, pp. 73–89.

Skyrms, Brian [1966] *Choice and Chance* Dickenson, Belmont, Calif.

Smiley, Timothy [1982] "The schematic fallacy" *Proceedings of the Aristotelian Society* 82 (1982/3), pp. 1–17.

Stalnaker, Robert L. [1968] "A theory of conditionals" *Studies in Logical Theory*, *American Philosophical Quarterly*, Monograph Series, no. 2, Blackwell, Oxford, pp. 98–112; reprinted in Harper, Stalnaker and Pearce [1981] pp. 41–55.

—— [1975] "Indicative conditionals" *Philosophia* 5, pp. 269–86; reprinted in Harper, Stalnaker and Pearce [1981] pp. 193–210.

—— [1980] "A defense of conditional excluded middle" in Harper, Stalnaker and Pearce [1981] pp. 67–104.

—— [1984] *Inquiry* MIT Press, Cambridge, Mass.

Strawson, P. F. [1950] "On referring" *Mind* 59, pp. 320–44; reprinted in Strawson [1971] pp. 1–27.

—— [1952] *Introduction to Logical Theory* Methuen, London.

—— [1971] *Logico-linguistic Papers* Methuen, London.

—— [1974] "Positions for quantifiers" in M. K. Munitz and P. Unger, eds, *Semantics and Philosophy* New York University Press, New York, pp. 63–79.

Tarski, A. [1937] "The concept of truth in formalized languages" in his *Logic, Semantics, Mathematics* Clarendon Press, Oxford, 1956, pp. 152–278.

Taylor, Barry [1985] *Modes of Occurrence* Basil Blackwell, Oxford.

Teichmann, Roger [1990] "'Actually'" *Analysis* 50, pp. 16–19.

Tennant, Neil [1978] *Natural Logic* Edinburgh University Press, Edinburgh.

Thomson, James [1967] "Is existence a predicate?" in P. F. Strawson, ed., *Philosophical Logic* Oxford University Press, Oxford, pp. 103–6.

Trollope, A. [1864] *Can you Forgive Her?* reprinted by Oxford University Press, Oxford, 1953.

Urbach, Peter [1988] "What is a law of nature? A Humean answer" *British Journal for the Philosophy of Science* 39, pp. 193–210.

Van Benthem, Johan [1983] "Correspondence theory" in Gabbay and Guenthner [1983] pp. 167–247.

—— and Doets, Kees [1983] "Higher-order logic" in Gabbay and Guenthner [1983] pp. 275–329.

Van Dijk, Teun, A. [1977] *Text and Context* Longman, London.

Van Fraassen, Bas [1977] "The only necessity is verbal necessity" *Journal of Philosophy* 84, pp. 71–85.

Van Inwagen, Peter [1985] "Plantinga on trans-world identity" in James Tomberlin and Peter van Inwagen, eds, *Alvin Plantinga: A Profile* Reidel, Dordrecht, pp. 101–20.

—— [1986] "Two concepts of possible worlds" in P. French, T. Uehling and H. Wettstein, eds, *Midwest Studies in Philosophy*, vol. 11: *Studies in Essentialism* University of Minnesota Press, Minneapolis, pp. 185–213.

Wiggins, David [1976] "The *de re* 'must': a note on the logical form of essentialist claims" in Evans and McDowell, eds [1976] pp. 285–312.

—— [1980a] "'Most' and 'all': some comments on a familiar programme and on the logical form of quantified sentences" in M. de B. Platts, ed., *Reference, Truth and Reality* Routledge and Kegan Paul, pp. 318–46.

—— [1980b] *Sameness and Substance* Basil Blackwell, Oxford.

—— [1985] "Verbs and adverbs, and some other modes of grammatical combination" *Proceedings of the Aristotelian Society* 86 (1985/6), pp. 273–304.

Wittgenstein, L. [1921] *Tractatus Logico-Philosophicus* (original German edn); translation by D. F. Pears and B. F. McGuiness, Routledge and Kegan Paul, London, 1961.

—— [1932] "Elementary propositions", section 4 of appendix to Part I of *Philosophical Grammar*, ed. Rush Rhees, tr. Anthony Kenny, Basil Blackwell, Oxford, 1974, pp. 210–14.

Wright, Crispin [1986] "Inventing logical necessity" in Jeremy Butterfield, ed., *Language, Mind and Logic* Cambridge University Press, Cambridge.

—— [1989] "Necessity, caution and scepticism" *Proceedings of the Aristotelian Society*, Supplementary Volume 63, pp. 203–238.

Yourgrau, Palle [1987] "The dead" *Journal of Philosophy* 84, pp. 84–101.

Index

Note: 'G' followed by a number refers to Glossary